D1329593

Wireless Spectrum Management

Wireless Spectrum Management

Amit K. Maitra

McGraw-Hill

New York Chicago San Francisco Lisbon London Madrid
Mexico City Milan New Delhi San Juan Seoul
Singapore Sydney Toronto

The McGraw-Hill Companies

CIP Data is on file with the Library of Congress

1 2 3 4 5 6 7 8 9 0 DOC/DOC 0 1 0 9 8 7 6 5 4

ISBN 0-07-140987-4

The sponsoring editor for this book was Stephen S. Chapman and the production supervisor was Sherri Souffrance. It was set in Century Schoolbook by International Typesetting and Composition. The art director for the cover was Anthony Landi.

Printed and bound by RR Donnelley.

This book is printed on recycled, acid-free paper containing a minimum of 50% recycled, de-inked fiber.

To uncle Krishna
For teaching me that
The Rule of Law is the
Highest Philosophy of the Land

Contents

Preface

This book explains basic concepts of spectrum management and planning for rational use of spectrum resources for radiocommunications and other applications vital for the prosperity, culture, and security of society. The author wants readers to gain an appreciation of the complexities of developing national spectrum policies and understand the depth of technical engineering analyses, including application of advanced mathematical methods and computer tools, needed to

1. Ensure that spectrum policies, rules, and regulations required for proper spectrum management nationally and internationally are technically valid
2. Derive the necessary technical facts that will lead to resolution of spectrum issues and problems
3. Provide a technical and engineering basis for future spectrum planning and standards
4. Provide new ways to adopt new spectrum efficient technologies so the national, state, and local governments can use the spectrum efficiently and effectively

With wireless phones allowing portable high-speed Internet connections, the United States and the rest of the world are experiencing the benefits of a new generation of personal mobile communications. During the last 20 years, the United States wireless industry has grown from virtually nothing to 100 million subscribers and growth is continuing at a rate of 25 to 30 percent annually. Globally, there are over 470 million wireless subscribers, and industry analysts predict that in less than 5 years the number will grow to approximately 1.3 billion. The wireless industry has developed leading technologies for current and future systems that have helped establish better communication in more places throughout the world, saving time, money, and lives. The next generation of wireless technology holds even greater promise, with broadband to hand-held devices and "mobile-commerce" (m-commerce) becoming a reality today. Currently, an international effort is underway to make it possible for the next generation of wireless phones to work anywhere in the world.

Former US President Bill Clinton signed a memorandum dated October 13, 2000 that provided the guiding principles for the development of third-generation (3G) wireless systems, including government and industry cooperation needed to develop recommendations and plans for identifying spectrum for 3G wireless systems. On October 20, 2000, the National Telecommunications and Information Administration (NTIA), an agency of the U.S. Department of Commerce that governs spectrum allocation for federal use, released a plan outlining the activities that would lead to selection of spectrum for 3G. NTIA and the Federal Communications Commission (FCC), an independent agency that has broad regulatory authority over nonfederal spectrum allocations, released interim reports on November 15, 2000.

According to these reports, the federal government has agreed to transfer the 1710 to1755 MHz band to the FCC in 2004 on a mixed-use basis as long as federal operations in particular areas are protected. The federal government has exclusive allocation of the 1755 to 1850 MHz band for fixed and mobile services, with the U.S. Department of Defense (DoD) being the predominant user. Other agencies also operate extensive fixed and mobile systems in this band. Major systems in the band include: (1) tracking, telemetry, and control for federal space systems (1761 to 1842 MHz band); (2) military tactical radio relay radios; (3) air combat training systems; (4) conventional fixed microwave systems (medium capacity); and other systems including precision guided munitions, video data links (high resolution), and land mobile video.

According to the interim report produced by NTIA, the interference from 3G transmitters to the U.S. Defense Department's satellite receivers might create a problem in sharing and recommended further analysis. The report stated that uncoordinated sharing was not desirable or feasible due to extensive use by federal government and projected build out of 3G systems. However, sharing could be an option if restrictions in space and/or time prove feasible, 3G operators reimburse federal operators to relocate or the conventional fixed systems are retuned, and major DoD functions are not impacted.

Two major private sector services use the 2500 to 2690 MHz band: (1) the multipoint distribution system (MDS) and (2) instructional television fixed services (ITFS). MDS and ITFS supply high-speed access to the Internet; broadband services to rural areas; video services for education, health, and other institutions; and a variety of analog and digital one and two-way services. The FCC interim report stated that, with few exceptions, 3G systems could not share frequencies with incumbent systems without extensive interference and that segmenting 2500 to 2690 MHz band would raise technical and economic difficulties for incumbents.

Under the President's Executive Memorandum on Advanced Mobile Communications/Third Generation Wireless Systems, NTIA, in conjunction with the U.S. State Department and U.S. diplomatic posts, initiated an international outreach initiative focusing on spectrum needs assessments to support third-generation wireless technologies. Third-generation wireless systems will provide mobile, high-speed access to the Internet and new telecommunications services anytime and anywhere. To maximize the effective and efficient use of this unique, ubiquitous natural resource and ensure that global roaming and economies of scale develop, international harmonization of spectrum allocation must be achieved.

Several goals are associated with the international outreach effort. First, NTIA, in cooperation with the U.S. Department of State, seeks to explain the President's Memorandum, particularly the need to consider requirements of incumbent operators in current spectrum bands, to foreign governments. These requirements may be reallocated, but many other nations do not face this issue. Second, NTIA wants to review the status of 3G developments in the rest of the world, particularly with countries where

1. 3G systems are preoperational
2. Regulatory and standards structures are already in place
3. People are using mobile devices, rather than computers, to access the Internet

National Telecommunications and Information Administration's outreach effort has gained momentum, with talks being held with Japan, Hong Kong, and China, and joint explorations and other cooperative activities being planned with Canada, the European Union, and other nations in the Americas to ensure efficient use of spectrum while providing new opportunities to access the Internet.

The foregoing remarks reveal that there are numerous forces affecting wireless spectrum management that have led countries to investigate, experiment with, and implement alternative ways of managing spectrum resource. Against this backdrop, Wireless Spectrum Management

1. Provides a survey of scientific, business, and industrial applications of spectrum, including spectrum management tasks
2. Identifies methods of conducting engineering analysis to derive the necessary technical facts that lead to resolution of spectrum issues and problems, thereby maximizing the value of spectrum to society

3. Provides basic information on governmental institutions and processes involved in developing spectrum plans and policies—both in the United States and abroad
4. Identifies some of the crucial choices to be made to improve deficiencies in spectrum management practices
5. Summarizes advantages and disadvantages of each

Amit K. Maitra

Acknowledgments

My indebtedness extends to many who have contributed to my thinking, research, and ability to write this book. They include Professor Dr. R. G. Struzak, ITU Regulation Board; Dr. Wim van Driel, Chairman Observatoire de Paris, Committee on Radio Astronomy Frequencies (CRAF); Dr. Titus Spoelstra, Frequency Manager/Secretary, CRAF; Dr. Olov Carlsson, Director, Spectrum Management, AeroteckTelub, Sweden; Dr. Joseph N. Pelton, Director, Space and Advanced Communications Research Institute, The George Washington University; Dr. Edwin C. Jones, Vanderbilt University; Dr. Raymond A. Greenwald, Johns Hopkins Applied Physics Laboratory; Dr. Peter F. Guerrero, Director, Physical Infrastructure Issues; U.S. General Accounting Office, and Christie Brown of Agilent Technologies. Their research, insights, and analytic clarity in their thinking about wireless spectrum management issues have directly contributed to this book. I would also like to acknowledge long-standing debts to the European Science Foundation, AeroteckTelub, and Agilent Technologies for sharing their research findings as well as granting me permission to reproduce in full or in part various texts, charts, tables, and figures. I could never have written this book without the backing of Dr. Titus Spoelstra, Dr. Joseph Pelton, Dr. Olov Carlsson, Dr. Peter F. Guerrero, and Dr. Edwin C. Jones, and the support of these institutions.

Steve Chapman, my editor at McGraw-Hill, has worked with me on this book for longer than either he or I anticipated. His patience and wisdom are greatly appreciated. I would like to acknowledge the value of our collaboration—and its resonance—in writing this book.

Finally, I thank my wife, Julie Binder Maitra, for tolerance of my preoccupation over the months of writing, and her loving support and help in reacting to drafts.

Wireless Spectrum Management

Chapter

1

Introduction

The ocean, the air, and space are common natural resources shared by the nations of the world. So is the radio frequency spectrum—no individual or government owns it; neither is its development or use controlled by any one country or group of countries. Rather, the world community has the collective responsibility of ensuring the wise and equitable use of this vital international resource.

The radio spectrum spans nearly 300 billion frequencies; however, it is the 1 percent of frequencies below 3.1 GHz, where lies the concentration of 90 percent of the radio spectrum's use.[1] The crowding of this region takes place because a wide range of wireless communication and entertainment services, including mobile phones, radio and television broadcasting, and numerous satellite communication systems, are made possible by these frequencies.[2]

Spectrum allocation, a process adopted both internationally and domestically, allocates frequencies among the above types of uses and users of wireless services. The process is designed to prevent radio congestion, which can lead to interference. Interference is caused by radio signals of two or more users interacting and disrupting the transmission and reception of messages. Spectrum allocation slices the radio spectrum into bands of frequencies that are then chosen for use by particular types of radio services or classes of users, such as broadcast television and satellites.

During the last 50 years, the world has witnessed numerous scientific and technological breakthroughs. New wireless technology has made available more usable frequencies, decreased the potential for interference, and enhanced the efficiency of transmission through various techniques, such as reduced spectrum needs to send information. All these new technology developments are certainly decreasing congestion within the radio spectrum; however, competition for additional spectrum is also rising. A case in point is the current terrorist threats that have made wireless services indispensable in practically all countries at all levels of government for national security, public safety, and other functions. The consumer market for wireless services is also witnessing phenomenal growth. For example, mobile phone service in the United States greatly exceeded the industry's original growth predictions, as it jumped from 16 million subscribers in 1994 to an estimated 160 million in 2003.[3]

Wireless Spectrum Management focuses on frequency engineering and management functions as part of planning, research, development, and operational activities involving use of the radio frequency spectrum. The book offers methods, tools, and techniques for analyzing the technical characteristics of newly planned U.S. and international radio communication services, technologies, and applications which may be intended for operation in the same frequency bands. In the United States, the methods occasionally result in lengthy negotiations between the Federal Communications Commission (FCC) and National Telecommunications and Information Administration (NTIA) over how to resolve some allocation issues.[4] Since nearly all of the usable radio spectrum has been allocated already, accommodating more services and users often involves redefining spectrum allocations. Therefore, a major portion of the book, namely Chaps. 5 to 8, underscores the importance of conducting the analyses in accordance with requirements and procedures of national and International Telecommunications Union (ITU) rules and regulations, while making efficient use of new techniques in the propagation prediction and utilization of radio frequencies.

One method, spectrum "sharing," enables more than one user to transmit radio signals on the same frequency band. In a shared allocation, a distinction is made as to which user has "primary" or priority use of a frequency and which user has "secondary" status, meaning it must defer to the primary user. Users may also be designated as "co-primary," in which the first operator to obtain authority to use the spectrum has priority to use the frequency over another primary operator. In instances where spectrum is shared between federal and nonfederal users—currently constituting 56 percent of the spectrum in the 0 to 3.1 GHz range[5]—FCC and NTIA must ensure that the status assigned to users (primary/secondary or co-primary) meet users' radio needs, and that users abide by rules applicable to their designated status.

Another method to accommodate new users and technologies is "bandclearing," or reclassifying a band of spectrum from one set of radio services and users to another, which requires moving previously authorized users to a different band. Bandclearing decisions affecting either only nonfederal or only federal users are managed within FCC or NTIA respectively, albeit sometimes with difficulty. However, bandclearing decisions that involve radio services of both types of users pose a greater challenge. Specifically, they require coordination between FCC and NTIA to ensure that moving existing users to a new frequency band is feasible and not otherwise disruptive to their radio operation needs.[6] While many such bandclearing decisions have been made throughout radio history, these negotiations can become protracted. For example, a hotly debated issue is how to accommodate third-generation wireless services.[7] FCC announced that the relationship between FCC and NTIA on spectrum management became more structured following the enactment of legislative provisions mandating the reallocation of spectrum from federal to nonfederal government use.[8]

To address the protracted nature of some spectrum bandclearing efforts, many industry and government officials who participated in the 2002 NTIA-sponsored Spectrum Management and Policy Summit[9] have suggested establishing a third party—such as an outside panel or commission, an office within the Executive branch, or an interagency group[10]—to arbitrate or resolve differences between FCC and NTIA.[11] In some other countries, decisions are made within one agency or within interagency mechanisms that exist for resolving contentious bandclearing issues. For example, the United Kingdom differs from the U.S. spectrum management structure in that a formal standing committee, co-chaired by officials from the Radiocommunications Agency and the Ministry of Defense, has the authority to resolve contentious spectrum issues.[12]

Another proposed mechanism is the preparation of a national spectrum plan to better manage the allocation process. The Omnibus Budget Reconciliation Act of 1993 required NTIA and FCC to conduct joint spectrum planning sessions.[13] The National Defense Authorization Act of 2000 included a requirement for FCC and NTIA to review and assess the progress toward implementing a national spectrum plan.[14] According to the top officials from FCC and NTIA, neither requirement has yet been fully implemented; however, their stated intention is to implement these directives.[15]

Wireless Spectrum Management is an extraordinary book in that it

1. Covers virtually *all* spectrum-dependent communications-electronic systems, from satellite communications, radar surveillance, point-to-point microwave, and air-to-air and land- or sea-based ballistic missile systems to conventional and trunked land mobile radio, very high frequency (VHF) and ultra-high frequency (UHF) air traffic control systems, telemetry, weapon systems, and the use of ultra-wide band (UWB), with technical factors affecting the properties and availability of key radio bands.

2. Provides and interprets the *hard-to-find information* on the properties and availability of key radio bands to encourage the rational use and conservation of the spectrum/orbit resources.
3. Focuses on *engineering analysis.*
4. Points to *frequency management* and *frequency allocation* approaches, tools, and methods that improve spectrum engineering, monitoring and enforcement mechanisms.
5. Discusses the *prevailing concept of spectrum management*, including all activities related to regulations, planning, allocation, assignment, use, and control of the radio frequency spectrum.
6. Discusses *different administrations*'[16] regulatory structures, including their assignment processes, the amount of flexibility allowed spectrum users, the existence of secondary markets, and their rules regarding interference.

From this encyclopedic work, the author summarizes alternative approaches to spectrum management used around the world and identifies similarities and differences between these approaches and those used in the United States. For example, 4.4 Appendix: Spectrum Use Summary in Chap. 4 describes how U.S. government agencies and private industry planning to develop, modify, or use communications systems must ensure that new proposed systems are compatible with existing frequency assignments and equipment operations and are in compliance with national and international frequency allocation tables. The table cites Office of Management and Budget (OMB) Circular No. A-11,[17] which states that estimates for development or procurement of major communication-electronics systems, including all systems employing satellite communications (space) techniques, will be submitted only after NTIA has certified that the radio frequency required for these systems is available. This implies that government agencies cannot legally obligate funds for procurement of spectrum-dependent communications-electronic systems unless spectrum support is assured. There are engineering tasks, namely technical examination, coordination, frequency planning, optimization, EMC[18] analyses, monitoring, etc., that need to be performed to obtain such support. In a narrow sense, *Wireless Spectrum Management* is about these frequency engineering and management functions to be performed as part of planning, research, development, and operational activities involving use of the radio frequency spectrum.

The point to underscore is that "spectrum" is a general term that encompasses the spatial and temporal properties of any medium, including fiber optic cable, coaxial cable, and ambient air. The term assumes a new degree of importance when one restricts its usage to the more common application of communications through wireless media. In today's environment, wireless communications technologies are assuming a greater importance in most nations and are viewed as a critical means to expanding economic opportunity. To the author, the current environment serves as the point of departure for *Wireless Spectrum Management.*

In recent annual World Radiocommunication Conferences (WRCs), more than 2000 delegates from over 150 countries have participated to ensure that their existing uses of the spectrum for their radiocommunications would be protected and that their future requirements for the spectrum would be satisfied. The point to underscore here is "future requirements."

Newer radio communication services, technologies, and applications are emerging on the horizon with tremendous speed. Third generation cellular mobile systems—International Mobile Telephony (IMT)-2000[19] and Universal Mobile Telephony System (UMTS)—are some of the technologies already available today. IMT-2000/UMTS systems provide for an integration of terrestrial and satellite technologies and are capable of providing a wide range of services, such as multimedia, videoconferencing and high-speed internet access. Some future IMT-2000/UMTS applications may require transmission at very high data rates where the user is stationary, and for such applications, it may be possible to utilize frequency bands above 3 GHz. These technologies are likely to be available in the long term.

The author had the opportunity to talk to members of many delegations to listen to their views on the many issues being addressed at the WRC.[20] It was apparent that all countries had definite views on obtaining additional spectrum for implementing IMT-2000 and future generations of advanced communications.

Following extensive analyses and international talks, the United States has had difficulty in identifying any single band that could likely be used on a global basis by IMT-2000 and other advanced communications technologies.[21] The current U.S. proposal calls for the identification of spectrum in several bands for consideration by countries around the world for implementation of IMT-2000 and other advanced communications technologies.[22] The U.S. proposal acknowledges that many countries, including the United States, are studying some of the spectrum (698–960 MHz, 1525–1559 MHz, 1610–1660.5 MHz, 1710–2025 MHz, 2110–2200 MHz, 2483.5–2690 MHz bands) to determine its feasibility within the national boundaries of that particular country.[23] There has been a call for adopting a second new resolution addressing national studies and ITU-R studies related to the 698–960 MHz, 1710–1885 MHz, and 2500–2690 MHz bands.[24] ITU studies will review aspects to facilitate the implementation of these new technologies, thereby offering the U.S. and other countries a new opportunity whereby all countries can come together and attempt a consensus on new procedures to ensure compatible operation.[25]

A major portion of *Wireless Spectrum Management*, namely Chaps. 5 to 8, underscores the importance of conducting the analyses in accordance with requirements and procedures of national and ITU rules and regulations. Asian, European, and other countries are interested in national rules and regulations, and within ITU, all countries try to accomplish spectrum management with engineering analysis at a level that is thoroughly familiar with and immediately responsive to the requirements of the ITU-R Study Groups (which study technical questions relating to radiocommunication issues and adopt recommendations) and sufficiently close to WRC treaty status to facilitate referral of issues requiring consideration by these offices. *Wireless Spectrum Management* is timely in that it addresses the fundamental needs for accomplishing spectrum management in consonance with ITU rules and regulations *and* in coordination with sound engineering practices that promise more effective and prudent use of radio frequency spectrum.

Wireless Spectrum Management is organized as follows:

Chapter	Title	Scope
Chapter 1	Introduction	The context: ▪ Technical principles ▪ Organization ▪ Competition
Chapter 2	Definitions	What is spectrum?
Chapter 3	Physical Characteristics	▪ Various segments[26] ▪ Dimensions of the spectrum
Chapter 4	Applications	▪ Active ▪ Passive
Chapter 5	Spectrum Management	▪ Demand ▪ Objectives ▪ Tasks: Policy, administration, and engineering
Chapter 6	Management Process	▪ International spectrum management ▪ U.S. spectrum management ▪ Telecom and broadcasting by regions
Chapter 7	Observations on the present national and world-wide spectrum management and its remedies	▪ What we should do today ▪ What should we not do today ▪ How are we operating scientific radio projects after 25 years? ▪ What management tools do scientists have? ▪ With what harm can we live? ▪ Passive versus active: the sharing problem

1.1 Audience

Intended for senior executives, such as the chairman and chief executive of a radio communications agency; vice president, industry relations and standards for wireless; managing director, defense spectrum management; director of major new radio-based technology; professor in the IT, telecommunications, or broadcasting field; and consultants with projects in the broadcast, communications and information sectors, *Wireless Spectrum Management* underscores the importance of conducting the analyses in accordance with requirements and procedures of national and ITU rules and regulations, while making efficient use of new techniques in the propagation, prediction, and utilization of radio frequencies. The book offers insights into methods, tools, and techniques for analyzing the technical characteristics of newly planned U.S. and international radio communication services, technologies, and applications which may be intended for operation in the same frequency bands.

1.2 Endnotes

[1]Radio waves are a form of electromagnetic radiation that propagates in space as the result of particle oscillations. The number of oscillations per second is called frequency, which is measured in units of hertz. The terms "kilohertz" refers to thousands of hertz and "gigahertz" to billions of hertz. The radio spectrum comprises a range of frequencies from 3 kHz to around 300 GHz.

[2]Today, a host of additional new technologies and services, such as digital audio broadcasting, high-definition television, personal communications services, satellite and microwave communications, and even baby monitors, are vying with existing radio-based applications for a slice of the valuable, but crowded, radio spectrum.

[3]"U.S. Mobile Phone Penetration Reaches 53% of Total Population in December 2002," Kensei News & Information Services, San Francisco, CA, February 11, 2003.

[4]Management of the spectrum for the private sector, including state and local governments, is the responsibility of the FCC. NTIA is responsible for managing the federal government's use of the radio spectrum. Chapter 6 provides more details on the principal responsibilities of the FCC and NTIA for developing and articulating U.S. domestic and international telecommunications policy.

[5]NTIA also reported that 42 percent of the shared allocations between federal and nonfederal users in the 0 to 3.1 GHz range are shared on a "co-primary" basis.

[6]The Strom Thurmond National Defense Authorization Act for the Fiscal Year 1999, P.L. 105-251, October 17, 1998, authorized federal entities to accept compensation payments when they relocate or modify their frequency use to accommodate nonfederal users of the spectrum. The National Defense Authorization Act for Fiscal Year 2000, P.L. 106-65, October 5, 1999, specified a number of conditions that have to be met if spectrum in which Department of Defense (DoD) is the primary user is surrendered. The act requires NTIA, in consultation with FCC, identify and make available to DoD for its primary use, if necessary, an alternate band(s) of frequency as replacement for the band surrendered. Further, if such band(s) of frequency are to be surrendered, the Secretaries of Defense and Commerce and the Chairman of the Joint Chiefs of Staff must jointly certify to relevant congressional committees that such alternative band(s) provide comparable technical characteristics to restore essential military capability.

[7]For more information on spectrum use decisions for third-generation wireless services, see *Defense Spectrum Management: More Analysis Needed to Support Use Decisions for the 1755–1850 MHz Band* (GAO-01-795, August 20, 2001).

[8]Omnibus Budget Reconciliation Act, P.L. 103-66, August 10, 1993, mandated that bands of frequencies not less than 200 MHz be transferred from use of the federal government to nonfederal users. NTIA was directed to make a report on the identification and recommendation for reallocation of frequency bands; utilize specific criteria in making recommendations; issue a preliminary report upon which public comment on proposed reallocations would be solicited; obtain analyses and comment from FCC on reallocations; and transfer frequency bands within specified time frames. It required FCC to gradually allocate and assign frequencies over the course of 10 years. The Balanced Budget Act, P.L. 105-33, August 5, 1997, imposed a stricter deadline for NTIA to identify for reallocation and FCC to reallocate, auction, and assign licenses by September 2002 for an additional 20 MHz of spectrum. (Eight megahertz of spectrum was subsequently reclaimed per congressional direction. Refer to section 1062 of the National Defense Authorization Act for Fiscal Year 2000, P.L. 106-65, October 5, 1999.) For further details, refer to Peter F. Guerrero, *Telecommunications: History and Current Issues Related to Radio Spectrum Management*, GAO-02-814T, June 11, 2002.

[9]Spectrum Management and Policy Summit sponsored by the National Telecommunications and Information Administration (NTIA), http://www.commerce.gov/opa/photo/NTIA/2002/April_04/gallery.html

[10]"Commerce Secretary Evans kicks off two-day 'Spectrum Summit' exploring solutions to challenges in managing nation's airwaves," http://www.commerce.gov/opa/photo/NTIA/2002/April_04/gallery.html.

[11]"FCC Chairman Michael K. Powell and Assistant Secretary Nancy J. Victory met formally to plan and coordinate the efforts of the FCC and NTIA to improve U.S. spectrum policy," NTIA Press Release, December 10, 2002, http://www.ntia.doc.gov/ntiahome/press/2002/spectrumleadership_12102002.htm.

[12]Guerrero, *ibid.* 8.

[13]47 U.S.C. § 922, *ibid.*

[14]P.L. 106-65, 113 Stat. 767 (1999), *ibid.*

[15]Guerrero, *ibid.*

[16]The ITU Radio Regulation defines an Administration as "any governmental department or service responsible for discharging the obligations undertaken in the Constitution of the International Telecommunication Union, in the Convention of the International Telecommunication Union and in the Administrative Regulations," ITU, 1993, *Final Acts of the Additional Plenipotentiary Conference (Geneva, 1992)—Constitution and Convention of the International Telecommunication Union*, International Telecommunication Union, Geneva, as referenced in "Science and Spectrum Management," European Science Foundation (ESF) Committee on Radio Astronomy Frequencies (CRAF) 2002.

[17]OMB Circular A-11 requires that certification of spectrum support be obtained prior to the submission of annual budget estimates to the OMB. This applies to the appropriation of funds for either the development, procurement, or modification of CNS or supporting systems or equipment that require use of the radio spectrum.

[18]Electromagnetic compatibility (EMC) involves understanding and the effects of both radiated and conducted electrical interference on equipment and biological systems. The effects are quantified as either emission levels (at the source) or immunity thresholds (at the victim). The paths that link the two entities are studied in detail.

[19]International Mobile Telecommunications-2000 (IMT-2000) is the global standard for third generation (3G) wireless communications, defined by a set of interdependent ITU Recommendations. IMT-2000 provides a framework for worldwide wireless access by linking the diverse systems of terrestrial and/or satellite based networks. It will exploit the potential synergy between digital mobile telecommunications technologies and systems for fixed and mobile wireless access systems.

[20]NTIA "Spectrum Management & Policy Summit," *Ibid* 9. Also refer to the Federal Communication Commission's 2003 World Radiocommunication Conference (WRC-03) Homepage: http://www.fcc.gov/wrc-03/

[21]Informal Working Group 1: IMT-2000 and Terrestrial Wireless Interactive Multimedia, http://www.fcc.gov/wrc-03/iwg_1.html

[22]United States of America, DRAFT PROPOSAL FOR THE WORK OF THE CONFERENCE, http://www.fcc.gov/wrc-03/files/docs/us_proposals/uspl_1_33.pdf

[23]"Spectrum Management Policies and the World Radiocommunications Conference," prepared testimony of the Honorable Gregory L. Rohde, Assistant Secretary for Communications and Information, U.S. Department of Commerce before the Subcommittee on Telecommunications, Trade, and Consumer Protection, Committee on Commerce, U.S. House of Representatives, Washington, D.C., July 19, 2000, National Communications System, Telecommunications Speech Service, Volume III, Number 46.

[24]World Radiocommunication Conference (WRC-03), Geneva, Switzerland, 9 June - 4 July 2003, http://www.fcc.gov/wrc-03/

[25]International Telecommunications Union activities on IMT-2000 comprise international standardization, including frequency spectrum and technical specifications for radio and network components, tariffs and billing, technical assistance and studies on regulatory and policy aspects.

[26]No discussion of the dimensions of the spectrum is complete without reference to satellites/orbits.

Chapter 2

Definitions

2.1 What Is the Spectrum?

There is no simple answer to the question, What is the spectrum?[1] Many have defined the spectrum as an ensemble and an open medium comprising the:

- Electric and magnetic fields that produce (electromagnetic) waves moving through space at different frequencies.
- Set of all possible frequencies called the "electromagnetic spectrum."
- Subset of frequencies between 3000 Hz and 300 GHz known as the "radio spectrum." Refer to Fig. 2-1.
- Radio waves not requiring a medium *per se*, for these radio waves can travel through a vacuum (e.g., outer space).

Jean-Baptiste Fourier (1768–1830) originally introduced the spectrum as an abstract mathematical idea for solving differential equations.[2] There were initial doubts, difficulties, and resistance to the idea. Peter Dirichlet (1805–1859) and George Riemann (1826–1866) helped resolve the doubts and transformed the idea into a powerful tool to be used in many branches of theoretical sciences, signal processing, communications, and computer technology. In more recent times, with the development and advancement of radio engineering, the concept of radio frequency spectrum has found use in our everyday world and RF spectrum analyzers have become basic instruments in radio laboratories.

The spectrum of radio waves carries energy and messages through space at no cost and at the speed of light, thereby making the spectrum a valuable resource from which everybody profits. Its chief attractiveness is free access from any place and at any time. Compounding the

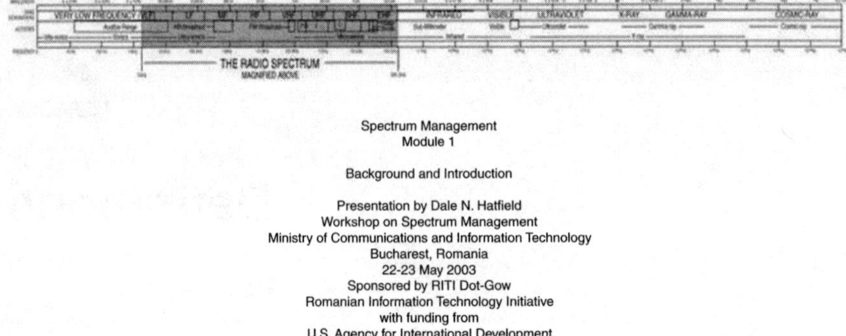

Figure 2-1 Electromagnetic spectrum.

benefits of the spectrum as a natural resource is its natural association with another abstract notion: invisible lines in space,the satellite orbits.[3] Table 2-1 delineates the nature of the spectrum resource.

Struzak[5] describes the spectrum in the context of wireless communications, illustrating that it has had more than one meaning over time, as shown in Table 2-2.

According to Struzak, the three elements that added new dimensions to the concept of spectrum include:

- Market demand fostered by the wireless service providers and equipment manufacturers
- International wireless services and their progress and expansion
- Cross-border radio interference, generating risks.

Table 2-3 and Fig. 2-2 identify the ever-increasing demand and uses of the spectrum at different frequencies and their properties. Later in this chapter, we will also discuss in general terms how several of these uses can cause radio interference.

All governments signed an international treaty which verifies that

> ...radio frequencies and the geostationary-satellite orbit are limited natural resources [...] that must be used [...] so that countries and group of countries may have equitable access to both...[6]

TABLE 2-1 Nature of the Spectrum Resource[4]

- A unique natural resource
- A national and international resource
- Infinitely renewable
- Like air or water, it can be polluted
- Scarcity of the resource—economic value

TABLE 2-2 Changes in the Concept of Spectrum over Time

Year	Approach: Spectrum is...	Major event
1822	...an abstract concept of no practical value	Concept of spectrum (Fourier)
1873		Concept of radio waves (Maxwell)
1888	...a physical object of no practical value	First experiments with radio waves (Hertz)
1895	...an inexhaustible natural resource from which everybody can profit freely	First experiments with wireless communications (Marconi, Popov)
1901		First transatlantic wireless transmission (Marconi)
1903	...a shared resource	First International Radiotelegraphic Conference in Berlin to coordinate the uses made of the spectrum
1910		First aviation radio
1921		First broadcasting network
1925	...a scarce resource that requires conservation	"...no more spectrum available..." declares a U.S. Secretary of Commerce
1927	...allocated to separate services	Creation of CCIR to study questions related to radio communications and of International Frequency Allocation Table that covers 10 kHz to 60 MHz
1932	...regulated by an intergovernmental telecommunication organization	Integration of radio, telegraph and telephone regulatory activities in the framework of the International Telecommunication Union (ITU)
1939		First commercial television
1947	...registered and controlled internationally	Creation of International Frequency Registration Board (IFRB) and International Frequency List. Spectrum use is to be coordinated worldwide.
1949		First public paging system
1949	...a "common heritage of mankind"	The ITU became the United Nations' specialized agency for telecommunications

(Continued)

TABLE 2-2 Changes in the Concept of Spectrum over Time (*Continued*)

Year	Approach: Spectrum is...	Major event
1957		First artificial Earth satellite
1958		First data transmission via satellite
1963		First World Space Radiocommunication Conference included the geostationary satellite orbit (GSO) into spectrum concept
1965		First commercial communication satellite
1978		First wireless public phone
1989	...a sellable commodity in some countries	Creation of tradeable rights in radio frequencies in New Zealand
1993		GSM (now Global System for Mobile communications) in commercial use
1995		Record spectrum auctions in United States: 258M USD/MHz
1995		Major ITU reform. The ITU membership open to nongovernmental entities. The IFRB replaced by part-time Radio Regulations Board (RRB).
>2000	Poised to be a disruptive force. Technology and standards dynamically share spectrum, using newer technologies: Low-power, wideband spread spectrum underlay Cognitive/agile/software-defined radios Mesh networks	Global information infrastructure (GII); universal mobile telecommunications systems (GMPCS, IMT-2000, UMTS); constellations of low-orbiting satellites (LEO); digital high-definition television (DTV, HDTV); digital audio broadcasting (DAB); orthogonal multiple carrier modulation (OMCM) systems.

Radio waves and satellite orbits represent a common heritage of humanity not to be operated by any one nation alone. Radio waves and satellite orbits are subject to misuse and pollution by man-made radio noise and interference that decrease its utility.

The remainder of this chapter discusses the interference, pollution, and the natural conflict issues surrounding the competing technologies/applications, users, the haves and the have-nots, and the managers and the users.

TABLE 2-3 Market Demand For the Spectrum Resource

Frequency	Wavelength	Interesting properties	Typical uses
10 kHz	30 km (20 mi)	Waves penetrate significant distance into water	Communication with submarines
100 kHz	3 km (2 mi)		Navigation
1000 kHz (1 MHz)	300 m (1000 ft)		AM broadcasting
10 MHz	30 m (100 ft)	Ionospheric reflection	CB radio, HF broadcasting
100 MHz	3 m (10 ft)		FM broadcasting TV broadcasting
1000 MHz (1 GHz)	30 cm (1 ft)		Cellular radio, top of UHF TV band
10 GHz	3 cm (1 in)	Higher ranges affected by intense rain	Satellite TV, point-to-point communications, radars

SOURCE: Hatfield, *Ibid.*

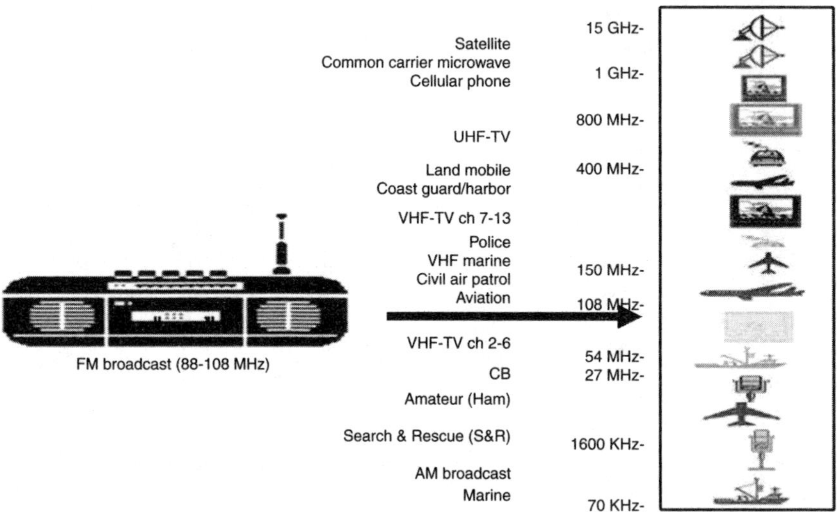

Figure 2-2 Uses of the spectrum resource at different frequencies. (*Source: Ibid.*)

2.2 Interference

Our commonsense understanding of radio spectrum capacity is based on everyday experience with radio technologies that have remained largely unchanged since the beginning of the last century. Radio and TV receivers handle one signal at a time, which must be of much higher amplitude than the noise floor and any other signal near the same frequency (refer to Fig. 2-3).

Interference is when there is a signal at or near the same frequency and with amplitude that approaches or exceeds the amplitude of the signal that a dumb receiver is attempting to receive. Figure 2-4 illustrates how two signals can overlap, giving rise to interference.

Interference does not happen just because two signals intersect. Rather, when a receiver fails to differentiate between the signal it is interested in and another irrelevant signal, we say there is interference.

The ITU Radio Regulations provide definitions for interference that play an important role in spectrum management. Table 2-4 lists these definitions with appropriate comments.

It is beyond the scope of this book to document all the concerns about interference. However, astronomers' worries about interference are generally indicative of the type of concerns about interference that various groups raise. The cosmic radio signals that astronomers try to detect from distant stars and galaxies are typically billions of times weaker[8] than the radio station signals picked up by home radios

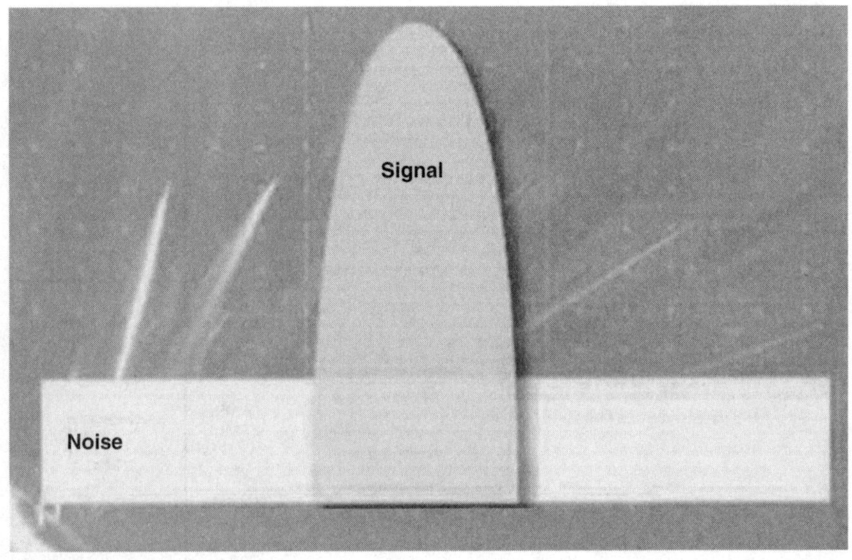

Figure 2-3 Legacy radio signal. (*Source:* Robert J. Berger, "Open Spectrum: A Path to Ubiquitous Connectivity," *The Wireless Revolution,* Vol. 1, No. 3, May 2003.)

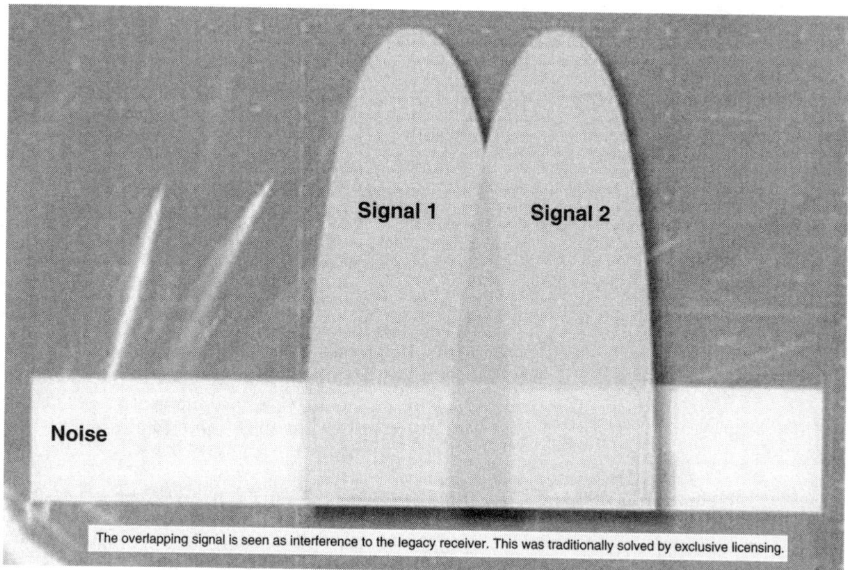

Figure 2-4 Overlap = Interference. (*Source: Ibid.*)

in the AM and FM bands.[9] Given the faintness of the cosmic radio signals and the great sensitivities of radio telescopes, radio astronomers must have some segments of the radio spectrum free from man-made transmissions.[10]

Technology advancement, particularly in telecommunications, is making it difficult for the astronomers to be free from an increasingly hostile environment characterized by radio frequency interference, or **RFI**. Today, we are witnessing the proliferation of mobile transmitters, cellular telephones, personal communications devices, and so on. These devices provide legitimate uses, but intensify the demand for spectrum and oftentimes interfere with one another. Also, airborne and satellite transmitters are growing in number, making the astronomers even more vulnerable.

2.2.1 What kinds of signals interfere with radio astronomy?

Radio frequencies, based on international agreement, are divided up into bands, designated for different types of uses. For example, AM radio stations are all within a certain range of frequencies that is different from the band of frequencies in which FM stations are located. Likewise, TV stations and police two-way radios have different frequencies, thereby preventing one type of station from interfering with stations of another type.

TABLE 2-4 Definitions and Criteria for Interference[7]

Topic	Definition
Interference	The effect of unwanted energy owing to one or a combination of emissions, radiations, or inductions upon reception in a radiocommunication system, or loss of information which could be extracted in the absence of such unwanted energy (ITU Radio Regulations, Article S1.166).
Harmful interference	Interference which endangers the functioning of a radionavigation service or of other safety services or seriously degrades, obstructs, or repeatedly interrupts a radiocommunication service operating in accordance with the ITU Radio Regulations (ITU Radio Regulations, Article S1.169).
Harmful Interference to the radio astronomy service	Interference levels are considered to be harmful to the Radio Astronomy Service when the rms fluctuations of the system noise increase at the receiver output by 10 percent owing to the presence of interference (ITU-R *Handbook on Radio Astronomy*). Note that a continuous process as well as an intermittent interference can be harmful.
Protection against interference	*Radio astronomy service:* For the purpose of resolving cases of harmful interference, the radio astronomy service shall be treated as a radiocommunication service. However, protection from services in other bands shall be afforded to the radio astronomy service only to the extent that such services are afforded protection from each other (ITU Radio Regulations, Article S4.6). *Space research (passive) service and earth exploration-satellite (passive) service:* For the purpose of resolving cases of harmful interference, the space research (passive) service and earth exploration-satellite (passive) service shall be afforded protection from different services in other bands only to the extent that such services are afforded protection from each other (ITU Radio Regulations, Article S4.7).

Radio astronomy has been allocated a number of frequency bands. The receiving equipment that the radio astronomers use is extremely sensitive and there is less need for *transmitting,* which is generally prohibited in radio astronomy bands. However, transmitters that use frequencies near those assigned to radio astronomy can cause interference to radio telescopes. This occurs when the transmitter's output is

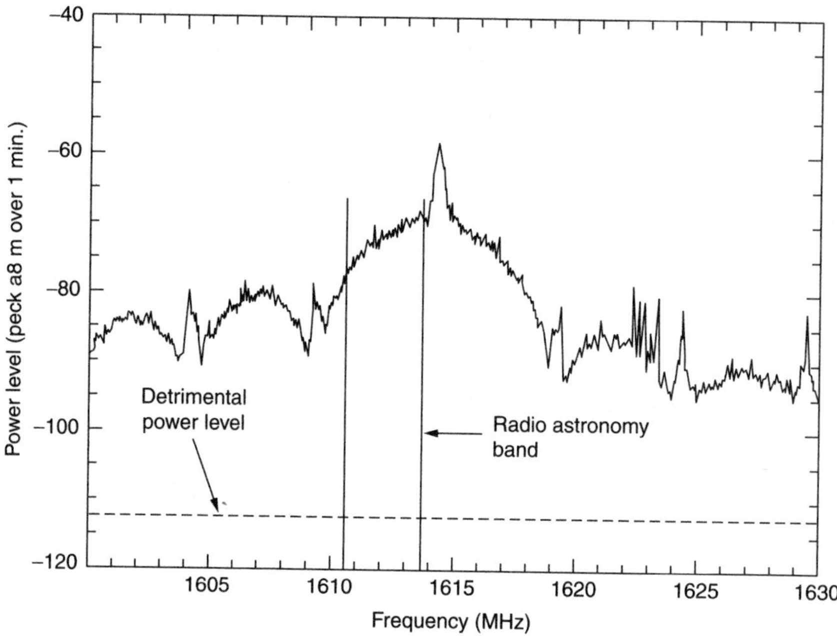

Figure 2-5 "Spillover" into a radio astronomy band by a too-broad transmitter. (*Source:* preserving the Common Heritage of All Humanity, *NRAO_http://www.aoc.nrao.edu/intro/rfi/rfimain.html.*)

unduly "broad" and spills over into the radio astronomy frequencies, or when the transmitter emits frequencies outside its intended range. Other interference may also arise because of radio transmitters unintentionally emitting signals at multiples of their intended frequency.

The graph in Fig. 2-5 illustrates interference to radio astronomy from a satellite transmitter with a very broad signal that causes it to spill into the band of frequencies allocated to radio astronomy by international agreement. The radio astronomy band is between the two vertical lines. The horizontal line is the power level that, according to the ITU, is detrimental to radio astronomy; signals above the horizontal line would exceed this detrimental level. Herein is the problem: the transmitter is spreading radio emissions over a very wide range of frequencies, far beyond its ITU-authorized range.

The use of radio for devices such as cellular telephones, wireless computer networks, garage door openers, and a whole host of other uses are rapidly increasing, as are the threats to radio astronomy from inadequately engineered transmitters. Transmitters in orbiting Earth satellites pose a major threat. Those transmitters are located overhead, precisely where radio astronomers must aim their telescopes to study the universe. To show the potential impact of such transmitters,

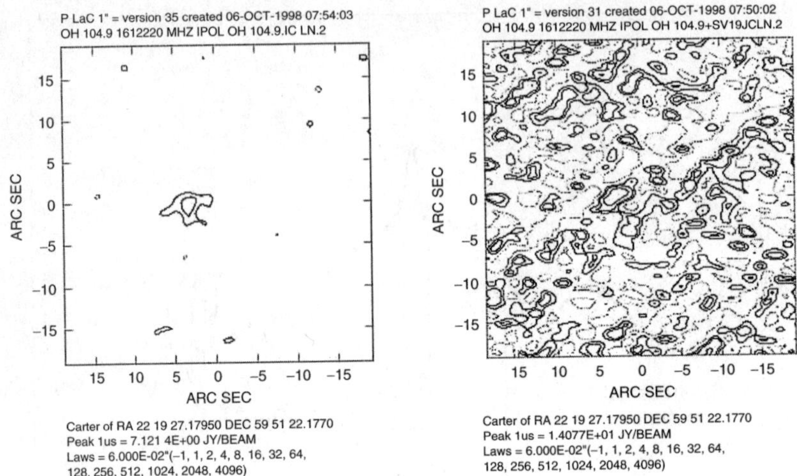

Figure 2-6 Effect of Radio Interference on Astronomical observations. (*Source: Preserving the Common Heritage of All Humanity, NRAO_http://www.aoc.nrao.edu/intro/rfi/rfimain.html.*)

Fig. 2-6 captures two images of OH/IR Star at 1612 MHz, one with no satellite present (left) and the second one with a satellite at 22° from the star (right).

In addition, many types of equipment not normally considered to be radio transmitters, particularly computers or systems incorporating microprocessors, emit undesirable radio signals.

Technology readily available can eliminate or drastically reduce the unwanted signals or interferences that pose problems for particular disciplines. Four basic approaches can minimize RFI effects:

- Eliminate the radiating source
- Shield either the source or the receiver
- Separate distance
- Improve circuit design.[11]

RFI will be an increasing concern in the future. Much is being done today to address this issue.[12]

2.3 Spectrum Pollution

In recent years, the radio spectrum has become a lucrative resource.[13] Extensive investments are being made in terrestrial use of radio and space applications to provide wireless communication services and information infrastructures.

For both terrestrial uses of radio and space applications, one persistent problem is spectrum pollution, as spectrum impurity degrades spectrum efficiency and reduces spectrum availability dramatically. Electrical or electronic appliances generate radio spectrum noise as a by-product of their desired function. Over time, these unwanted emissions add to natural background noise until services operating at very low desired signal levels are gradually swamped by noise.

Many international studies are being conducted, and standards are being developed to address unwanted emissions from some devices and other appliances' susceptibility to them. As these standards are developed, particular attention must be paid to active users of the radio spectrum, as well as the requirements of passive radiocommunication services.[14]

The remote sensing community uses both passive and active techniques for studies of the Earth's surface and the atmosphere:

- The ozone hole.
- The health of the rain forests.
- The distribution of natural resources.
- Pollution and global and local weather patterns.

On the other hand, scientific users, namely astronomers, use only "passive" systems in certain spectral bands to detect natural emissions. Although astronomers operate in allocated spectral bands set aside for their purposes, they are struggling to keep these spectral windows free from active users in adjacent or even distant bands.[15] Often, scientific spectrum users find themselves on the defensive against commercial users who have dramatically different operational standards and conceptions of spectrum efficiency. Some commercial operators have defective systems which complicate the matter. The defects could originate from:

a. Malfunction.

b. Ignorance in the design and construction phase.

c. Plain bad system design.

These defects generate harmful interference or some other negative impact on other systems.[16] Also, a complicating factor is that these owners and operators of defective systems do not pay adequate attention to system quality if, in their view, it is not commercially justified. The ITU-R Radio Regulations are inadequate in that they do not provide guidance to the administrations in managing such situations.

Chapter 5 details on how the radiocommunication sector of the ITU-R, part of the United Nations, regulates radio spectrum. At biannual ITU-R World Radiocommunication Conferences (WRCs), governments, commercial enterprises, and international organizations participate in discussions leading to the finalization of spectrum allocations. "Full Members" of the ITU are national administrations that make the final decisions.

In recent years, however, the intense participation of commercial sector members has had a dramatic impact on the ITU-R. Governments nowadays have less money to spend, thereby forcing the ITU to rely on financial contributions from commercial sector members. In return, commercial sector members are granted more direct participation in allocation processes, which may erode the objectivity of the ITU-R. Given the pressure of commercial entities, the ITU may allow them a larger part of the decision-making process in the future. That is a cause for concern for the passive services.[17]

The passive Radio Astronomy and Earth Exploration Services maintain different protection requirements than active services. Inter-service protection criteria within the ITU-R Radio Regulations are based on an "equal for all" principle, thereby making the provisions for protecting science services inadequate.[18]

The majority of spectrum users are commercial telecommunication entities. It costs money for the active spectrum services to protect the passive services. If these entities do not view something as commercially justified, they simply do not do anything,[19] which amounts to not achieving the "equal interference for all" criterion. However, protection of the more vulnerable passive services should amount to protecting the "active services" from each other. That being the case, the commercial entities should consider this concept as "the cost of doing business."[20]

Passive users are also concerned about the rapid changes in the space services as digital and spread spectrum techniques are being applied to provide seamless communication systems.

The new systems in the space services require uplinks, downlinks, feeder links and inter-satellite links to operate. These link transmissions represent the single-most important interference threat for other spectrum users. As for radio astronomy and remote sensing operations in adjacent or nearby bands, they could mean *disaster*. About 75 percent of the radio astronomy bands are already being threatened by adjacent or nearby space service bands. While some space transmission systems are specifically designed to protect radio astronomy bands, there are others which cause or will cause serious damage.[21] The following three examples[22] explain how things go wrong for the passive users:

1. GLONASS is a Russian global positioning system providing a service similar to the U.S. Global Positioning System (GPS). According to International Civil Aviation Organization (ICAO), both 24-satellite systems are useful as the Global Navigation Satellite System for aviation use. However, faulty system design of GLONASS has resulted in severe interference for passive users in the 1610.6- to 1613.8-MHz

spectrum band. After extensive discussions with the radio astronomers and several administrations, the GLONASS system is presently undergoing modification to eliminate unwanted emissions in the radio astronomy band and also to make spectrum available for Mobile Satellite System operators in the 1610- to 1626.5-MHz band. But this interference problem will remain until the year 2005.

2. IRIDIUM, originally envisioned as a 66-satellite constellation, was designed by Motorola to provide a seamless but expensive global telecommunication system. In 1992 when IRIDIUM was presented and spectrum was allocated, Motorola promised that radio astronomy operations in the 1610.6 to 1613.8 MHz band would be protected from the IRIDIUM downlink in an adjacent band. Later evidence suggested that the IRIDIUM could not possibly protect the radio astronomy spectrum (RAS) band except during 4 to 5 h of night-time low traffic periods. IRIDIUM carried no final stage filters to eliminate unwanted emissions into the neighboring RAS band.[23]

   There were further attempts at coordination between IRIDIUM and the radio astronomy community in various countries, but IRIDIUM could only admit that it would not interfere more than the present GLONASS system.[24] This state of events could not promote spectrum efficiency. Instead, the spectrum was polluted and IRIDIUM provided no clear technical solutions for protection of the RAS operations as demanded by the ITU-R Radio Regulations.[25]

3. ASTRA-D, operated from Luxembourg in a Fixed Satellite Service band of 10.7 to 11.7 GHz without permission from the ITU-R, was a direct TV broadcasting satellite. Its inadequate ground-based system testing resulted in a technical flaw that caused the broadcasting signals to spill over into the 10.6- to 10.7-GHz radio astronomy band. However, the system provided popular TV channels to a large audience; thus, nobody dared to turn off the satellite. Instead some administrations had made arrangements for radio astronomy to operate in a shifted band to avoid the unwanted signals of ASTRA-D.

The above three cases provide ample evidence of spectrum pollution, which will trouble all spectrum users sooner or later, because there are practical limits to increasing the number of users in the spectrum.[26] The ITU-R must come to the realization that "equal for all" does not protect the scientific users. Cohen *et al.* in their article, "What Should We Do about Radio Interference?", produce a point-counterpoint discussion.[27] They observe that radio astronomers search for "faint" radio static from cosmic objects. The signals that the astronomers are searching for often get lost in the "din" of terrestrial radio communications, particularly the requirements of cellular phones and other modern communications using satellites, raising concerns about the spectrum pollution.[28]

2.4 Natural Conflict

2.4.1 The haves and the have-nots

At a recent World Radio Conferences, more and more HF spectrum from the fixed service had been reallocated to broadcasting, aeronautical mobile, and maritime mobile services. From the developing countries' standpoint, HF is cost-effective for their domestic radiocommunications, such as for national broadcasting, mobile and fixed point-to-point radiocommunications. This creates a conflict over allocating the HF internationally: The developing countries want to retain it for their domestic radio communication needs, while the developed world wants to use the band for international broadcasting (Voice of America, Radio Moscow) and long-distance mobile communication.

2.4.2 Competing technologies/applications

In October 2000, President Clinton directed the Department of Commerce and the Federal Communications Commission (FCC) to develop a plan for selecting frequencies for use by third-generation (3G) wireless systems. He also tasked the two agencies to prepare a report on existing spectrum uses and the potential for reallocating or sharing frequency bands to clear space for wireless services. This situation exhibited how competing technologies and applications evoke natural conflicts in the spectrum policy matters.

President Clinton made the announcement at about the same time his Council of Economic Advisors released a white paper on the economic importance of 3G technology. Martin Baily, the chairman of the President's Council of Economic Advisors, said that the spectrum devoted to 3G services could lead to development of business and industries akin to those that were generated by the development of the computer chip and the Silicon Valley technology boom.

The council's report about the potential economic impact of 3G technology stated that even in year 2000 timeframe, the existing wireless systems were already producing "annual consumer benefit" in the range of \$53 to \$111 billion. The promised new frequencies were expected to produce similar benefits, the report added.

The United States had 189 MHz of bandwidth allocated to mobile services. Greg Rhode, the then assistant secretary of commerce for communications and information, mentioned in another press conference that an additional 160 MHz was needed to support new broadband wireless services.

Rhode acknowledged that there was a major challenge before the Commerce Department and the FCC because a key block of frequencies identified for potential 3G uses were extensively used by WorldCom

Inc. and Sprint Corp. to provide fixed wireless services that bypassed local phone companies for "last mile" telecommunications services to businesses and consumers. Public and Catholic school systems as well as colleges and universities also relied heavily on those frequencies to broadcast instructional TV programming.

The Department of Defense (DoD) heavily used another block of frequencies also targeted by the Clinton administration for potential use by 3G wireless services. That block was used for satellite operations and supporting field communications by deployed Army units.

The FCC in its report released on November 15, 2000, said 3G wireless services, targeted for the same piece of spectrum used by schools and others for high-speed Internet connections, would "raise technical and economic difficulties for the incumbents."

Patrick Gossman, the then chairman of the Detroit-based National ITFS Association and director of University Television at Wayne State University, also in Detroit, said it would be "criminal" to reallocate the spectrum and make the multichannel multipoint distribution service (MMDS) and instructional television fixed service (ITFS) operators move to another band. The FCC more than 40 years ago licensed the spectrum to the ITFS, which had 120 MHz of spectrum allocated nationwide, and MMDS, which occupied 66 MHz, to provide one-way television services. The two-way data services were added later. Commercial and educational operations were interleaved throughout the 2500- to 2690-MHz band.

Gossman added that ITFS/MMDS represented one technology that could easily bridge the digital divide: "We can go into a rural area, set up a single (antenna) and provide broadband service to a 35-mi radius. This is the only technology that can bring high-speed Internet to a number of people not served by other carriers." ITFS licenses covered all areas of the country "except for the middle of the desert," he said.

Gossman and others like Jack Keithley, vice president of federal regulatory affairs at Sprint, which had invested more than $1 billion in MMDS, voiced the concerns of the educational institutions and broadband fixed-wireless operators. They both disputed the efficacy of moving ITFS and MMDS operators to another portion of the spectrum. In Gossman's opinion, pushing ITFS and MMDS to higher bandwidths would cut down the service area because higher frequencies travel shorter distances than lower frequencies. It would also increase the costs of covering the same area.

Keithley believed that one solution for third-generation wireless services was for mobile service providers to develop broadband services by more efficiently using the spectrum they already owned.

While these debates were taking place, the FCC was promoting the use of ultrawideband wireless technology as potentially providing "enormous benefits for public safety, consumers, and businesses.[29] In May 2000, it adopted a proposal[30] to consider use of ultrawideband (UWB) technology on an unlicensed basis. But the commission said it would require further testing to ensure that UWB signals don't interfere with services such as the satellite-based Global Positioning System.[31]

Ultrawideband showed the potential to provide short-range, high-speed wireless data transmissions that could make access to Web pages over the air as fast as a wired connection would be. The technology would achieve that by spreading signals over a broad swath of the frequency spectrum, including the portion of the band used by GPS satellites.

Ralph Petroff, chairman of Time Domain Inc., a Huntsville, Alabama–based company that had championed UWB, said at the May 2000 NetWorld/Interop conference in Las Vegas that the new technology had the potential to deliver "megabits of information at microwatts" of power. Petroff argued that UWB could "ease the spectrum crunch" in the United States for wireless users, but not until the federal government finishes tests on the interference issue.[32]

The FCC, the Department of Transportation, and the National Telecommunications Information Administration conducted the UWB tests in this country.[33] Based on these test results, the Bush administration in conjunction with the FCC, approved the use of UWB technology, which enables broadband connections and assists in the performance of critical safety services. During 2002, the Department of Commerce worked closely with the FCC to authorize mechanisms to accommodate UWB wireless technology without causing serious impact to critical radio communications services.[34]

2.4.3 Competing users

The preceding discussion highlights the natural conflict that exists between competing users.[35] President Clinton articulated his concern about the United States losing its market share in "the industries of the 21st century" if it didn't exploit the potential of the 3G technology. His concern was immediately translated into an action plan for selecting frequencies that could be used by 3G wireless systems. He directed the Department of Commerce and the FCC to assess spectrum uses that existed and then identify the potential for reallocating or sharing frequency bands to clear space for wireless services.[36]

In addition, Clinton wanted the FCC to auction 3G broadband wireless licenses no later than September 30, 2002.[37] That was not an easy task for the Commerce Department or the FCC to manage.[38] WorldCom Inc. and Sprint Corp. used a key block of frequencies identified for potential 3G uses to provide fixed wireless services that bypassed local phone companies for "last mile" telecommunications services to businesses and consumers. Public and Catholic school systems as well as colleges and universities also relied heavily on those frequencies to broadcast instructional TV programming.[39]

Educational institutions, as referenced under the section on competing technologies/applications, were operating fixed wireless services, ITFS, in the 2-GHz band. ITFS backers inundated the FCC with comments contending that their use of those frequencies was essential to bridging the "digital divide" that limits the technology access of poor people.

In February 2001, Jim Leutze, Chancellor of the University of North Carolina at Wilmington, wrote to the commission that continued access "to superior ITFS frequencies is a crucial piece of how we will overcome the digital divide and bring new opportunities to children, workers and families who live in the rural south." He noted that shifting ITFS to new frequencies would derail a statewide network in North Carolina "that has been in the planning stages for more than seven years."[40]

A consortium of 28 educational groups—including the American Association of State Colleges and Universities, the National Education Association, and the National Association of Independent Schools—also argued against any reallocation of the spectrum used by the ITFS.[41]

The groups stated in their filing with the FCC that the need for additional spectrum for 3G services remained largely unproven. They recommended that the commission should examine the prevailing spectrum usage and needs of various providers before determining whether any additional spectrum was required.

The educational institutions also contacted Sprint Corp. and WorldCom Inc. to bolster their campaign to keep their spectrum positions. This alliance was supported by the CTIA, a Washington-based trade group. In its filing with the FCC, the CTIA noted that a substantial amount of spectrum would need to be allocated to ensure that carriers could offer advanced wireless services competitively. The trade group added that rapid growth of the Internet and mobile data services warranted additional allocations of spectrum for 3G networks.[42]

The fixed wireless carriers adamantly opposed reallocation of the 2.1 GHz and 2.5 to 2.7 GHz frequency bands that provided "last-mile" services to corporate and home users. Sprint noted that it had spent billions of dollars to acquire frequencies in markets across the U.S. to bypass local wireless carriers, and that its investment would be threatened by a reallocation of the spectrum.[43] Further, Sprint pointed out that the FCC "actively encouraged (Sprint) and other fixed wireless carriers to provide last-mile Internet and other wideband services over the 2-GHz bands." Any reallocation of those frequencies to 3G services would constitute "an arbitrary departure from established commission policy."[44]

Sprint also argued that "interference concerns" would make it impossible for fixed wireless and 3G mobile services to share the same spectrum and that forcing fixed wireless carriers to switch to another frequency band would require "relocation of many transmitters and customer receivers."[45]

The DoD heavily used another block of frequencies that the Clinton administration targeted for potential use by 3G wireless services. These frequencies were used for satellite operations and supporting field communications by deployed Army units. Linton Wells, Principal Deputy Assistant Secretary of Defense for Command, Control, Communications and Intelligence in the Clinton administration, reminded listeners at a press conference that these frequencies were used by equipment supporting U.S. forces engaged in peacekeeping missions in Kosovo and Bosnia.[46]

The FCC Chairman William Kennard recognized the needs of various user communities and said that competing commercial and government interests for the increasingly valuable airwaves needed to be resolved.[47]

Drawing on what Kennard said, Craig Mathias, an analyst at Far Point Group in Ashland, Massachusetts, argued that the FCC should think thoroughly before reallocating spectrum frequencies from either ITFS or fixed wireless services to 3G networks. On one hand, he noted, spectrum is "the stuff of life" for wireless carriers. The spectrum needed to support 3G mobile network services had to be reallocated because all usable frequency bands were already occupied. On the other hand, fixed wireless had proven to be a viable alternative to high-speed wired services. The battle "isn't about engineering," Mathias added. "It's about politics and lawyers."[48]

2.4.4 Managers and users

The preceding two sections provide a classic example of the spectrum users' debate in today's complex world. We reviewed concerns about the reallocation of spectrum frequencies to make room for the 3G mobile wireless services. The Cellular Telecommunications & Internet Association (CTIA), which sponsored the 2001 Wireless Conference, devoted significant energy to organizing and honing the users' views on spectrum policy. In advance of the conference, users comprising mobile and fixed wireless service providers and educational institutions had been engaging in a quiet, but intense, lobbying effort to influence the FCC.

In general, the U.S. government tries to meet demand for new radio services by allocating unused spectrum, improving the efficiency of currently used spectrum, and moving users to new bands to clear existing bands for new applications. These methods are considered on a case-by-case basis. The manager in these deliberations is the FCC, an independent regulatory agency (i.e., it is not part of the executive branch).

Its five Commissioners are appointed for five-year terms by the President, with the advice and consent of the Senate.[49] The agency has exclusive jurisdiction over nonfederal government spectrum management issues.[50] The FCC carries out its spectrum management responsibilities through procedures set forth in the Communications Act of 1934 and through more general statutes governing administrative procedures used by federal agencies. These procedures, commonly referred to as rulemaking proceedings, require the agency to notify the public of proposed actions; to allow opportunities for public comment; to provide reasoned, written decisions based on the public record; and to permit appeals of those decisions to the federal court system.

The Communications Act of 1934 authorizes the FCC to regulate nonfederal government use of the radio spectrum in the public interest but gives the president the authority to assign radio frequencies for federal government use. The president has designated the Secretary of Commerce to undertake this responsibility. Under the Secretary of Commerce, the Administrator of the National Telecommunications and Information Administration (NTIA) coordinates the federal government's use of its portion of the radio spectrum with the advice of the Interdepartmental Radio Advisory Committee (IRAC). The procedures used by the NTIA and IRAC are engineering intensive, with formal management coordination among affected agencies. Secrecy is often warranted, especially when national security considerations are on the coordination agenda. It is not the purpose of this section to review the NTIA's spectrum management approach;[51] rather, this section will focus primarily on the procedures the FCC uses in managing nonfederal government use of the spectrum by the FCC.

In a filing submitted to the FCC in February 2001, the CTIA aligned with mobile providers that planned to offer 3G services on frequencies already used by other companies and organizations.[52] As noted previously, fixed wireless carriers such as Sprint Corp. and WorldCom Inc. and educational institutions ranging from local school districts to universities came together to protect their existing spectrum positions. The actions of all these groups displayed the classical dimensions of the conflict between the spectrum manager and the users.

The spectrum needed to support 3G mobile network services had to be reallocated because all usable frequency bands were already occupied. The FCC, in coordination with the NTIA, also looked at a frequency band occupied by the U.S. Department of Defense. The Pentagon had not made any public statement about the possible reallocation of its spectrum. The conflict *among* users was unavoidable for the simple reason that the spectrum is "the stuff of life" for wireless carriers.

The FCC cannot make everyone equal. Instead, its responsibility is to maximize public interest from their vantage point. Redistribution of the bundle of spectrum rights is where Congress has limited the agency's discretion to act and where some of the most heated spectrum battles are waged.[53]

FCC Commissioner Kathleen Abernathy talked about the key battleground in the spectrum debate and addressed the issue of deciding who gets the rights.[54] She posed two questions:

1. What is FCC licensing?
2. What should be the FCC's goal?

To answer her first question, the Commissioner stated that FCC licensing was a way of government distributing a good and sanctioning its appropriate use.

As regards the second question, the Commissioner said that the FCC's management responsibility should be to maximize the efficiency of commercial spectrum use by promptly getting as many rights as possible into the marketplace, while protecting licensed uses from harmful interference.

She talked about two effective paradigms of rights distribution mechanisms:

1. Private property rights[55]
2. "Commons."

The Commissioner discussed the "commons" model by noting that it allows some goods to be enjoyed by all, as long as certain government-sanctioned norms are adhered to. She used the analogy of real estate to describe the idea of "commons." For example, land is largely distributed by a market-based private property mechanism, yet the use of the roads that connect various private lands is sanctioned as a common. As long as users obey certain government imposed norms (i.e., obeying speed limits, driving a safe vehicle, having reasonable eyesight, carrying insurance, and so on.), they can freely use the common.

In light of these opposing viewpoints of spectrum policy, Commissioner Abernathy asked: What is a manager to do? She makes two points. The first, which paves the way for the second, is that the distribution of rights to spectrum can be analyzed as a continuum from a full property rights model to a pure commons model. Secondly, users and the manager are well served by utilizing both the property-like rights approach and the commons model to fully maximize the value of spectrum, while minimizing conflicts.

From Commissioner Abernathy's remarks, it is clear that spectrum resource commands a thoughtful and deliberate approach to its management. The manager's task is to be creative in its policies and focus on maximizing spectral use and minimizing conflicts between the manager and the users and among users.

2.5 Endnotes

[1]Dedric Carter, Andrew Garcia, David Pearah (all from *Massachusetts Institute of Technology),* Stuart Buck, Donna Dutcher, Devendra Kumar, and Andres Rodriguez (all from *Harvard Law School*) define the term "spectrum" in terms of temporal and spatial opportunities to transmit information. They offer a multilane highway as a natural and useful analogy to the nature of spectrum. They view a single car (or, in some instances, a group of cars) traveling along the highway as a signal. The car can move along the highway following the empty spaces both within and between lanes. Each lane in the highway, to them, represents a different frequency, and they liken each car in a specific lane to be in a distinct temporal slot. Refer to "Spread Spectrum: Regulation in Light of Changing Technologies," paper prepared as part of the course requirements for MIT 6.805/STS085: Ethics and Law on the Electronic Frontier, and Harvard Law School: The Law of Cyberspace—Social Protocols, Fall 1998.

[2]*Ibid.*

[3]Chap. 3 of this book discusses the physical characteristics of the spectrum and its various segments and dimensions. No discussion of the dimensions of the spectrum is complete without reference to satellites/orbits. The author wishes to cover this topic more exhaustively in a future edition.

[4]Hatfield, *Ibid.*

[5]Struzak, R., "Access to Spectrum/Orbit Resources and Principles of Spectrum Management,"
http://www.ictp.trieste.it/~radionet/2000_school/lectures/struzak/AcceSpctrOrbICTP.htm

[6]As referenced in Struzak, *Ibid.* 5.

[7]As quoted in "CRAF Handbook for Frequency Management, Science and Spectrum Management," *European Science Foundation*, February 2002, pp. 58–59.

[8]The radio signals arriving on Earth from astronomical objects are millions (or billions) of times weaker than the signals used by communication systems. For example, a cellular telephone located on the moon would produce a signal on earth that radio astronomers consider quite strong.

[9]For definition of these bands, refer to Chap. 5.

[10]Because the cosmic radio sources are so weak, they are easily masked by man-made interference. Possibly even worse than complete masking, weaker interfering signals can contaminate the data collected by radio telescopes, potentially leading astronomers to erroneous interpretations.

[11]Mostia, William (Bill) L., Jr., "Understanding and Preventing Radio Frequency Interference," from the pages of Control Engineering, Reed Business Information, a division of Reed Elsevier Inc., 2004,
http://www.manufacturing.net/ctl/index.asp?layout=articlePrint&articleID=CA188299

[12]A good reference in this area is the book: *Noise Reduction Techniques in Electronic Systems* by Henry W. Ott, referenced in *Ibid.*

[13]Willem A. Baan, WRC-97 and the Passive Spectrum Users: Protecting the Radio Windows to the Universe, NAIC/AO Newsletter, November 1997,
http://www.naic.edu/ about/newslett/nov97/number23-7.html

[14]In chap. 4, the active and passive uses of spectrum are clearly defined. Active use refers to radio communication service in which both a man-made transmitter and receiver are used, while the passive use refers to "receive-only" radiocommunication service.

[15]Pankonin, V., "Protecting Radio Windows for Astronomy," *Sky & Telescope,* April 1981, p. 308.

[16]Roth, J., "Will the Sun Set on Radio Astronomy?" *Sky & Telescope,* April 1997, p. 40. Explains the "pollution" of the radio spectrum by human activities.

[17]*Ibid.*, 13.

[18]*Ibid.*

[19]*Ibid.*, 3.

[20]Likewise, the ITU_R should modernize and improve its protection standards of the passive services as it introduces and regulates the new space-based telecommunication industries. Also, refer to *Ibid.*, 15.

[21]*Ibid.*, 15.

[22]*Ibid.*, 13.

[23]An important point to understand is that a transmitter generates both "desired" and "unwanted" signals. The unwanted signals are not essential for the transfer of information. Immediately outside the bandwidth of the desired signal, there is the:

"Out-of-band emission" and beyond that,

"Spurious emission"

Spurious emission is defined as emission on a frequency or frequencies which are outside the necessary bandwidth (which, for a given class of emission, refers to the width of the frequency band just sufficient to ensure the transmission of information at the rate and with the quality required under specified conditions and the level of which may be reduced without affecting the corresponding transmission of information).

Spurious emissions include: harmonic (go to: http://www.atis.org/tg2k/ harmonic.html) emissions; parasitic emissions; intermodulation (go to: http://www.atis.org/tg2k/ intermodulation.html) products and frequency conversion products, but exclude out-of-band emissions.

With filters and good engineering practices, one can reduce the unwanted emissions; but, there are practicable limits, both economic and technical. In practice, one cannot always make a transmitter operate within some well-defined sharp spectrum edges.

[24]Feder, T. "Europe's Radio Astronomers Score in Spectrum Battle," *Physics Today*, October 1998, p. 75.

[25]Refer to the ITU-R Recommendations of SM series.

[26]Cohen, J. "Radio Pollution: The Invisible Threat to Radio Astronomy," *Astronomy & Geophysics*, December 1999, Vol. 40, issue 6, p. 8.

[27]Cohen, N. and Clegg, A. "What Should We Do about Radio Interference?" *Mercury*, September/October 1995, p. 10.

[28]Carpenter, S. "Lost Space: Rising Din Threatens Radio Astronomy," *Science News*, September 11, 1999, Vol. 156, p. 168. Also, refer to a short news article on a problem in India: "Cell Phones Threaten Radio Telescope," *Science*, Vol. 278, p. 1569 (28 November 1997).

[29]Brewin, Bob, "FCC Studying New Ultra-wideband Wireless Technology," *Computerworld*, May 15, 2000.

[30]Refer to New Public Safety Applications and Broadband Internet Access Among Uses Envisioned by FCC Consideration of Ultrawideband Technology, an unofficial announcement of Commission action, FCC News, May 10, 2000.

[31]For the final test results, refer to NTIA Report 01-384: Measurements to Determine Potential Interference to GPS Receivers from Ultrawideband Transmission Systems, http://its.bldrdoc.gov/pub/pubs.html.

[32]Those tests continue at NTIA's laboratory located at Boulder, CO. See for instance, the NTIA Report 01-389: Addendum to NTIA Report 01-384: Measurements to Determine Potential Interference to GPS Receivers from Ultrawideband Transmission Systems, http://its.bldrdoc.gov/pub/pubs.html or the NTIA Report TR-03-402: Measurements to Determine Potential Interference to Public Safety Radio Receivers from Ultrawideband Transmission Systems, http://its.bldrdoc.gov/pub/pubs.html

[33]For further updates on these tests, refer to Chap. 3, Section on UWB.

[34]Refer to Fact Sheet on Spectrum Management: Taking Action to Improve Spectrum Management, Presidential Action, News and Policies, June 2003, http://www.whitehouse.gov/news/releases/2003/06/20030605-5.html

[35]A high-stakes battle for radio-frequency spectrum that pits companies looking to build high-speed mobile networks against fixed wireless carriers and other spectrum occupants became a major topic of discussion at a wireless industry conference in Las Vegas. See Brewin, Bob., "Spectrum Battle to be in Spotlight at Wireless Conference," *Computerworld*, March 19, 2001.

[36]Brewin, Bob, "Clinton Puts 3G Wireless Plans on Fast Track," *Computerworld*, October 13, 2000.

[37]*Ibid.*

[38]An update: The Bush Administration has identified new spectrum for advanced third-generation (3G) wireless services and technologies for consumers. In July 2002, the Department of Commerce released a plan in concert with the FCC and the Department of Defense to make 90 MHz of spectrum available for 3G wireless services while accommodating critically important spectrum requirements for national security.

[39]Brewin, Bob, "3G Proposals Could Broaden Digital Divide: School, Broadband Wireless Operators Decry Spectrum Plan," *Computerworld*, November 27, 2000.

[40]Brewin, *Ibid.* 35.

[41]*Ibid.*

[42]*Ibid.*

[43]*Ibid.*

[44]*Ibid.*

[45]*Ibid.*

[46]Brewin, *Ibid.* 36.

[47]Refer to Telecommunications: History and Current Issues Related to Radio Spectrum Management, Testimony of Peter F. Guerrero before the Committee on Commerce, Science, and Transportation, US Senate, Tuesday, June 11, 2002.

[48]Brewin, *Ibid.* 35.

[49]Chapter 6 provides a more in-depth look at the FCC and its arrangements and limitations.

[50]The FCC does share certain regulatory functions with agencies of the individual states.

[51]Chapter 6 describes more fully the spectrum management procedures used by the NTIA and IRAC.

[52]Brewin, Bob, "Spectrum Battle to be in Spotlight at Wireless," *Computerworld*, March 19, 2001.

[53]Remarks of FCC Commissioner Kathleen Q. Abernathy, "My Vision of the Future of American Spectrum Policy," Before the CATO Institute's Sixth Annual Technology and Society Conference, Washington, DC, November 14, 2002, http://www.fcc.gov/Speeches/ Abernathy/2002/spkqa228.html

[54]*Ibid.*

[55]The property rights paradigm is best described in terms of real estate. Land distribution is done through market-based mechanisms, and then government steps in to ensure appropriate use of that land through zoning, building permits, and liability rules. The rules provide protection against owners that may otherwise be able to externalize costs to other, often adjacent, land owners.

Chapter

3

Physical Characteristics

3.1 Radiocommunications Theory

This chapter will discuss the characteristics and properties of radio waves and take a detailed look at the electromagnetic spectrum. Since the first radio transmissions between St. John's, Newfoundland, and Cornwall, England, radiocommunications have been continually improving. Many sophisticated systems are currently used to communicate with space vehicles millions of miles from earth and nuclear submarines cruising the ocean's depths. Further research into powerful radio transmissions to probe deep into the universe is also taking place with a view to establishing contact with alien life. What we are discovering is that radiocommunications are a unique and reliable form of communications that play a vital role in military, commercial, space, and other scientific advances.

3.2 Radio-Wave Creation and Propagation

We are all familiar with many methods of communication. These methods may be simple and direct or highly technical. For example, people engaged in conversation, either directly or by telephone, are using a common and simple means of exchanging ideas. Communications can, therefore, be defined as a means or system by which our thoughts, opinions, information, and intelligence are exchanged with others.

In the past 100 years or so, we have witnessed a remarkable evolutionary process which made possible the discovery of radio waves[1] and the invention of the numerous tubes, transistors, resistors, and other components that enable transmission and reception. The invention of radio cannot be attributed to any one person; neither can it be traced back to a specific date. However, there are certain individuals who should be noted for making important initial contributions to its development.

3.2.1 1865 to the present

The invention of the telegraph and telephone radically changed all previous systems of communication. The telegraph and telephone used electrical devices for both the sender and the receiver and a wire or cable as the medium for the transmission, thereby making it possible to communicate between any two points on the face of the earth that could be bridged by a cable or wire.

The development of the *wireless* or *radiocommunications* was the next important milestone in the process of message transmission. By using the air as a transmission medium rather than a wire or cable, the wireless proved to be superior to the telegraph and telephone.

In 1865, James Clerk Maxwell, a Scottish physicist, made a scientific observation that any electrical or magnetic disturbance created in free space could be propagated (transmitted) through space as an electromagnetic wave. He also observed that this electromagnetic wave was one in which the disturbance would be at right angles to the direction of travel, and the speed of such waves would be approximately 186,000 mi/s, that is, the speed of light. Maxwell suggested that such waves could be created by setting up electrical vibrations in a wire capable of conducting electricity. These observations were revolutionary and have proven to be correct.

For instance, Heinrich Hertz, a German scientist, performed a number of experiments during the latter part of the nineteenth century. He confirmed that electromagnetic waves could in fact be produced while verifying and validating that these waves were invisible and moved at the speed of light. Hertz performed more experiments to show that light waves are only a few thousandths of an inch long, whereas electromagnetic waves vary in length from millimeters to thousands of miles long. There were other scientists from many parts of the world, who performed simultaneous experiments with the propagation of electromagnetic waves. Figure 3-1 highlights the work of one such scientist, Sir J. C. Bose.

The cumulative work of all these scientists eventually contributed to the development of radio. By 1895, Guglielmo Marconi, an Italian scientist and inventor, developed a working radio telegraph system. In 1901, he succeeded in transmitting the letter S (three dots in Morse code…) across the Atlantic Ocean, thereby starting transoceanic radiocommunications.

Radiocommunication depends on a number of basic characteristics: frequency, amplitude, and wavelength. Frequency is defined as the number of cycles[2] that a radio wave completes in 1 s. It is the most common description of a radiocommunication signal. Hertz (Hz) represents 1 cycle per second[3] and is the international unit of frequency measurement. Amplitude refers to a measure of the value of a radio wave, measured in volts. Refer to Fig. 3-2.

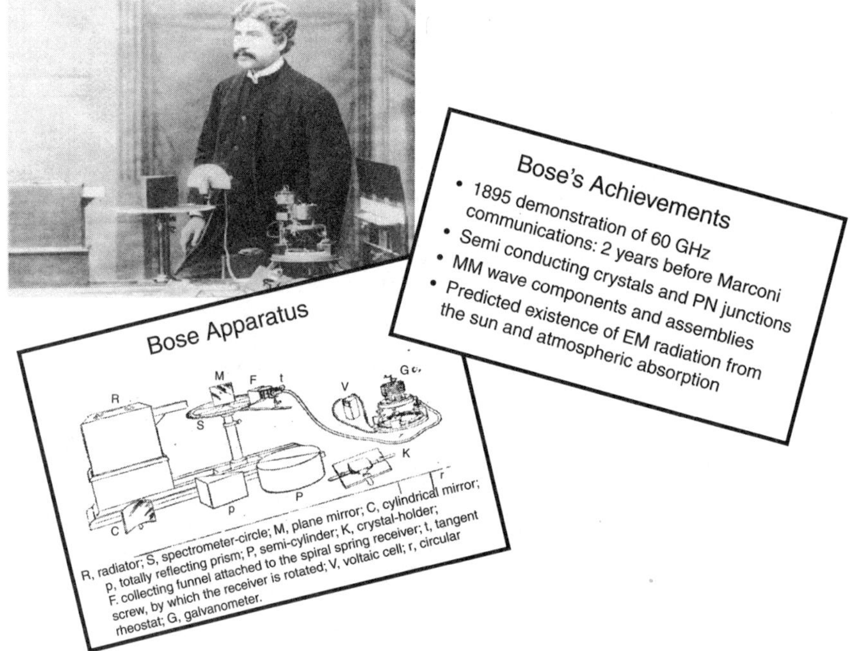

Figure 3-1 Bose's achievements. (*Source:* Wayne Pleasant, Telaxis Communications, IEEE T802.16-02/02a, *http://www.ieee802.org/802_tutorials/march02/T80216-02_02a.pdf.*)

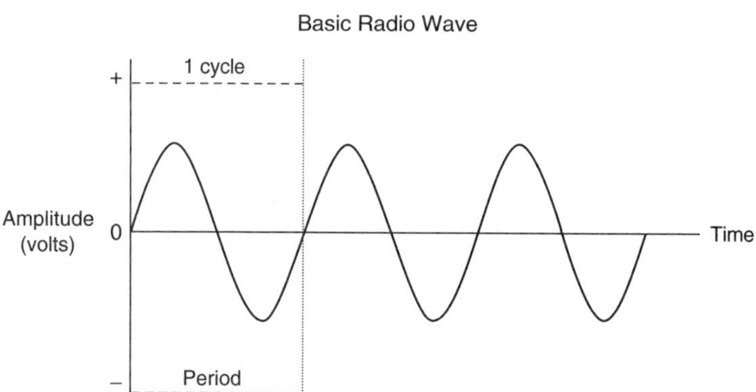

Figure 3-2 Cycle of a pure radio wave. (*Source:* Office of Technology Assessment, based on Harry Mileaf (ed.), *Electronics One,* revised 2d ed. (Rochelle Park, NJ: Hayden Book Company, 1976) p. 1–10 as referenced in *The 1992 World Administrative Radio Conference: Issues for U.S. International Spectrum Policy,* November 1991, OTA-BP-TCT-76, NTIS order. #PB92-157601.)

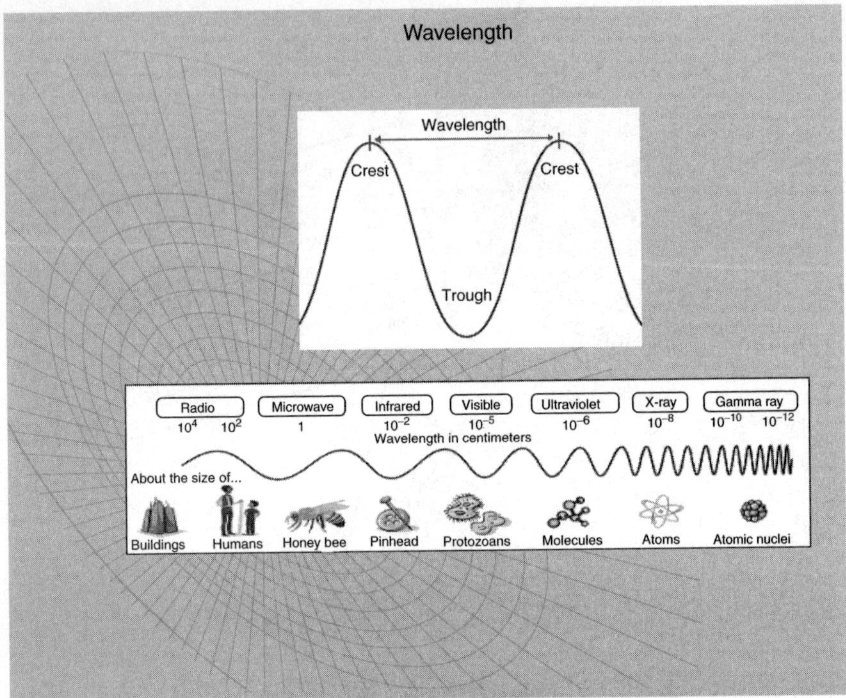

Figure 3-3 Frequency of a continuous wave. (*Source: http://imagers.gsfc.nasa.gov/ems/waves3.html*)

Wavelength refers to the distance between successive peaks of a continuous radio wave, as illustrated in Fig. 3-3. This distance is expressed in feet or meters.[4]

The radio spectrum is divided into several *bands* that are identified by their frequencies or wavelengths, or by several types of descriptive names attached to various portions of the spectrum. For instance, one method refers to relative position in the spectrum: very low frequency (VLF), high frequency (HF), very high frequency (VHF), superhigh frequency (SHF), and so on. Figure 3-4 shows the frequency ranges.[5] Each range of frequencies behaves differently and performs different functions, as indicated in Table 3-1.[6]

The letter designations, such as L-band, S-band, and K-band, refer to another method developed in World War II to maintain secrecy about the actual frequencies employed by radar and other electronic devices. These letter designations do not provide precise measures of frequency because the different segments of the electronics and telecommunications industries define the band limits differently. The International Telecommunication Union (ITU) classifies frequencies by

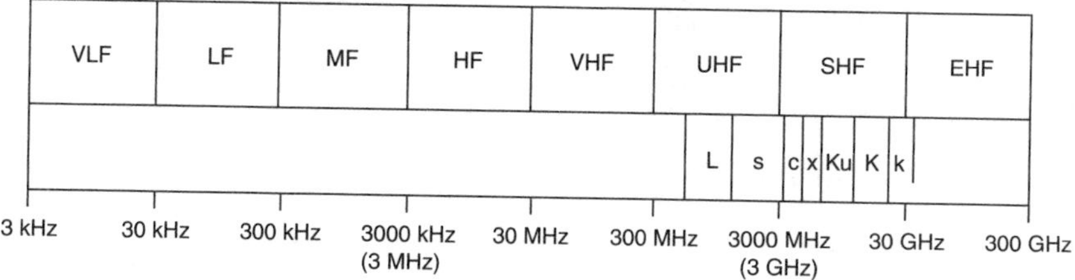

Figure 3-4 Frequency band designations. (*Source:* Office of Technology Assessment, 1991, based on Richard G. Gould, "Allocation of the Radio Frequency Spectrum," OTA contractor report, Aug. 10, 1990, as referenced in *The 1992 World Administractive Radio Conference: Issues for U.S. International Spectrum Policy*, November 1991, OTA-BP-TCT-76.)

TABLE 3-1 Radio Frequency Bands and Uses

Name	Frequency range	Examples of services
Very low frequency (VLF)	3–30 kHz	Marine navigation
Low frequency (LF)	30–300 kHz	Marine and aeronautical navigation equipment
Medium frequency (MF)	300–3,000 kHz	AM radio broadcast, LORAN maritime navigation, long-distance aeronautical and maritime navigation
High frequency (HF)	3–30 MHz	Shortwave broadcast, amateur radio, CB radio
Very high frequency (VHF)	30–300 MHz	Private radio land mobile services such as police, fire, and taxi dispatch; TV channels (2 through 13); FM broadcasting; cordless phones; baby monitors
Ultrahigh frequency (UHF)	300–3,000 MHz	UHF TV channels; cellular phones; common carrier point-to-point microwave transmission used by long-distance phone companies; satellite mobile services
Superhigh frequency (SHF)	3–30 GHz	Radar, point-to-point microwave, and satellite communication
Extremely high frequency (EHF)	>30 GHz	Satellite communications and space research.

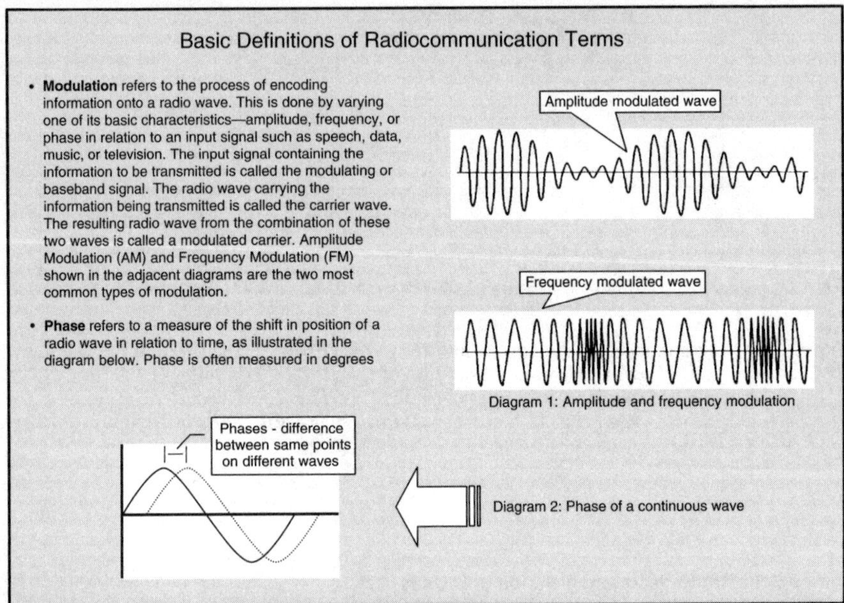

Figure 3-5 Basic definitions of radiocommunication terms. (*Sources:* Harry Mileaf (ed.), *"Electronics One,"* revised 2nd ed., Rochelle Park, NJ: Hayden Book Company, 1976; U.S. Congress Office of Technology Assessment, *"The Big Picture: HDTV & High Resolution Systems,"* OTA-BP-CIT-64, Washington, DC: U.S. Governmrnt Printing Office, 1990; William Stallings, *"Data and Computer Communications,"* New York, NY: McMillan Publishing, 1985, as referenced in *The 1992 World Administrative Radio Conference: Issue for U.S. International Spectrum Policy*, November 1991, OTA-BP-TCT-76.)

yet another method: Band numbers, such as Band 1 and Band 2. We are also very familiar with two frequency bands by the services which use them: amplitude modulation (AM) and frequency modulation (FM) radio broadcast bands. AM and FM refer to two common formats for analog transmission in which the message or information to be transmitted is impressed onto (modulates) a radio carrier wave,[7] causing some property of the carrier—the amplitude, frequency or phase (see Fig. 3-5)—to vary in proportion to the information being sent.

3.2.2 Various segments

The physical properties of radio waves and their transmission characteristics discussed in the preceding sections determine where particular radio frequencies are better suited to certain kinds of communications services. Table 3-2 shows the spectrum, encompassing the bands designated as:

- Extremely low frequency
- Super low frequency

TABLE 3-2 International Band Designators

Designation	Description	Frequency range
ELF	Extremely low frequency	3–30 Hz
SLF	Super low frequency	30–300 Hz
ULF	Ultra low frequency	300–3000 Hz
VLF	Very low frequency	3–30 kHz
LF	Low frequency	30–300 kHz
MF	Medium frequency	300–3000 kHz
HF	High frequency	3–30 MHz
VHF	Very high frequency	30–300 MHz
UHF	Ultrahigh frequency	300–3000 MHz
SHF	Super high frequency	3–30 GHz
EHF	Extremely high frequency	30–300 GHz

- Ultra low frequency
- Very low frequency
- Low frequency
- Medium frequency
- High frequency
- Very high frequency
- Ultra high frequency
- Super high frequency
- Extremely high frequency

In what follows, these frequency bands, some of their uses and the factors affecting transmission of radio signals in them are discussed. The technical content is thorough for introductory purposes; however, for those looking for a much more advanced treatment of the frequency bands, the author recommends a companion book.

3.2.3 Extremely low frequency

The extremely low frequency (ELF) range is from 3 to 30 Hz, capable of transmitting signals 5000 mi or more. ELF propagates through the earth's substrate and produces high-power sounds that can penetrate ocean depths to several hundred feet. ELF communications systems require enormous transmit antennas, called *transducers*, which cover thousands of acres, operating at very high transmitting powers—in the 100-MW range. These "antennas" transfer the transmitted RF energy to the earth and vice versa. ELF covers greater distance range than that of any other terrestrial communications system, without being affected greatly by atmospheric disturbances.

Extremely low frequency operates in the range of audible sound, but it is capable of only very low transmission rates. This slow data rate makes ELF transmissions impractical for normal character message transmission and impossible to use with current communications security (COMSEC) devices. The primary use of this area of the frequency spectrum is, therefore, underwater communications, mainly to communicate with submerged submarines in message formats that are only one or two characters in length and are transmitted by interrupted continuous wave (ICW).

3.2.4 Super low frequency

There are dozens of articles with wrong frequency designations. The International Telecommunications Union Radio Regulations provide frequency range designations that are accepted as the frequency band standard shown in Fig. 3-6. The super low frequency band is 30 to 300 Hz.

3.2.5 Ultra low frequency[8]

Electromagnetic waves observed with incredibly long wavelengths are classified as ultra low frequency (ULF) waves. The frequency range of ULF is from 300 MHz to 3000 Hz, that is, from the lowest the magnetospheric[9] cavity can support up to the various ion gyrofrequencies.[10] Pulsation frequency is considered to be ultra low when it is lower than

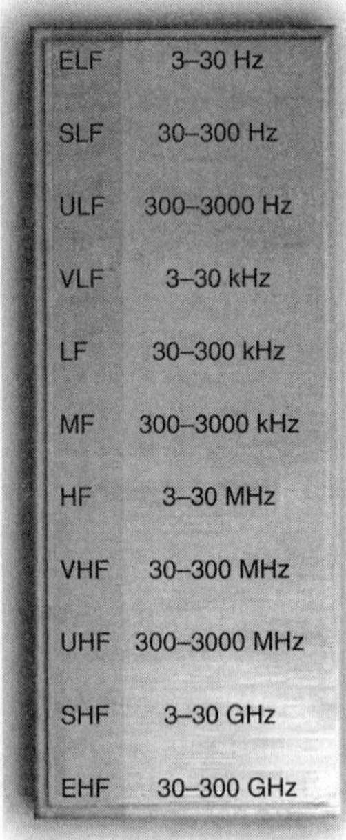

ELF 3–30 Hz
SLF 30–300 Hz
ULF 300–3000 Hz
VLF 3–30 kHz
LF 30–300 kHz
MF 300–3000 kHz
HF 3–30 MHz
VHF 30–300 MHz
UHF 300–3000 MHz
SHF 3–30 GHz
EHF 30–300 GHz

Figure 3-6 ITU radio regulations frequency range designations.

the natural frequencies of the plasma,[11] like plasma frequency and the ion gyrofrequency. Balfour Stewart first observed geomagnetic pulsations in 1859 and published his findings in 1861.[12]

A typical classification scheme for the ULF waves is based on the period of the pulsation.[13]

Pulsation Classes

	Continuous pulsations					Irregular pulsations	
	Pc 1	Pc 2	Pc 3	Pc 4	Pc 5	Pi 1	Pi 2
T (s)	0.2–5	5–10	10–45	45–150	150–600	1–40	40–150
F	0.2–5 Hz	0.1–0.2 Hz	22–100 MHz	7–22 MHz	2–7 MHz	0.025–1 Hz	2–25 MHz

People are interested in the sounds produced by ULF waves. For example, ultra low frequency waves are now widely used in underwater acoustic to get information on propagation parameters (acoustic and elastic).[14] Because ULF waves are almost nonaffected by absorption during their underwater propagation, they can be used to make long-range detection.[15] There is also research underway to determine what affect ULF may have on people's health[16] and on women's menstrual cycles.

3.2.6 Very low frequency, low frequency, and medium frequency

Table 3-1 groups very low frequency (VLF), low frequency (LF), and medium frequency (MF) that fall in the 3-to 3000- kHz range together. In this portion of the spectrum, radio signals are transmitted in the form of *ground-waves* that travel along the surface of the Earth, following its curvature, as depicted in Figs. 3-7, 3-8, and 3-9.

Much of the energy of the ground-waves is lost to the Earth, as they travel along its surface. Ground-waves travel greater distance over water than over land, but high power is needed for long-distance communication throughout this portion of the spectrum.

1. *Ground-waves* (surface waves) travel close to the earth. When these are transmitted over the earth, they take either a direct path, a surface path, or a ground-reflected path to the receiver. Refer to Fig. 3-8.

   If the conductivity which is a measure of the ability of the earth to conduct electric current or the efficiency with which a current is passed is good, the surface path may be more useful for communications from one ground station to another when lower frequencies are used. Type of soil and water in the propagation path (i.e., a radio wave traveling from one point to another) determine the earth's conductivity. Soil with poor conductivity quickly attenuates (weakens) radio signals. If a ground wave was transmitted over sea water, the

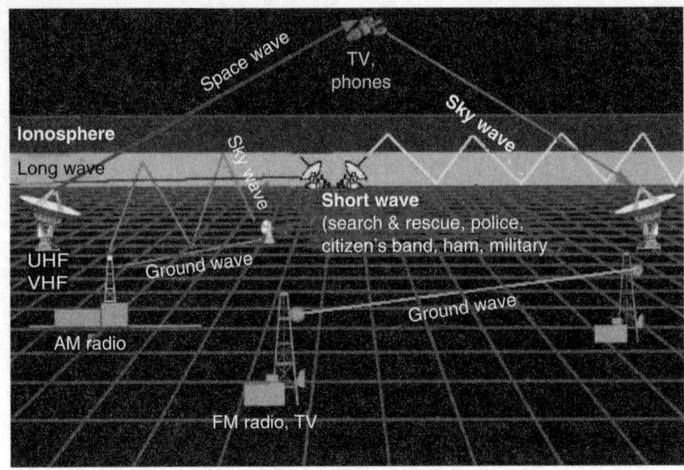

Figure 3-7 Radio wave propagation path. (*Source: Windows to the Universe, http://www.windows.ucar.edu/*, University Corporation for Atmospheric Research (UCAR). © 1995–1999, 2000 The Regents of the University of Michigan; © 2000–03 University Corporation for Atmospheric Research.)

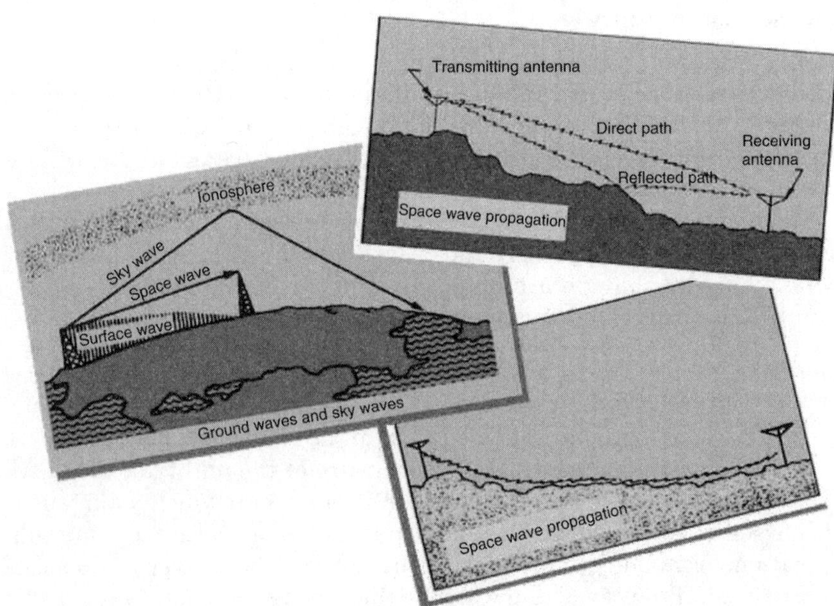

Figure 3-8 Radio wave transmission. (*Source: http://www.infodotinc.com/neets/book10/40c.htm.*)

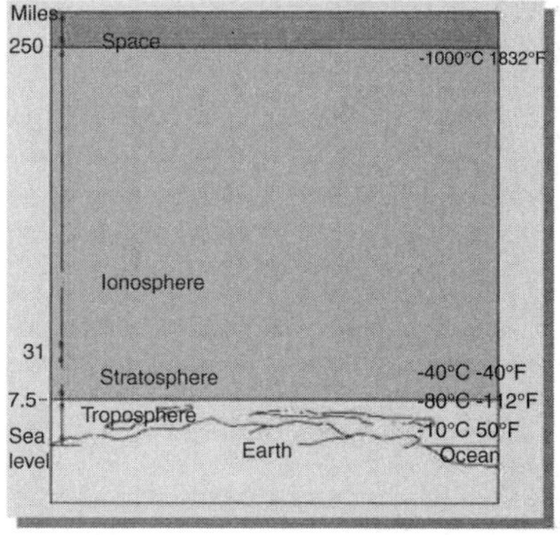

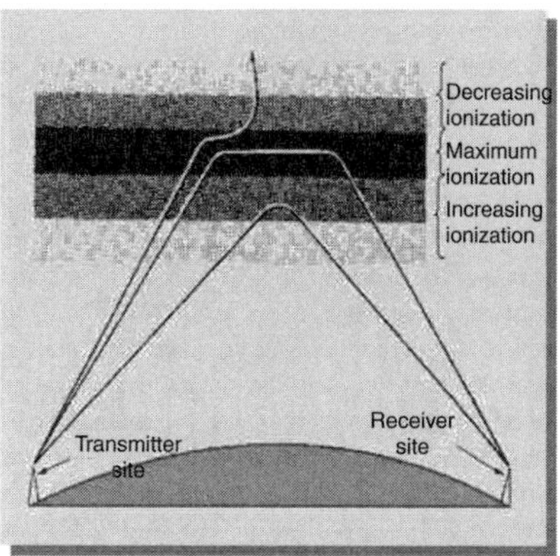

Figure 3-9 Ionospheric density and radio waves. (*Source: http://www.infodotinc.com/neets/book10/40c.htm.*)

direct path would only travel the short line-of-sight distance, but the surface path might travel up to 700 mi. Table 3-3 lists the relative conductivity of different surface conditions.

2. *Sky waves* travel upward and are redirected by atmospheric properties back to the earth. High above the earth these radio waves meet the ionosphere, consisting of layers of gases ionized by the ultraviolet rays of the sun. Passing through these ionized layers, radio waves are bent from their original course. Sky-wave communications become possible when the bending of the waves is great enough to return them to earth, as shown in Fig. 3-9. Sky-wave transmissions are very effective for long-distance communications in the high-frequency range (3 to 30 MHz).
3. *Direct waves* travel through the air in a straight line (line-of-sight or LOS) from the transmitting antenna to the receiving antenna and continue to travel in a straight line until they are interrupted by an object, as shown in Fig. 3-10, or weaken over a great distance.

TABLE 3-3 Surface Conductivity

Surface	Relative conductivity
Sea water	Good
Flat, loamy soil	Fair
Large bodies of fresh water	Fair
Rocky terrain	Poor
Desert	Poor
Jungle	Unusable

SOURCE: *http://www.infodotinc.com/neets/book10/40c.htm.*

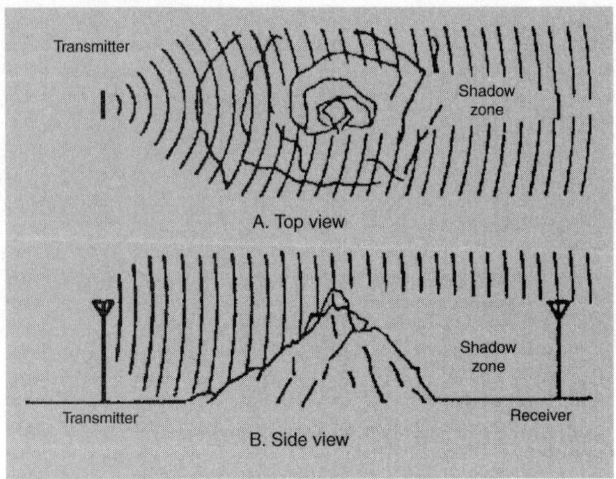

Figure 3-10 Interruption of direct waves. (*Source: Ibid.*)

The average distance of direct-wave communications is limited by the height of the transmit or receive antenna. At frequencies greater than 30 MHz (VHF and above) with antennas at ground level, a direct wave is normally limited to under 20 mi. This is due to the curvature of the earth, as depicted in Fig. 3-11.

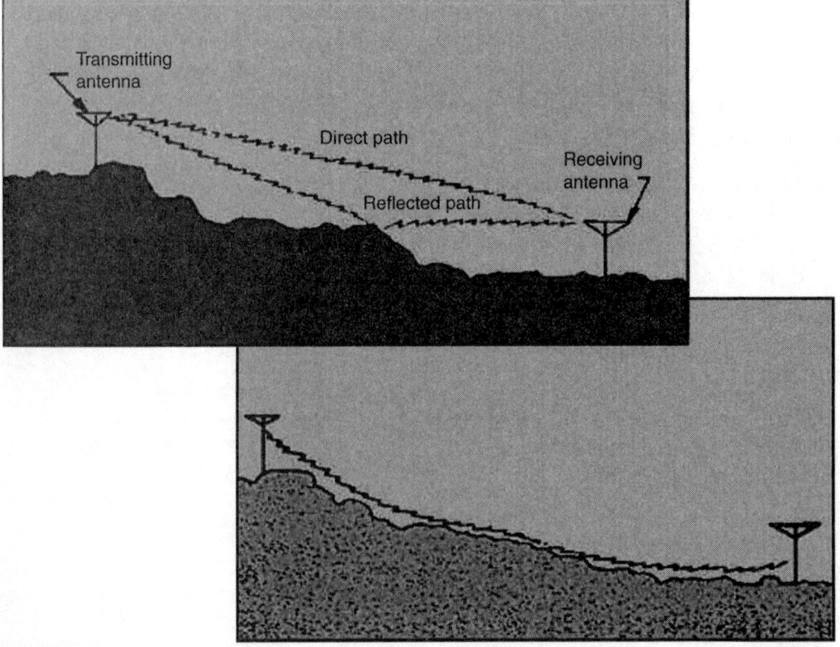

Figure 3-11 Direct waves. (*Source: Ibid.*)

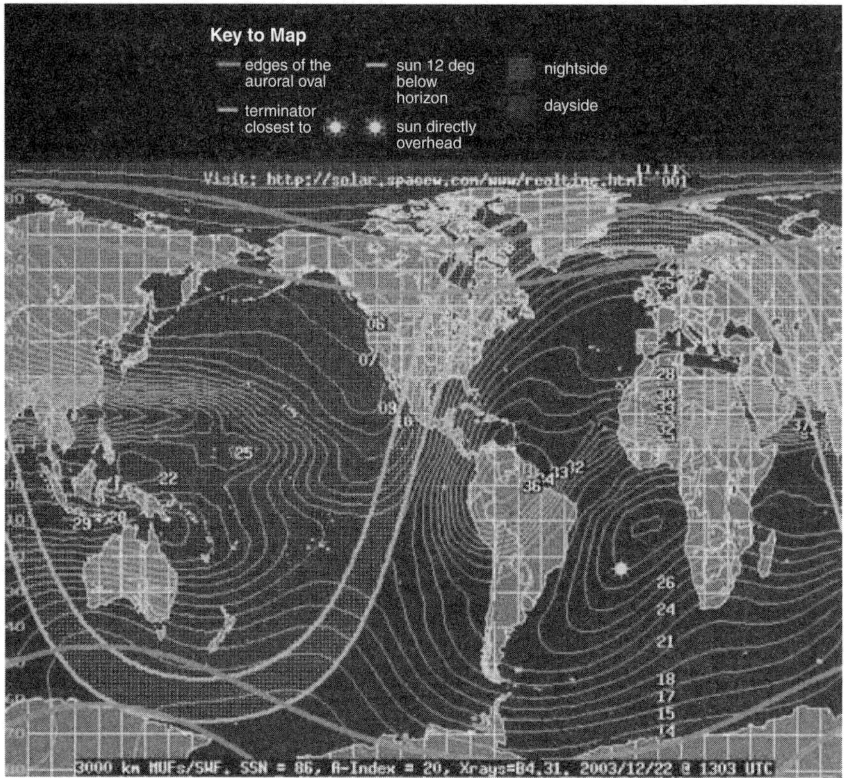

Source: Solar Terrestrial Dispatch. Contact C Oler@Solar.Stanford.Edu.

Figure 3-12 Contour map of the maximum useable radio frequency. (*Source: Windows to the Universe, http://www.windows.ucar.edu/*, University Corporation for Atmospheric Research (UCAR). © 1995–1999, 2000 The Regents of the University of Michigan; © 2000–03 University Corporation for Atmospheric Research. All Rights Reserved.)

If and when the height of either antenna is increased, the distance between the antennas can also be increased. Direct waves become usable for long-range UHF or SHF satellite communications or VHF/UHF communications with aircraft by eliminating obstructions.

Figure 3-12 is a global contour map, showing:

1. The highest frequency that will reflect from the Earth's ionization layer for a 3000-km path length. Read the maximum usable frequency (MUF) value at the halfway point of the radio path (1500 km). Higher frequencies will punch through the ionosphere and travel into space, as depicted in Fig. 3-13.
2. The current location of the auroral oval. Signals traveling through the auroral oval will most probably be degraded.

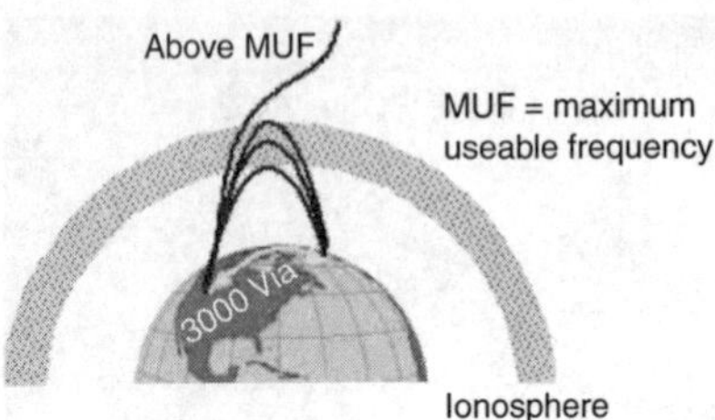

Figure 3-13 Frequencies higher than the maximum useable frequency. (*Source: http://www.infodotinc.com/neets/book10/40e.htm.*)

3. The location of the sunrise/sunset terminator and the regions where the sun is less than 12° below the horizon (called the gray-line corridor). In the gray-line corridor, the lowest altitude ions (that degrade the signal) are rapidly lost as the sun sets, but the high altitude ions (that reflect the signal) are still plentiful. These are particularly good conditions for short-wave radio signal propagation.

At the lower end of this portion of the spectrum, transmissions are better suited for low data rate communications with submarines and for navigation. The maritime mobile service has been allocated this band for communication with ships at sea. Table 3-4 lists the appropriate range and conditions under which VLF, LF, and MF become usable.

3.2.7 High frequency

In the HF range, propagation of a sky-wave supplements the ground-wave, as depicted in Fig. 3-14.

The ground-wave dies out at about 100 mi; however, the sky-wave can be bent back to Earth from layers of ionized particles in the atmosphere (the ionosphere), as illustrated in Fig. 3-15. As the signal returns to Earth, it may be reflected again just like the previous time.

The signal can make several *bounces* while traveling around the Earth, making long-distance communication possible.

There are a few factors associated with these bounces: The bending or refraction of a radio wave is caused by an abrupt change in the velocity of the upper part of a radio wave as it enters a new medium. The amount of refraction that occurs depends on the:

1. Density of ionization of the layer.[20] Refer to Fig. 3-16 and the following description on Simplified Model of Propagation in the Ionosphere.

TABLE 3-4 Very Low, Low, and Medium Frequencies, 3–3000 kHz

Frequency range	Use
	Designation: Very Low Frequency
3–30 kHz	Spans 5000 mi or more and penetrates vegetation and water. Used mainly for navigation and to communicate by low-speed secure teletypewriter[17] with submarines at sea when they are submerged at shallow depths (about 10 ft). Very Low Frequency transmitters are normally shore-based, but certain command and control (C2) aircraft such as airborne command posts may have a VLF capability, using long trailing wire antennas and transmitters powered in the 100 to 200 kW range.
	Designation: Low Frequency
30–300 kHz	Spans distances of 1000 to 5000 mi. Used for medium-distance communications, particularly with submarines and surface ships at sea, and for navigation.[18] Also, used for conducting efficient airborne operations. Penetrates vegetation and water; however, it is less effective than ELF or VLF.
	Designation: Medium Frequency
300–3000 kHz	Propagates by groundwave, skywave, and direct wave. Spans from 100 to 1000 mi by groundwave and from 1000 to 3000 mi by skywave, depending on the transmitter output power and the atmospheric conditions. Main uses include medium-distance communications, radio navigation, and amplitude modulation (AM) broadcasting.[19]

2. Frequency of the radio wave.[21] Refer to Fig. 3-16 and the descriptive material included on simplified model of propagation in the atmosphere.
3. Angle at which the wave enters the layer.[22] Refer to Fig. 3-16 and descriptive material included on simplified model of propagation in the atmosphere.

The physics of ionospheric radio wave propagation is a vast and complex topic. A basic understanding of ionosphere properties is paramount to understanding HF radio propagation over distance. The insert provides a simplified overview of important ionosphere mechanics used in propagation research and discusses limitations caused by the complex dynamics of the ionosphere.[23] The conclusions are based on observations from recent HF radio testing that used the PTC-II[24] HF digital controller to derive HF radio skip distance over a range of HF frequencies and validate the predictions made by the equation of motion discussed under the simplified model of propagation in the ionosphere.

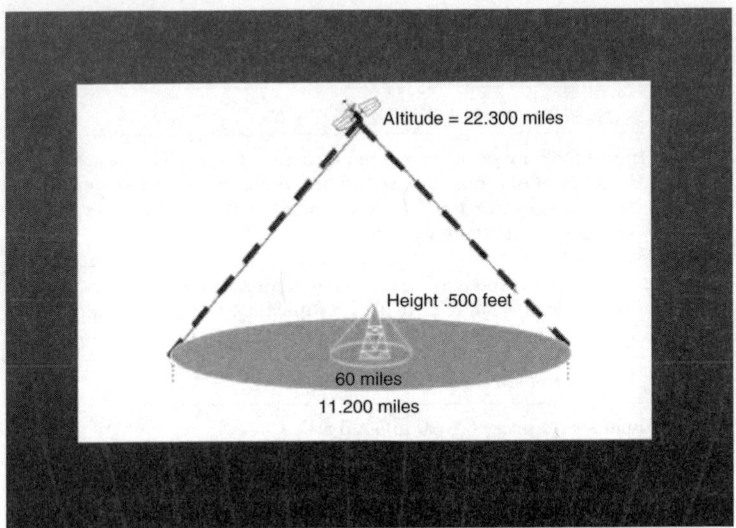

Figure 3-14 Terrestrial and satellite transmission ranges. (*Source:* Office of Technology Assessment, 1991, based on Richard G. Gould, *"Allocation of the Radio Frequency Spectrum,"* OTA Contractor Report, August 10, 1990.)

Supporting data for these conclusions are included. In the simplified model of propagation in the ionosphere presented below, Gaussian units are utilized and all vectors are denoted in *bold*.

The simplest approach to describe radio wave propagation is to explain the index of refraction given by $\eta = (\mu\varepsilon)^{1/2}$, where μ = magnetic permeability ($1.25664 \times 10^{-6}\ Hm^{-1}$) and ε = dielectric constant, the ratio of the strength of an electric field in a vacuum to that in the ionosphere.

The index expresses the relationship between the angles of incidence and refraction through Snell's law:

$$\frac{\mathrm{Sin}\,i}{\mathrm{Sin}\,r} = \frac{\eta'}{\eta}$$

This is shown graphically in Fig. 3-17, which shows an incident wave **k** striking a plane interface between different media, giving rise to a reflected wave **k″** and a refracted wave **k′**.

Since magnetic permeability is constant in the ionosphere, the goal is to work out the dielectric constant from the equation of motion.

Let us consider the simple problem of a tenuous electron plasma of uniform density trapped in a strong, static, and uniform magnetic induction $\mathbf{B_o}$. If we assume that the transverse radio waves propagate

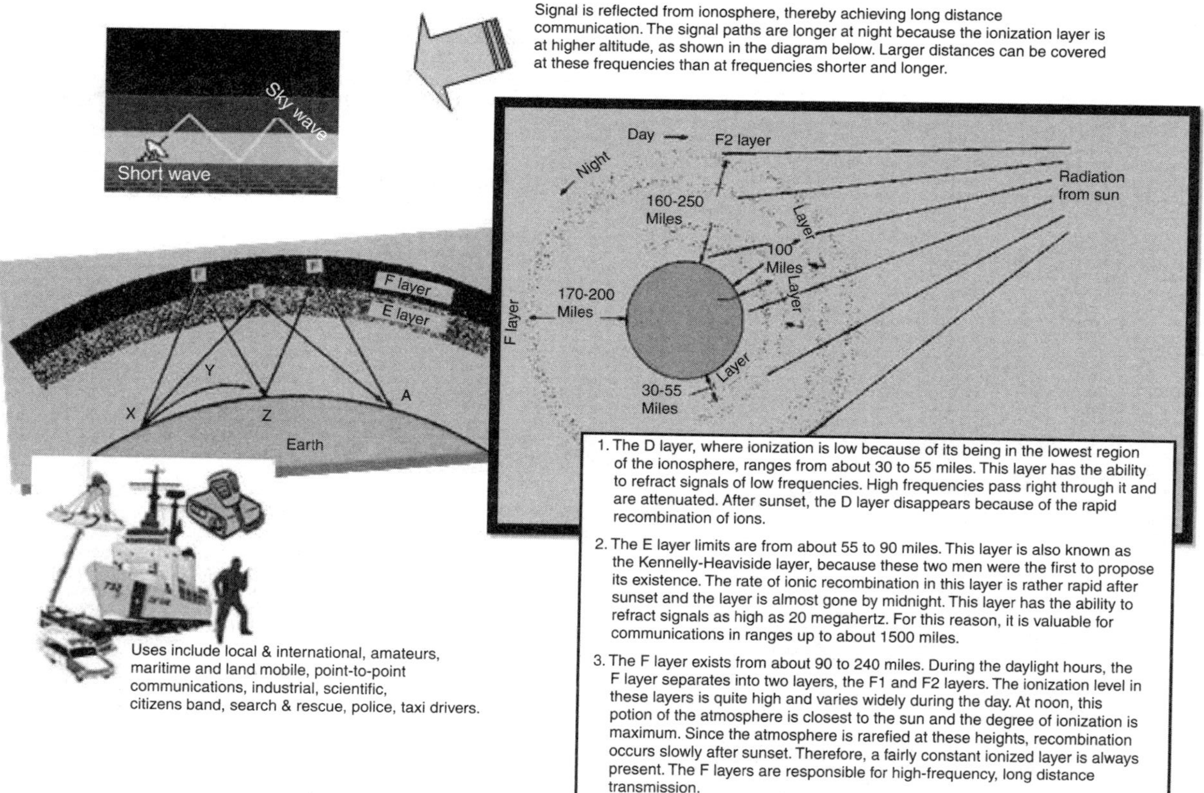

Short wave
Sky wave
Signal is reflected from ionosphere, thereby achieving long distance communication. The signal paths are longer at night because the ionization layer is at higher altitude, as shown in the diagram below. Larger distances can be covered at these frequencies than at frequencies shorter and longer.
F layer
E layer
X
Y
Z
A
Earth
Uses include local & international, amateurs, maritime and land mobile, point-to-point communications, industrial, scientific, citizens band, search & rescue, police, taxi drivers.
Day
Night
F2 layer
160-250 Miles
Layer
100 Miles
Layer
170-200 Miles
F layer
30-55 Miles
Layer
Radiation from sun
1. The D layer, where ionization is low because of its being in the lowest region of the ionosphere, ranges from about 30 to 55 miles. This layer has the ability to refract signals of low frequencies. High frequencies pass right through it and are attenuated. After sunset, the D layer disappears because of the rapid recombination of ions.
2. The E layer limits are from about 55 to 90 miles. This layer is also known as the Kennelly-Heaviside layer, because these two men were the first to propose its existence. The rate of ionic recombination in this layer is rather rapid after sunset and the layer is almost gone by midnight. This layer has the ability to refract signals as high as 20 megahertz. For this reason, it is valuable for communications in ranges up to about 1500 miles.
3. The F layer exists from about 90 to 240 miles. During the daylight hours, the F layer separates into two layers, the F1 and F2 layers. The ionization level in these layers is quite high and varies widely during the day. At noon, this potion of the atmosphere is closest to the sun and the degree of ionization is maximum. Since the atmosphere is rarefied at these heights, recombination occurs slowly after sunset. Therefore, a fairly constant ionized layer is always present. The F layers are responsible for high-frequency, long distance transmission.

Figure 3-15 Sky-wave bent back to Earth. (*Source:* Windows to the Universe, *Ibid. and http:www.windows.ucar.edu.*)

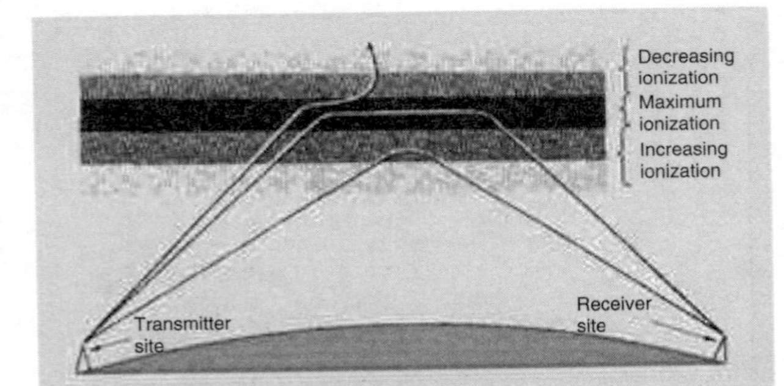

Ionospheric Density and Its Effect on Radio Waves

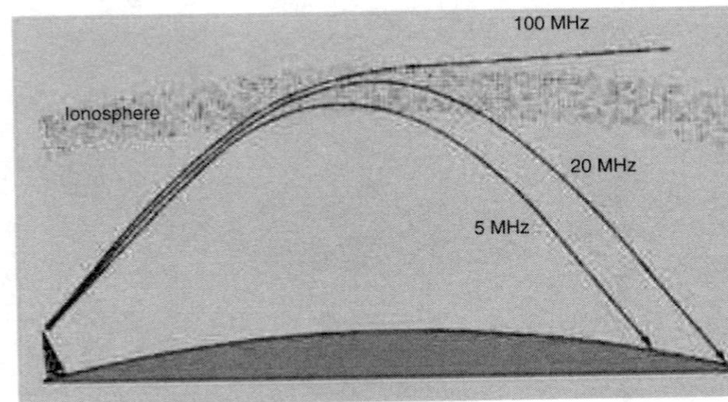

Frequency Versus Refraction and Distance: In this illustration, three separate waves of different frequencies are entering an ionospheric layer at the same angle. The 5-magahertz wave is refracted quite sharply. The 20-magahertz wave is refracted less sharply and returned to Earth at a greater distance. The 100-megahertz wave is greater than the *Critical Frequency* for the ionized layer and is not refracted.

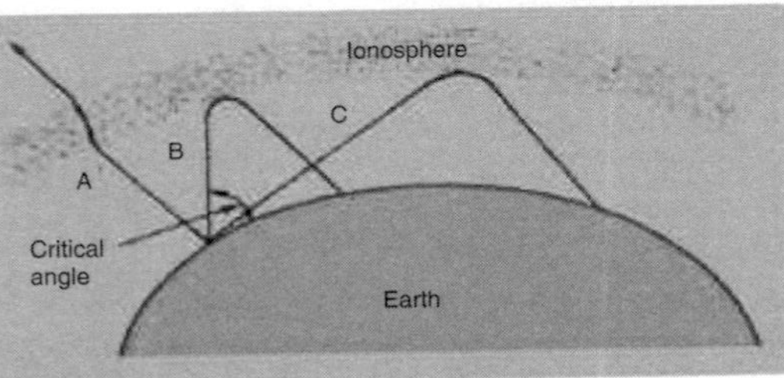

Different Incident Angles of Radio Waves

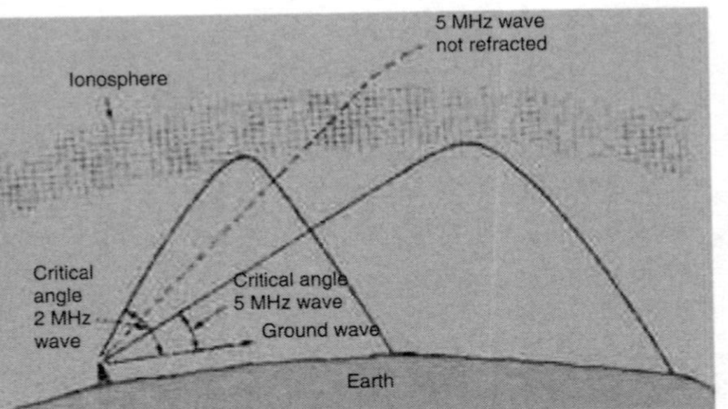

Effects of Frequency on the Critical Angle. This diagram illustrates that if and when the frequency of the radio wave is increased, the critical angle must be reduced for refraction to occur. The 2-megahertz wave strikes the layer at the critical angle for that frequency and is refracted back to Earth. The 5-megahertz wave (broken line) strikes the ionosphere at a lesser angle, penetrates the layer but then it passes into space. When the angle is lowered from the vertical, the 5-magahertz wave reaches its critical angle, making refraction back to Earth, possible.

Figure 3-16 Determining factors for refraction in radio waves. (*Source: http://www.infodotinc.com/neets/book10/40c.htm.*)

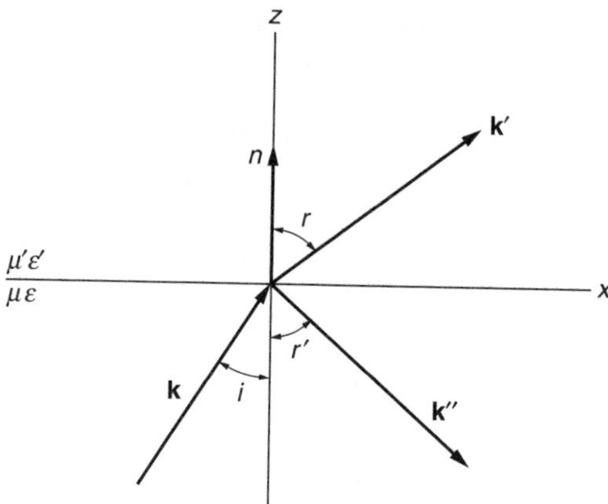

Figure 3-17 Relationship between the angles of incidence and refraction.

parallel to $\mathbf{B}_o$, the equation of motion for electrons trapped in this ionospheric plasma is given by

$$m\frac{d^2\mathbf{x}}{dt^2} - \frac{e}{c}\mathbf{B}_o \times \frac{d\boldsymbol{x}}{dt} = -e\mathbf{E}e^{-iwt}$$

where the influence of the **B** field of the transverse wave is negligible compared to the static induction $\mathbf{B}_o$ and the electron charge is given by $-e$. It is customary to describe the electric field component of the radio waves as circularly polarized, which implies

$$\mathbf{E} = (\varepsilon_1 \pm i\varepsilon_2)\mathrm{E}.$$

Likewise, **x** will have a similar expression.

$\mathbf{B}_o$ is orthogonal to both $\boldsymbol{\varepsilon}_1$ and $\boldsymbol{\varepsilon}_2$; therefore, the cross product in the equation of motion has components only in the directions $\boldsymbol{\varepsilon}_1$ and $\boldsymbol{\varepsilon}_2$, and the transverse components decouple. This leads to a steady-state solution given by

$$\mathbf{x} = \frac{e\mathbf{E}}{m\omega}(\omega \pm \omega_B)$$

where ω_B is the frequency of precession of a charged particle in a magnetic field:

$$\omega_B = \frac{e\mathrm{Bo}}{mc} \approx 6 \times 10^6\ \mathrm{s}^{-1}$$

(in the earth's magnetic field Bo = 0.3 G). The frequency dependence of the steady-state solution can be determined by transforming the equation of motion to a coordinate system precessing with frequency ω_B about the direction of $\mathbf{B}_o$. If the static magnetic field is negligible, the force on the electrons has an effective frequency ($\omega \pm \omega_B$), depending on the sign of the circular polarization.

The steady-state solution implies a dipole moment for each electron and yields, for a bulk sample, the dielectric constant of the ionosphere

$$\varepsilon^{\pm} = 1 - \left\{ \frac{\omega_p^2}{\omega(\omega \pm \omega_B)} \right\}$$

where the upper sign corresponds to a positive helicity wave (left-handed circular polarization in optics notation), while the lower sign corresponds to negative helicity. Furthermore, ω is the frequency of the radio wave of interest, and ω_p is the plasma frequency of the ionosphere specified by

$$\omega_p^2 = \frac{4\pi \mathrm{NZ} e^2}{m}$$

where NZ is the density of electrons per unit volume. For propagation anti-parallel to the magnetic field $\mathbf{B}_o$, the signs are reversed. Furthermore, for propagation in directions other than anti-parallel to the static field $\mathbf{B}_o$, it is straightforward to show that, if terms of order ω_B^2 are negligible compared to ω^2 and $\omega\omega_B$, the dielectric constant is still given by $\varepsilon^{\pm}$ above.

In this simplified problem of ionospheric radio wave propagation, the essential characteristic of waves of right-handed and left-handed circular polarizations is observed: these waves propagate differently. In other words, the ionosphere has both birefringent and anisotropic scattering properties. This has an important implication for the Law of Reciprocity which states that in line of sight communications, two stations running the same power output observe the same signals regardless of their station antennas. In HF sky-waves, there is the possibility that one station will not hear another station and/or vice versa even if they have identical antennas and are running identical power outputs.

The density of free electrons in the earth's ionosphere ranges between 10^4 and 10^6 electrons/cm^3, and the corresponding plasma frequency ω_p is on the order of ~$6 \times 10^6 - 6 \times 10^7$ s^{-1}. These values and the precession frequency ω_B imply a wide interval in frequency within the HF spectrum where one state of circular polarization cannot propagate

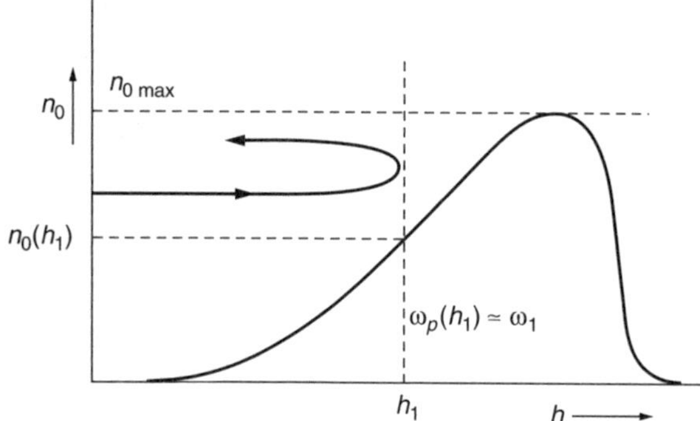

Figure 3-18 Variation of electron density versus vertical height above the earth's surface.

within the medium at all; instead, this wave is totally reflected back towards the earth. The other state of polarization is partially transmitted. Thus, when a linearly polarized wave is incident on a plasma, the reflected wave will be elliptically polarized, with its major axis generally rotated away from the direction of polarization of the incident wave.

The propagation of HF radio waves off the ionosphere is explicable in terms of these ideas, but the presence of several layers of plasma with densities and relative positions varying with height and time makes this problem considerably more complicated than this simplified model implies. Therefore, most ionospheric research groups seek to understand the variation of electron density versus vertical height above the earth's surface. The electron densities at various heights can be inferred by studying the reflection of radio pulses transmitted vertically upwards. Such studies show that the number n_0 of free electrons per unit volume slowly increases with height within a given layer of the ionosphere where it reaches a maximum before falling off abruptly with further increases in height as shown in Fig. 3-18.

A pulse of given frequency ω_1 enters a layer without reflection because of the slow change in n_0. When the density n_0 is large enough, $\omega_p(h_1) \sim \omega_1$, the dielectric constant vanishes, and the pulse is reflected vertically back to the earth. The electron density NZ required to induce this vertical reflection for a pulse of frequency ω is found by combining the equations above to give

$$\mathrm{NZ} = \left(\frac{m}{4\pi e^2}\right)(\omega^2 \pm \omega\omega_B).$$

By defining the minimum frequency at which a radio wave just penetrates a specific layer of ionization as the critical frequency f_c for that layer, the above equation can be rearranged into a more useful form. Two modes contributing to vertical reflections off a layer of ionization of electron density NZ, identified as ordinary mode and extraordinary mode, are expressed using meter, kilogram, and second (MKS) units and frequencies given in Hertz. The critical frequencies of these two modes are:

$$f_c = 8.98(\mathrm{NZ})^{1/2} \quad \text{for the ordinary mode}$$

$$f_c = 8.98(\mathrm{NZ})^{1/2} + 0.5 \times \mathrm{Bo}(e/m) \quad \text{for the extraordinary mode.}$$

By measuring the time interval between the initial transmission and reception of the reflected signal, the height h_1 corresponding to this electron density can be found. Therefore, by varying the frequency f and measuring the change in time intervals, the electron density can be determined as a function of height. If the frequency f is too high, the index of refraction remains finite, and no vertical reflection occurs. The minimum frequency at which vertical reflection disappears determines the maximum electron density in a given ionospheric layer. An example of an ionogram typically recorded at various ionospheric research stations around the world is shown in Fig. 3-19.

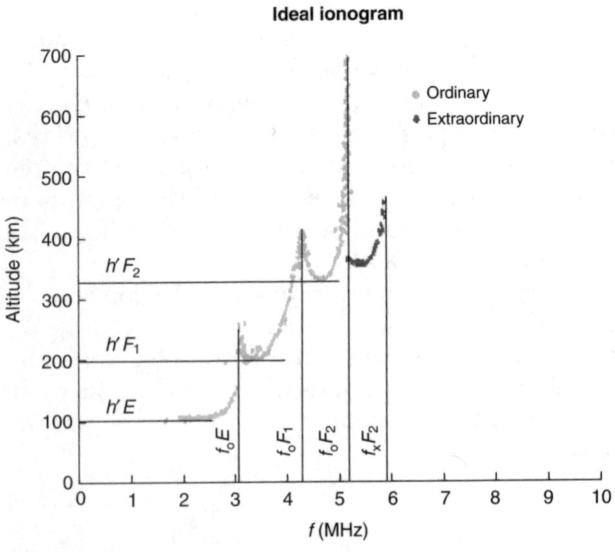

Figure 3-19 Ionogram typically recorded at various ionospheric research stations around the world.

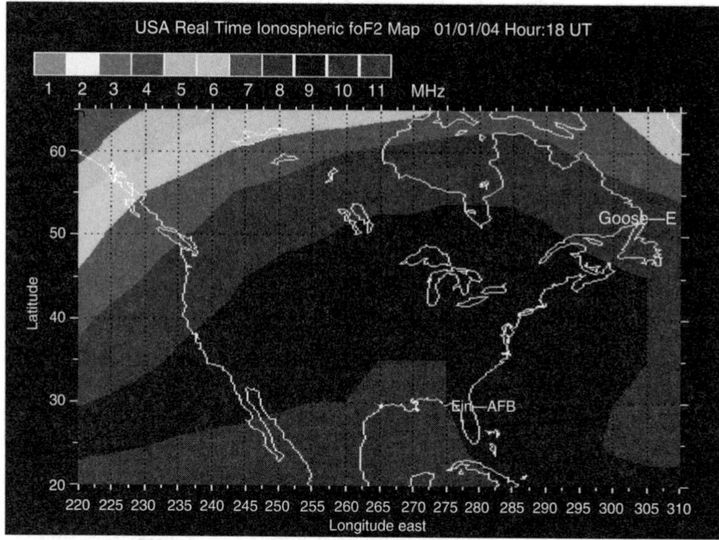

Figure 3-20 USA real time ionospheric f_0F_2 map.

The maximum frequency that vertically reflects off the F2-layer of the ionosphere at a given time and place is known as f_0F_2. The extraordinary mode reflecting off the F2-layer is known as f_xF_2. Similarly, the maximum frequencies that vertically reflect off the E-layer and F1-layer are known as f_0E and f_0F_1, respectively. Real-time f_0F_2 data are readily available from several government agencies in the United States and Australia. The two plots in Figs. 3-20 and 3-21 are real-time f_0F_2 plots of the United States and Australia that are provided by the Australian government.

Vertical reflections typically disappear at frequencies higher than the ~30 m range, so the more popular HF DX bands (10–20 m) cannot be adequately characterized by vertical ionograms. For these higher frequencies, oblique ionograms are utilized to yield empirical data on the refractive properties of the ionosphere and the virtual altitudes of these reflections; however, these data are limited, as the goal of most ionospheric research studies is aimed at determining the variation of electron density with height. Figure 3-22 illustrates the minimum ground distance needed (blind zone) to establish a digital communications link via reflection off the ionosphere for these higher frequencies.

The data show that as the frequency increases, the blind zone increases with no vertical reflections occurring at frequencies higher than ~7.5 MHz (SFI = 179) or ~14 MHz (SFI = 261). Finally, the data indicate that the blind zone decreases with higher solar flux (SFI) and increases with frequency.

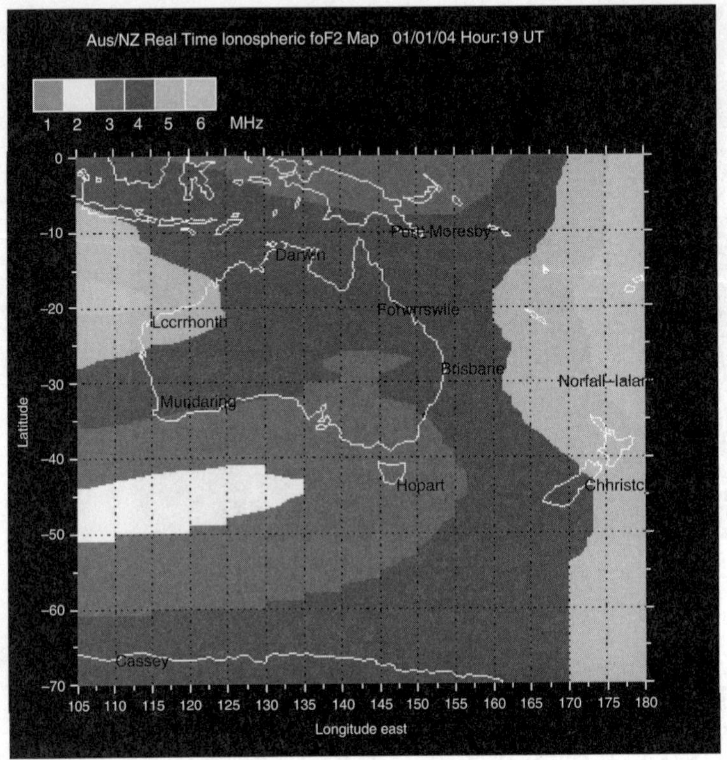

Figure 3-21 Australia/New Zealand real time ionospheric f_0F_2 map.

These data were obtained by connecting to a Nashville, TN, pactor base station from a mobile pactor station. Refer to Figs 3-23 and 3-24.

The frequency dependence of the dielectric constant $\varepsilon_{-}(\omega)$ at low frequencies is responsible for a peculiar magnetospheric propagation phenomenon known as *whistlers*. As $\omega \to 0$, $\varepsilon_{-}(\omega)$ tends to positive

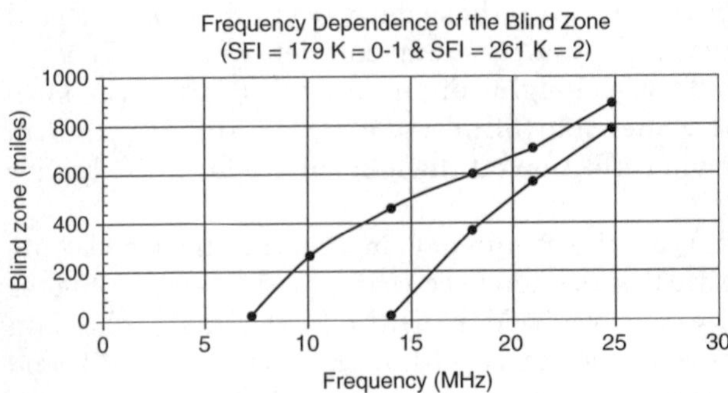

Figure 3-22 Frequency dependence of the Blind Zone.

Figure 3-23 The mobile Pactor-II station conducting a propagation experiment.[25] (*Source: http://ecjones.org/backscatter.html*, Courtesy of Dr. Edwin C. Jones, Vanderbilt University and Dr. Raymond A. Greenwald, Johns Hopkins Applied Physics Laboratory.)

infinity: $\varepsilon_{-} \sim \omega_p^2/\omega\omega_B$. Propagation occurs, with a wave number

$$k_{-} \approx \left(\frac{\omega_p}{c}\right)\left(\frac{\omega}{\omega_B}\right)^{1/2}$$

This corresponds to a highly dispersive medium. Energy transport

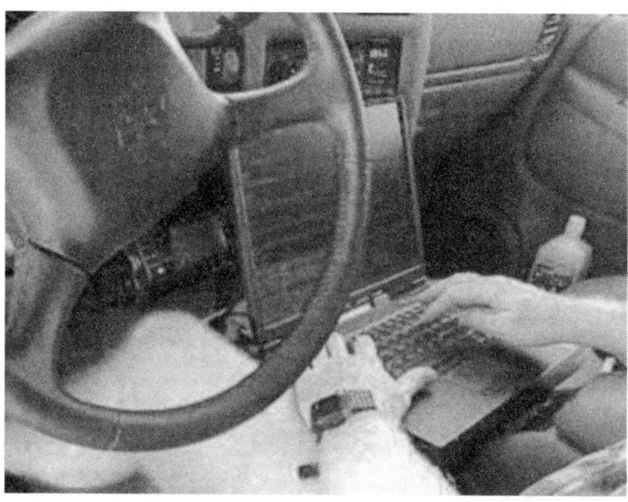

Figure 3-24 The mobile station consists of an Icom-706 transceiver, a SCS Pactor-II controller (not visible), and a computer terminal. (*Source: Ibid.*)

governed by the group velocity $v_g = [d\omega/dk]_0$ leads to a solution (for the group velocity) in the magnetosphere at the MF range characterized by:

$$v_g(\omega) \approx 2v_p(\omega) \approx 2c\left[\frac{(\omega\omega_B)^{1/2}}{\omega_p}\right]$$

This equation indicates that pulses of radiation at different frequencies travel at different speeds: in particular, the lower the frequency, the slower the speed. Whistlers occur when thunderstorms in one hemisphere generate a wide spectrum of frequencies, some of which propagate along the dipole lines of the earth's magnetic field, with the higher frequency components reaching the antipodal point first and the lower ones later. These whistlers generally occur at frequencies below 100 kHz, and when received by a radio receiver, the pulses sound like a whistle dropping in frequency. Assuming a distance of 10^4 km to a lightening discharge, the time interval for these whistlers occurs in seconds.

3.2.8 RF Skin effect

The skin effect *penetration depth* is the distance to which a radio wave can penetrate into a conductive medium (metal, salt water, ionosphere,and so forth) leaving only 37 percent ($1/e$) of its initial intensity. As predicted by Maxwell's equations below, RF energy decays exponentially when it encounters a conductive medium.
From Maxwell's equations,

$$\nabla \times \mathbf{H} = \frac{4\pi J + \dfrac{dD}{dt}}{c}$$

In an electron gas, $\mathbf{E} = \dfrac{\mathbf{J}}{\sigma}$; therefore, this becomes

$$\nabla \times \mathbf{H} = \frac{4\pi\sigma\mathbf{E} + \dfrac{dD}{dt}}{c}$$

where $\mathbf{H}$ = magnetic field in matter
$\mathbf{B}$ = magnetic induction
c = velocity of light
$\mathbf{J}$ = volume current
$\mathbf{D}$ = electric field displacement
σ = electrical conductivity
$\mathbf{E}$ = electric field

Faraday's law of induction states

$$\nabla \times \mathbf{E} = -\frac{\left(\frac{dB}{dt}\right)}{c}$$

Combining the above two equations gives

$$\nabla x(\nabla x\mathbf{E}) = \nabla(\nabla \cdot \mathbf{E}) - \nabla^2\mathbf{E} = -\nabla^2\mathbf{E} = -\frac{\left[\frac{d(\nabla x\mathbf{B})}{dt}\right]}{c}$$

$$\nabla^2 \mathbf{E}_0 + \left(\frac{4\pi\mu\sigma}{c^2}\right)\frac{d\mathbf{E}}{dt} = 0$$

Recognizing that this is the diffusion equation, we write

$$\mathbf{E}(t) = \mathbf{E}_0 e^{-iwt}$$

$$\nabla^2 \mathbf{E}_0 + t^2\mathbf{E}_0 = 0, \text{ then}$$

$$t^2 = \left(\frac{4\pi\sigma\mu\omega}{c^2}\right)i$$

$$t = \frac{(1+i)(2\pi\sigma\mu\omega)^{1/2}}{c} = \frac{(1+i)}{\delta}$$

In MKS units the skin depth δ, therefore, is given by

$$\delta = \frac{c}{(2\pi\sigma\mu\omega)^{1/2}}$$

where σ = electrical conductivity
μ = permeability
ω = angular frequency

In MKS units, the skin depth δ is given by

$$\delta = \left(\frac{2}{\mu\omega\sigma}\right)^{1/2}$$

Recognizing that the relative RF intensity at the skin depth is 37 percent of the incident intensity and that this is equivalent to 8.68 dB, the

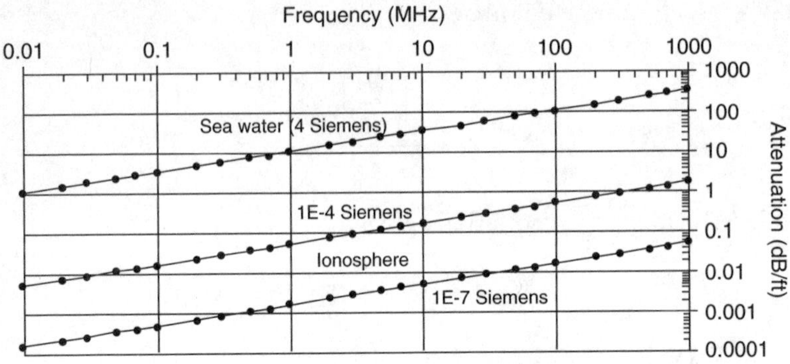

Figure 3-25 Attenuation of RF passing through conductive media.

attenuation of radio waves in dB/ft can be solved for various conductivities σ. Figure 3-25 indicates the attenuation of sea water ($\sigma = 4$ Siemens) shown in blue as a function of frequency. In addition, typical *ionospheric conductivities*[26] lie between 1×10^{-7} Siemens $< \sigma < 1 \times 10^{-4}$ Siemens; therefore, the attenuation per foot is shown as red and green for the upper (daytime) and lower (nighttime) conductivity limits for the ionosphere, respectively.

Figure 3-25 indicates that as the frequency increases, the RF attenuation per unit length increases. The RF attenuation also increases with electrical conductivity σ. The implications of these findings include the existence of discrete ranges of frequencies that can reflect off the ionosphere as well as the ability to communicate through sea water using extremely low frequencies.

Oftentimes, there are disturbances of the ionosphere, including sunspots, which could interfere with HF communications. The reliability of HF communications is, therefore, uncertain and so is its quality. For a detailed discussion on how the ionosphere is disturbed as it reacts to certain types of solar activity, refer to Appandix A.

Notwithstanding these shortfalls, HF is widely used for long-distance communications, short-wave broadcasting, over-the-horizon (OTH) radar, and amateur radio. HF transmitter power can range from as low as 2 W to above 100 kW, depending on the intended use.

In the HF range, two-way voice and data (record) communications are supported in various ways: point-to-point broadcast and air/ground/air operating modes using upper or lower sidebands. Besides long-range communications, HF is also widely used to supplement communications in tactical environments, if and when line-of-sight (LOS) radio is not possible or feasible.

Another HF mode is short-range near-vertical-incidence sky wave (NVIS) used with the NVIS antenna. When stations are separated by obstacles (such as mountains), and direct communication is not possible, a NVIS antenna can radiate an HF signal almost straight up for reflection down (over a mountain peak) to another station only a few miles away. NVIS operations are most effective at the lower HF frequencies (2 to 6 MHz).

High frequency can accommodate international maritime communications, voice, and teletypewriter operating modes and can operate in secure modes using a variety of available COMSEC devices. HF radios can be mounted in vehicles, ships, or aircraft and can be fixed, portable, or manpack configured.[27] Transmissions are normally in either the single sideband (SSB)[28] or independent sideband (ISB) mode.

High frequency sky-wave propagation can be easily intercepted, particularly the high-powered, long-haul systems. The HF part of the spectrum is the frequency band most susceptible to jamming. Electronic countermeasure (ECM) jammers far from the receiver can jam or disrupt HF skywave communications. Proper use of COMSEC devices and burst transmission techniques can, however, reduce this vulnerability. For applications within critical command and control systems, the HF communications must have an appropriate level of anti-jam protection.

From the developing countries' standpoint, HF is cost-effective for their domestic point-to-point systems.[29] This creates a conflict over allocating the HF internationally: The developing countries want to retain it for their domestic radiocommunication needs, while the developed world wants to use the band for international broadcasting (Voice of America, Radio Moscow) and long-distance mobile communication. For an overview of HF allocations in the United States, refer to Appendix B at the end of this chapter.

3.2.9 Very high, Ultrahigh, and Superhigh frequencies: 30 MHz to 30 GHz

Many important communication and entertainment services, including television broadcast signals, FM radio, and land mobile communications use this part of the spectrum.[30] Additionally, radiolocation service for long-range radars (12,350–2900 MHz), aircraft landing radar (around 9000 MHz), and point-to-point radio relay systems (various bands between 2000 and 8000 MHz) use these frequencies. In recent years, increasing use of frequencies in this band has been made by communication satellite.

Particularly valuable is the portion of the band between approximately 1 and 10 GHz. At the lower end, there is increasing cosmic and other background noise, whereas the upper end is affected by precipitation attenuation; however, in between, communications can be carried out well. Today, the 1- to 3-GHz band is especially sought after for mobile communications, including personal communications services (PCS), and for new broadcasting technologies such as digital audio broadcasting (DAB).

At 10 GHz and above, radio transmissions become increasingly difficult. Rain, snow, fog, clouds, and other forms of water in the signal's path cause greater attenuation of the radio signals. Nonetheless, development of the region above 10 GHz is taking place as a result of the crowding in the bands below 10 GHz. Frequencies above 10 GHz are relatively unused, offering extremely wide bandwidths. For example, the 3 to 30 MHz, HF band, is 27 MHz wide, making enough bandwidth available for about 9000 voice channels at 3 kHz each. The frequency range 3 to 30 GHz is 27,000 MHz wide, which can accommodate about 9 million voice channels. Table 3-5 compares the particulars of the VHF, UHF, and SHF.

TABLE 3-5 Very High, Ultrahigh, and Superhigh Frequencies: 30 MHz to 30 GHz

Frequency range	Use
	Designation: Very High Frequency
30–300 MHz	Signals propagate principally by LOS. Line-of-sight restrictions limit the ground range of VHF systems. However, LOS is an effective means of ground communication for distances up to 25–50 mi (depending on terrain and antenna height) without using a repeater. Long-range VHF transmissions possible through a series of short LOS hops. The higher the antenna, the greater is the possible LOS distance in each link. Links provide excellent circuit quality, comparable to cable systems with up to 99 percent reliability. Links can handle either analog or digital voice and data transmissions in single/multichannel modes. Data rates may vary from 45.5–75 bps (bits per second) for mobile VHF radio nets to 1.2–9.6 Kbps (thousand bits per second) per channel for LOS multichannel radio relay systems.
	Designation: Ultrahigh Frequency
300–3000 MHz	Main propagation methods include: ▪ Tropospheric scatter ▪ Satellite ▪ Air/Ground/Air ▪ Line-of-sight Ultrahigh frequency communications are flexible, making the distance range variable as follows: ▪ Line-of-sight—15–100 mi, terrain dependent, 300+ nautical miles (nmi) LOS from aircraft. ▪ Satellite—thousands of miles, depending on altitude, power, and antenna configuration. Transmitter power can range from a low of 10–100 W for LOS and satellite systems. Troposcatter systems operate in the 2500–10000-W range.

TABLE 3-5 Very High, Ultrahigh, and Superhigh Frequencies: 30 MHz to 30 GHz (*Continued*)

Frequency range	Use		
	Designation: Superhigh Frequency		
3–30 GHz	Used mainly for high-data-rate LOS microwave, multichannel radio relay, troposcatter, and satellite systems. Distances range from line-of-sight for terrestrial microwave links to thousands of miles for satellite connectivity:		
	▪ Line-of-sight	Ground-wave mode	40 mi (approximately).
	▪ Line-of-sight	Direct-wave mode (satellite)	Limited only by power and sensitivity (gain) of transmit and receive antennas.
		Troposcatter mode	Analog: ▪ 100 to 200 mi with 132 voice channels ▪ 200 to 300 mi with 72 voice channels ▪ Over 300 mi with 12/24 voice channels
	LOS microwave systems provide reliable, high-capacity, long-distance communications through radio relay sites, when properly engineered. Carrier signals permit large bandwidths, and are thereby able to handle significant amounts of data over multiplexed voice channels and television. High-speed data with rates of 2.4 Kbps and more (250 Kbps) are possible.		

3.2.10 Extremely high frequency

The extremely high frequency range is from 30 to 300 GHz. Military is conducting research and development to find appropriate application of this band. Two promising applications include: EHF satellite systems and millimeter-wave (MMW) transmissions.

The EHF Military Strategic Tactical and Relay (Milstar) satellite system provides worldwide coverage, with geosynchronous space segments in both equatorial and polar orbits. EHF satellite systems with cross-satellite linking have a global range; however, as EHF transmissions pass through the atmosphere, rain and other atmospheric conditions can cause attenuation.

Extremely high frequency systems can transmit secure voice and high-speed data at rates of up to 100 Mbps (million bits per second), operating either in the single-channel or multichannel mode. The extensive bandwidths are available in the EHF range, which permit as many as 600 channels per link, depending on the type of multiplexing equipment used. Additionally, EHF has the following advantages: increased capacity, jam resistance, electromagnetic pulse (EMP) protection, low power, narrow beam width, and excellent mobility.

3.3 No Borders

3.3.1 Ultra wideband

Ultra wideband (UWB) communicates data by varying the timing or phasing of extremely short radio frequency pulses. Available UWB techniques result in low-level emissions distributed across megahertz or gigahertz of spectrum, and therefore across the boundaries of numerous radio services.

Current UWB research is focused on 2 to 6 GHz and other bands including low VHF. The frequency range used by UWB may cause harmful interference to non-UWB services. The Federal Communications Commission (FCC) maintains that UWB devices can generally operate above 2 GHz without causing harmful interference to other radio services. FCC ruling permits such usage without restriction; however, FCC has *significant concerns* about UWB operation below 2 GHz, except for Ground Penetrating Radars (GPR) and *possibly through-wall imaging devices*. It is monitoring, for example, UWB/Global Positioning System (GPS)[31] tests that National Telecommunication and Information Administration (NTIA) and other researchers are conducting, and gathering explanation, notes, and clarifications on possible regulatory approaches.[32] Table 3-6 lists UWB technologies proven to be beneficial in a host of applications.

3.3.2 Spread spectrum[33]

Spread spectrum refers to a class of modulation techniques developed over the past 50 years. A signal is qualified as spread spectrum when the following criteria are met:

1. The transmitted signal bandwidth is greater than the minimal information bandwidth needed to successfully transmit the signal.
2. Some function other than the information itself if being employed to determine the resultant transmitted bandwidth.

Most commercial spread spectrum systems transmit an RF signal bandwidth that is one to two orders of magnitude greater than the bandwidth of the information being sent. A number of benefits are obtained from spreading the transmitted signal bandwidth. First, the spread spectrum signal is spread over a large bandwidth, thereby allowing it to coexist with narrow-band signals with only a slight increase to the noise floor in a given slice of spectrum. Because the spread-spectrum

TABLE 3-6 Ultrawideband Technology Applications

UWB technology	Applications	Description/particulars
UWB technologies use extremely narrow pulses with their concomitant ultra wide bandwidths, high repetition frequencies, and low duty cycles. UWB can offer privacy, low likelihood of interference, low cost, high capacity, and when used in radar and localizer applications, the power to make extremely precise measurements of distance.	Ground Penetrating Radar	Ground Penetrating Radars are wheeled, lawnmower-like instruments that reveal X-ray-like images of structures such as roads and bridges, for safety inspection. A GPR has to be within one meter of the ground and aimed directly into the ground to qualify as a permitted product.
	Potential medical uses	A mattress-installed breathing monitor to guard against Sudden Infant Death Syndrome, and heart monitors that act like an electrocardiogram except that they measure the heart's actual contractions instead of its electrical impulses.
	Other applications	Liquid level sensors for everything from water-conserving toilets to oil refinery tanks. Auto focusing cameras to calculate distances more accurately.
	Experimental microphones	Utilize radar instead of air to pick up sound from the vocal organs.
	UWB radars and covert communication devices	Interest is strong among defense, safety, and law enforcement communities. Companies manufacturing and selling these devices must limit the distribution and keep appropriate records of purchasers and users.

receiver looks over a large range of frequencies, it does not see the narrow-band frequency, making coexistence possible. Even if the spread-spectrum receiver detects the narrow-band signal, it does not recognize the signal because its transmission is not being made with the proper code sequence. Spread-spectrum modulations have a number of manifestations. The two popular forms are direct sequence and frequency hopping.

The popularity of direct sequence is based on the simplicity with which direct sequencing can be implemented. A pseudo-random noise generator creates a high-speed pseudo-noise code sequence, which is transmitted at a maximum bit rate called the *chip rate*. The pseudo-random code sequence is used to directly modulate the narrow-band carrier signal; thus, it directly sets the transmitted RF bandwidth. The chip rate has a direct correlation to the spread of the information. The information is demodulated at the receiving end by multiplying the signal by a locally generated version of the pseudo-random code sequence.

Another popular from of implementing spread spectrum is Frequency Hopping in which spreading takes place by hopping from frequency to frequency over a wide band. A hopping table generated with the help of a pseudo-random code sequence determines the specific order in which the hopping occurs. The rate of hopping is a function of the information rate. The pseudo-random noise sequence dictates the order of frequencies selected by the receiver. The transmitted spectrum of a frequency-hopping signal is different from that of a direct sequence signal; however, the data is spread out over a signal band larger than is necessary to carry it. In both cases, the resultant signal appears noise-like, and the receiver utilizes a similar technique to the one employed in transmitting in order to recover the original signal.

TABLE 3-7 Spread Spectrum Techniques and their Applications

Application/Equipment	Examples
	Band: 902–928 MHz
Government radiolocation	Military systems: high power air surveillance radars on aircraft carriers, tracking and telemetry radar systems used in aeronautical flight testing, systems that monitor the positions of missiles, drone and manned aircraft and land units, and perimeter protection devices for intrusion detection at military facilities.
Naval radars	Detection of sea skimmers, fast-moving targets over water.
Radar wind profiling	Weather forecasting, aviation warning, marine observations, and environmental studies.
Government fixed and mobile radio systems are in this band on a secondary basis to radiolocation.	Mobile and portable radios, the transmission of images seen by bomb disposal robots, and fixed systems for such purposes as control of power utilities and video links for monitoring entry points at national borders.
	Band: 902–928 MHz
Spread spectrum remote control of robot jockeys	One of the earliest commercial applications of spread spectrum communications technology.
	ISM frequency (915 ± –13 MHz)
ISM equipment	Industrial heaters that cure glue, inks, and rubber, welding equipment, food equipment such as bacon dryers and donut fryers, and medical instruments for magnetic resonance imaging, diathermy (tissue heating), microwave aided liposuction, and thermo-therapy for treatment of prostate disease.

Spread spectrum has many advantages. For example, spread-spectrum receivers can effectively ignore narrow-band transmissions, thereby making it possible to share the same frequency band with other users. These users can tolerate a significant degree of overlap without interference effects. In both mechanisms discussed above, a pseudo-random noise sequence has been employed—either to directly modulate the signal or to determine the order of frequencies in the hopping table. This pseudo-random signal makes the transmitted signal appear as noise; therefore, only receivers possessing the proper duplicate pseudo-random noise code sequence are able to recover the signal, thereby ensuring the privacy of point-to-point communications (or point to multi-point communications, as the case may be). The U.S. military has used the noise-like character of the transmitted signal that drastically reduces the probability of signal detection and interception to ensure secure communications.[34]

Industrial, scientific, and medical (ISM) devices are highest priority among nongovernmental uses. The FCC defines ISM equipment as "...appliances designed to generate and use radio frequency energy to perform some work other than telecommunications." It (FCC) permits spread-spectrum devices to transmit up to one watt of power in the ISM bands (specified as 902–928 MHz, 6.765–6.795 MHz, 13.553–13.567 MHz, 26.957–27.283 MHz, 40.66–40.70 MHz, 2400–2500 MHz, 5.725–5.875 GHz, 24–24.25 GHz, 61–61.5 GHz, 122–123 GHz, and 244–246 GHz), depending on the specific technology used.[35] Table 3-7 lists some of the equipment and applications.

3.4 Appendix A: Radio Wave Propagation[36]

The sun's electromagnetic radiation is a continuum that spans radio wavelengths through the infrared, visible, ultraviolet, x-ray, and beyond. Ultraviolet radiation, through a process termed photo ionization, interacts with upper atmospheric constituents to form an ionized layer called the ionosphere.

The ionosphere affects radio signals in different ways depending on their frequencies (see Fig. 3-26), which range from extremely low frequency to

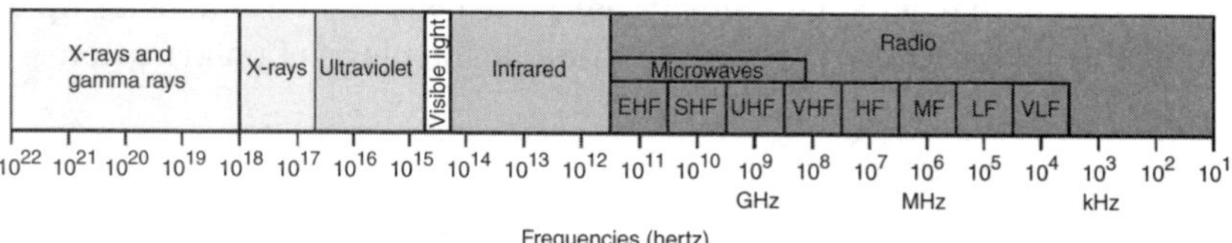

Figure 3-26 The electromagnetic spectrum includes x-rays, visible light, and radio waves.

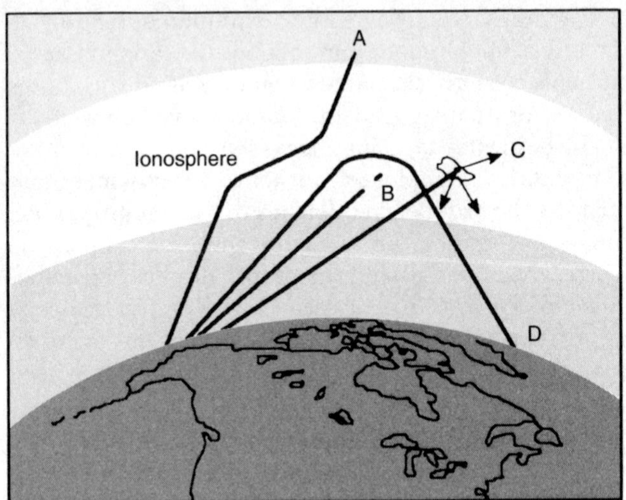

Figure 3-27 Radio waves that reach the ionosphere can go astray. A.Wave penetrates the ionospheric layer. B. Wave is absorbed by the layer. C. Wave is scattered in random directions by irregularities in the layer. D. Wave is reflected normally by the layer.

extremely high frequency. On frequencies below about 30 MHz, the ionosphere may act as an efficient reflector, allowing radiocommunication to distances of many thousands of kilometers. Radio signals on frequencies above 30 MHz usually penetrate the ionosphere and, therefore, are useful for ground-to-space communications. The ionosphere occasionally becomes disturbed as it reacts to certain types of solar activity. Solar flares are an example; these disturbances can affect radiocommunication in all latitudes. Frequencies between 2 and 30 MHz are adversely affected by increased absorption, whereas on higher frequencies (e.g., 30–100 MHz) unexpected radio reflections can result in radio interference.

Scattering of radio power by ionospheric irregularities produces fluctuating signals (scintillation), and propagation may take unexpected paths. TV and FM (on VHF) radio stations are affected little by solar activity, whereas HF ground-to-air, ship-to-shore, Voice of America, Radio Free Europe, and amateur radio are affected frequently. Figure 3-27 illustrates various ionospheric radio wave propagation effects. Some satellite systems, which employ linear polarization on frequencies up to 1 GHz, are affected by Faraday rotation of the plane of polarization.

3.4.1 Solar flare effects

A solar flare is a sudden energy release in the solar atmosphere from which electromagnetic radiation and, sometimes, energetic particles and bulk plasma are emitted (see Fig. 3-28).

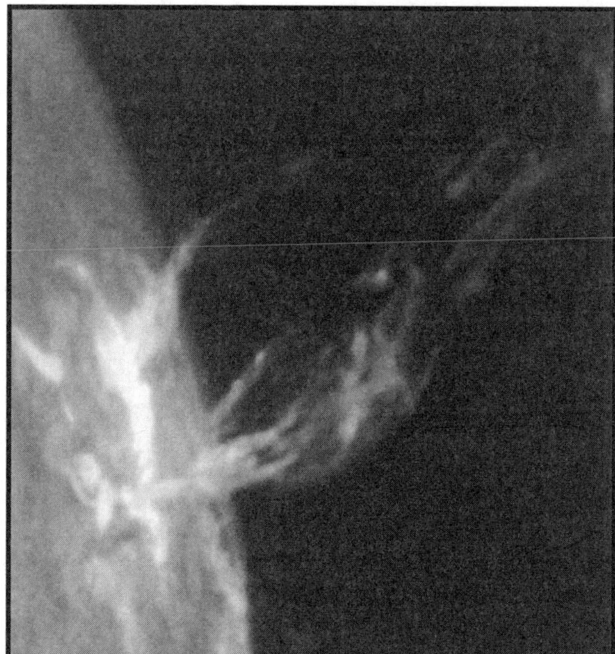

Figure 3-28 An eruption on the limb of the sun. This picture was taken in Hydrogen-α light (656.3 nm).

A sudden increase of x-ray emissions resulting from a flare causes a large increase in ionization in the lower regions of the ionosphere on the sunlit side of Earth. A sudden ionospheric disturbance (SID) of radio signals can ensue. An SID can affect very low frequencies (e.g., OMEGA) as a sudden phase anomaly (SPA) or a sudden enhancement of signal (SES). At HF, and sometimes at VHF, an SID may appear as a shortwave fade (SWF). This disturbance may last from minutes to hours, depending on the magnitude and duration of the flare.

Solar flares also create a wide spectrum of radio noise; at VHF (and under unusual conditions at HF) this noise may interfere directly with a wanted signal. The frequency with which a radio operator experiences solar flare effects will vary with the approximately 11-year sunspot cycle; more effects occur during solar maximum (when flare occurrence is high) than during solar minimum (when flare occurrence is very low). A radio operator can experience great difficulty in transmitting or receiving signals during solar flares.

3.4.2 Energetic particle effects

On rare occasions a solar flare will be accompanied by a stream of energetic particles (mostly protons and electrons). The more energetic

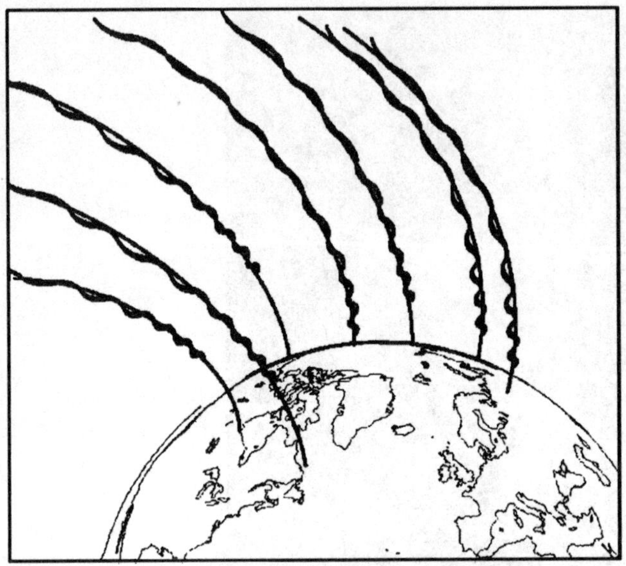

Figure 3-29 Solar energetic particles following Earth's magnetic field lines can penetrate the upper atmosphere near the magnetic poles, resulting in ionization and creating a polar cap absorption event.

protons, traveling at speeds approaching that of light, can reach Earth in as little as 30 minutes. These protons reach the upper atmosphere near the magnetic poles (see Fig. 3-29). The lower regions of the polar ionosphere then become heavily ionized, and severe HF and VHF signal absorption may occur. This is called a polar cap absorption (PCA) event. PCA events may last from days to weeks, depending on the size of the flare and how well the flare site is magnetically connected to Earth. Polar HF radio propagation often becomes impossible during these events.

3.4.3 Geomagnetic storm effects

Sufficiently large or long-lived solar flares and disappearing filaments (DSF) are sometimes accompanied by the ejection of large clouds of plasma (ionized gases) into interplanetary space. These plasma clouds are called coronal mass ejections (CMEs). A CME travels through the solar wind in interplanetary space and sometimes reaches Earth (see Fig. 3-30). This results in a worldwide disturbance of Earth's magnetic field, called a geomagnetic storm. Another type of solar activity, known as a coronal hole (CH), produces high-speed solar wind streams that buffet Earth's magnetic field (see Fig. 3-31); geomagnetic storms that may be accompanied by ionospheric disturbances can result.

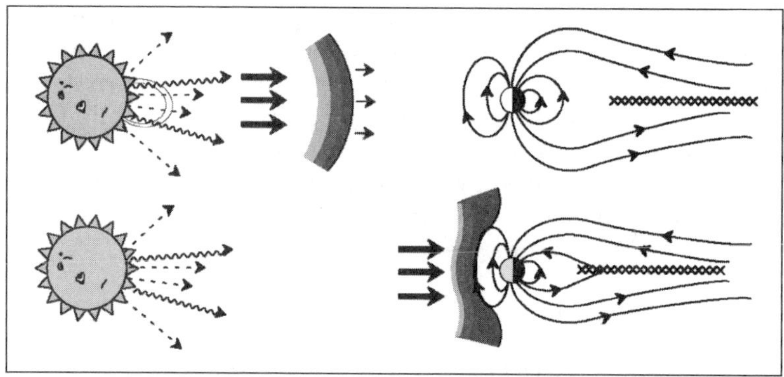

Figure 3-30 An ejection from the sun travels to Earth and distorts Earth's magnetic field, resulting in geomagnetic activity.

These ionospheric disturbances can have adverse effects on radio signals over the entire frequency spectrum, especially in auroral latitudes. In particular, HF radio operators attempting to communicate through the auroral zones (the regions of visible aurora, or "Northern Lights") during storms can experience rapid and deep signal fading due to the ionospheric irregularities that scatter the radio signal.

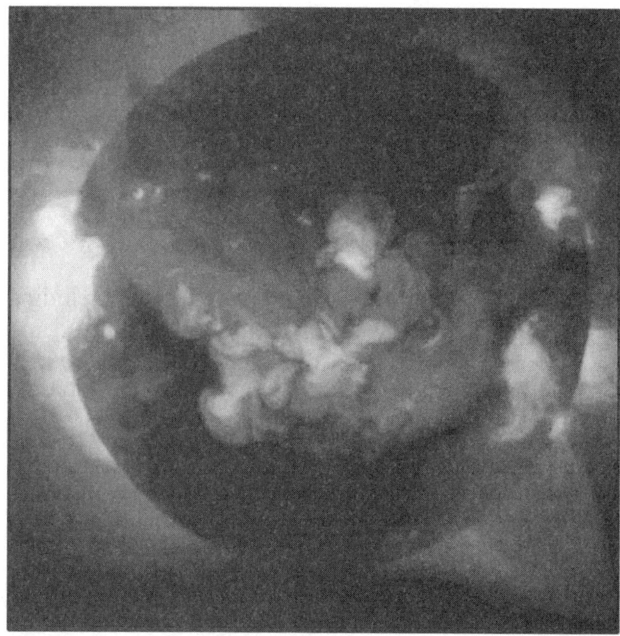

Figure 3-31 The sun as seen in x-rays. The darkest areas are coronal holes; bright areas overlie active regions.

Auroral absorption, multipathing, and nongreatcircle propagation effects combine to disrupt radiocommunication during ionospheric storm conditions. During large storms, the auroral irregularity zone moves equatorward. These irregularities can produce scintillations that adversely impact phase-sensitive systems on frequencies above 1 GHz (e.g., the Global Positioning System). Geomagnetic storms may last several days, and ionospheric effects may last a day or two longer.

Systems affected by solar or geomagnetic activity:

HF Communication

- Increased absorption
- Depressed Maximum Usable Frequencies (MUF)
- Increased Lowest Usable Frequencies (LUF)
- Increased fading and flutter

Surveillance Systems

- Radar energy scatter (auroral interference)
- Range errors
- Elevation angle errors
- Azimuth angle errors

Satellite Systems

- Faraday rotation
- Scintillation
- Loss of phase lock
- Radio Frequency Interferences (RFI)

Navigation Systems

- Position errors

3.5 Appendix B: HF Allocations in the United States[37]

In the United States, the HF band is divided into 101 subbands including four exclusive federal bands and 20 exclusive nonfederal bands. The fixed and mobile services share the four exclusive federal bands, a total of 1170 kHz or approximately 4 percent of the HF band. Of the 20 exclusive nonfederal bands, which total 4905 kHz or approximately 18 percent of the HF band, 10 are for the amateur services, 3 for the fixed service, 5 for the land mobile service, 1 for the mobile service, and 1 shared by the fixed and mobile services. The remaining 77 subbands are shared by federal and nonfederal users and amount to approximately 78 percent of the HF band.

Currently, approximately 30 federal agencies and numerous state and local government agencies and private sector entities hold HF assignments or licenses. Over 80,000 assignment records are reflected in the Government Master File (GMF) and the FCC Master Frequency List.

The largest holder of HF frequency assignments in the federal government is the Department of the Navy with over 16,000 assignments, followed by the U.S. Coast Guard (USCG) with approximately 8700 assignments. The Departments of the Army, Air Force, Justice, and the Federal Aviation Administration (FAA) follow with approximately 8200, 4700, 1800, and 1200, respectively. The private sector has over 26,000 frequency licenses held by various entities such as private citizens, businesses, utilities, state and local governments.

In terms of the total amount of HF spectrum allocated, the shared fixed and mobile allocated service is the largest followed by maritime mobile, amateur, fixed, broadcasting, aeronautical mobile, mobile, radio astronomy, and standard frequency and time signal services. The fixed and mobile services comprise 9094 kHz of spectrum in 22 subbands with over 34,000 assignments. The maritime mobile service has over 32,000 assignments in 12 subbands that total 4808 kHz. In 3550 kHz, the amateur service supports approximately 400,000 licensed amateur radio operators in the United States in 10 subbands. The fixed services are difficult to characterize because of close pairing with mobile assignments in multiple station configurations. In 15 subbands, the fixed service has over 8700 assignments in 3357 kHz of allocated spectrum. The broadcasting service has 2930 kHz of allocated spectrum in eight subbands containing over 3100 assignments. Broadcast use of the bands normally is on a scheduled basis and requires extensive global coordination to minimize interference on a given frequency. The aeronautical mobile service has 21 subbands and over 5900 assignments in 2026 kHz of allocated spectrum. The mobile service, for purposes of this report, comprises the one mobile and five land mobile bands with over 2500 assignments in 985 kHz of allocated spectrum. In 170 kHz of allocated spectrum, the radio astronomy service contains about 35 assignments, while the standard frequency and time signal service has over 2100 assignments in 80 kHz of allocated spectrum. Figure 3-32 depicts the U.S. HF band allocations and their relative sizes.

Note: This figure may suggest that an HF channel is the same or equal anywhere in the HF band. The lower portion of the HF band is very congested and has generally better propagation conditions than the upper portion which is less congested.

3.5.1 HF band usage in the United States

The HF spectrum is finite and cannot be expanded to accommodate other users or additional radio services. For radio services seeking expansion to satisfy their spectrum requirements, it is usually at the expense of another radio service. In the NTIA *Requirements Study*, the fixed service was the only significant radio service where no additional spectrum requirement was expressed in the HF band. This portion of

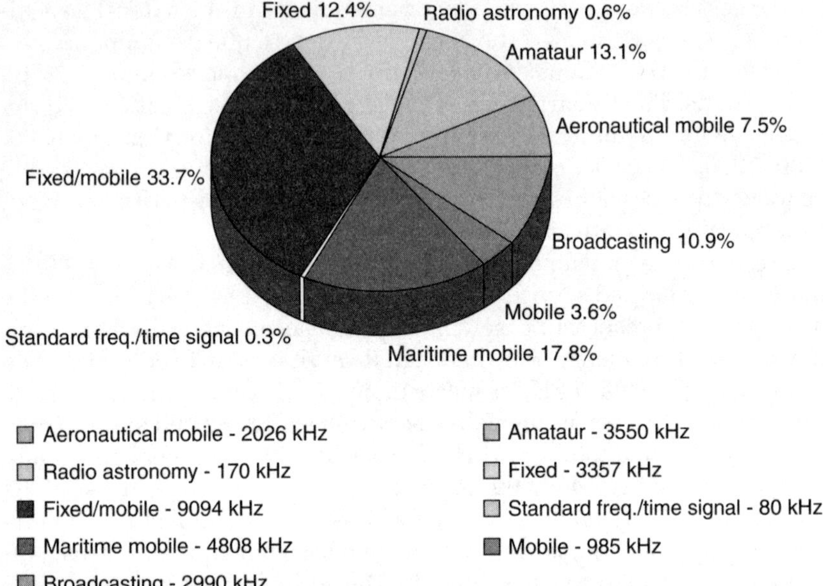

Figure 3-32 United States HF band allocations.

the study will examine the U.S. use of the HF spectrum and its spectrum flexibility. It is important for the reader to know that operations in the radio services of the HF band support numerous public benefits such as public safety, emergency medical assistance communications, weather observation reporting, environmental resource management. There are no current plans to auction any HF spectrum.

In the HF band, there are 15 subbands allocated to the fixed service, 12 of which are shared between the federal and nonfederal users. The remaining three are allocated for nonfederal use. An examination of this band revealed that the 12 shared subbands have over 8700 assignments in 3167 kHz of allocated spectrum. Further, a high degree of spectrum flexibility is evident in each frequency block. For instance, federal usage in the 5005 to 5060 kHz band shows 446 assignments for the fixed service, while 177 assignments are maritime mobile, 283 assignments for mobile, 29 for aeronautical mobile, 10 for experimental, and 4 for radio determination. Figure 3-33 graphically depicts the general distribution of radio services using the 5005 to 5060 kHz fixed band. Other fixed service bands in this group of 12 bands generally resemble the extent of usage by other radio services. Table 3-8 reflects the flexibility usage of the other 12 bands. The three fixed service bands allocated exclusively for nonfederal use contain approximately 100 assignments within 190 kHz of spectrum. Of the approximately 100 assignments, about 40 are

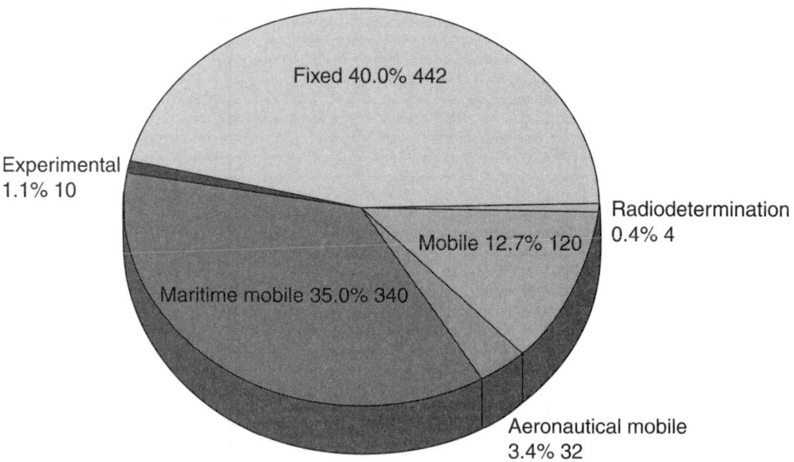

Figure 3-33 Assignments by radio services in the 5005–5060 kHz fixed service band.

held by federal agencies. As can be seen in the fixed service bands, a significant amount of sharing by other radio services takes place among federal users. Examination of the nonfederal licenses revealed little flexibility in the fixed service band usage where fixed operations and associated mobiles tended to be the licensed operations.

3.5.2 Aeronautical mobile service

This radio service is used primarily for air traffic control and has two categories: aeronautical mobile (R) and aeronautical mobile (OR) services. The aeronautical mobile (R) service is intended for communications with aircraft or between aircraft relating to safety and regularity of flight, primarily along national or international civil air routes. The aeronautical mobile (OR) service is for communications with aircraft or between aircraft outside national or international civil air routes. As a general rule, HF frequencies are not used for aeronautical mobile (R) communications in the domestic services when aircraft are within the airspace of the conterminous United States. The need to use HF was eliminated through successful implementation of VHF air traffic control communications. The use of the aeronautical mobile (OR) HF frequencies is primarily to satisfy military aeronautical communications requirements.

The aeronautical mobile service has 21 bands totaling 2026 kHz shared between the federal and nonfederal users. Eleven bands are designated as aeronautical mobile (R) and the remaining 10 are designated aeronautical mobile (OR). Federal and nonfederal aeronautical stations that operate in the aeronautical mobile (R) service within

TABLE 3-8 Fixed Service Bands and Other Radio Service Usage by the Federal Government

	Fixed radio service band (kHz)											
	5005–5060	9040–9500	9900–9995	11400–11650	12050–12230	15600–16360	17410–17550	18030–18068	18900–19680	19800–19990	21850–21924	22855–23000
Fixed	442	1093	239	580	832	1124	349	138	468	108	95	181
Maritime mobile	340	321	16	167	333	411	83	9	66	10	0	40
Aeronautical mobile	32	145	17	207	114	258	93	25	94	5	3	15
Mobile	120	128	15	114	171	130	48	13	88	0	0	1
Experimental	10	57	6	8	33	56	20	10	22	2	6	14
Broadcasting	0	0	0	4	0	0	0	0	0	0	0	0
Radiolocation	4	1	0	0	0	0	0	0	0	0	0	0

the United States and Possessions are normally authorized by the FAA and the FCC. Frequencies in the bands allocated exclusively to the (OR) service are used primarily for the satisfaction of military aeronautical requirements. The aeronautical mobile (R) bands have 1181 kHz of HF spectrum while the aeronautical mobile (OR) bands have 845 kHz of HF spectrum. There are over 5800 assignments in the HF aeronautical mobile bands: over 1100 in the (R) bands and over 4600 in the (OR) bands. Less flexibility is shown in the aeronautical mobile bands than the fixed service primarily because of the operational and safety requirements associated with national and international aviation.

3.5.3 Maritime mobile service

The maritime mobile bands are used primarily to communicate between and among fixed coast stations, ships and other offshore vessels. Specific uses include the command, control, and communications of U.S. Navy and USCG ships and vessels, distress and safety communications, treaty and law enforcement, drug interdiction, geological survey operations, and national marine fishery operations.

The maritime mobile service has 4808 kHz of spectrum allocated in the 12 shared HF bands. Stations use this spectrum either exclusively to the maritime mobile service or on a shared basis with other radio services. The federal agencies have approximately 13,000 assignments in the maritime mobile bands, while there are almost 10,000 assignments for nonfederal users. Several international plans detail the specific uses of certain maritime mobile HF frequencies. National planning for the maritime mobile bands closely follows international use.

3.5.4 Amateur services

Amateurs have been active in radio since its earliest days. The amateur service is the oldest radio service and predates regulation of radiocommunications. The amateur service allows its users to provide a unique service to the public while enjoying a popular, technical hobby. Radio amateurs have made significant contributions to the field of radio propagation, HF single-sideband radio, HF data communications systems, packet radio protocols and communications satellite design. Further, amateur radio continues to play an important role in disaster-relief communications where amateurs provide radiocommunications independent of the telephone network or other radio services, particularly in the first few days before relief agencies are at the scene and have set up disaster telecommunications services. It is estimated that there are in excess of 650,000 amateur radio operators in the United States and over 2.4 million worldwide.

In the HF band, the amateur operator has a choice of narrow frequency bands each with different propagation properties for long-distance communications. The amateur operator can follow the changing maximum usable frequency (MUF) as propagation conditions change in the amateur bands and still be able to communicate. Having a good selection of frequencies is critical to maintain reliable communications for both voice and data operations in the HF band. All U.S. amateur and amateur-satellite allocations in the HF band are allocated on a primary basis for exclusive nonfederal use only.

3.5.5 Broadcasting service

The use of HF frequencies for the broadcasting service in the United States includes: (i) operations by US-based broadcasters who are trying to reach non-U.S. audiences; (ii) operations by both US-based and foreign-based broadcasters who are trying to reach U.S. audiences. Broadcasting services that are intended to be transmitted across international borders are HF broadcast operations subject to the ITU Radio Regulations. For decades, nations have made increasing use of the electromagnetic spectrum to conduct public diplomacy by broadcasting speech and music directly to receivers throughout the world. There is a total of 2930 kHz of HF spectrum presently allocated in the United States for the broadcasting service within eight HF subbands. These bands are allocated on a primary basis to both federal and nonfederal users.

3.5.6 Other services

Rounding out the other radio services that have HF allocations are the standard frequency and time signal service, radio astronomy, land mobile, and on a secondary basis, the radiolocation service, with an allocation at 3230 to 3400 kHz. The standard frequency and time signal service is allocated on a shared, primary basis at 4995–5005, 9995–10,005, 14,990–15,010, 19,990–20,010 and 24,990–25,010 kHz. The radio astronomy service is allocated on a primary basis to federal and nonfederal users at 13,360–1,3410 and 25,550–25,670 kHz. The land mobile service is allocated exclusively to nonfederal users on a primary basis at 25,010–25,070, 25,210–25,330, 26,175–26,480, 27,410–27,540, and 29,700–29,800 kHz.

3.6 Endnotes

[1]There are three major force fields in nature: gravitational, electric, and magnetic. Radio is concerned with the electric and magnetic fields. The following relationships exist between the electric and magnetic fields:

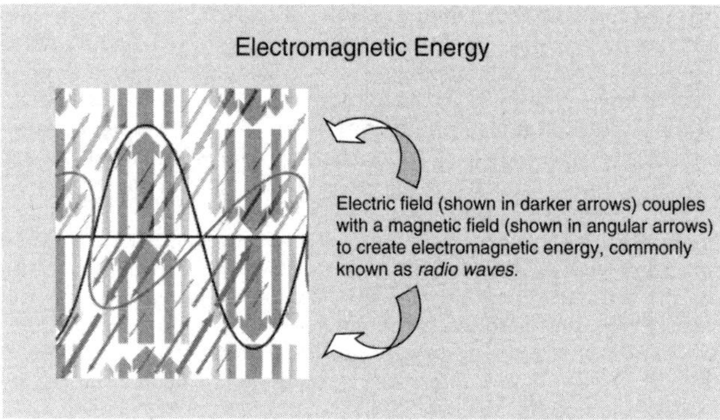

Electromagnetic Energy
Electric field (shown in darker arrows) couples with a magnetic field (shown in angular arrows) to create electromagnetic energy, commonly known as *radio waves.*

Figure 3-34 Radio waves are magnetic fields generated by continually changing electric fields. (*Source: http://imagers.gsfc.nasa.gov/ems/waves2.html*)

- A magnetic field is created when an electric field changes.
- An electric field is created when a magnetic field changes.

Radio waves are magnetic fields generated by continually changing electric fields, as illustrated in Fig. 3-34.

The magnetic fields created around a wire conducting a 60 cps (see Fig. 3-2) current are very small. They extend only a short distance from the wire and collapse with each current reversal. As the current increases in frequency, however, the magnetic fields do not collapse nearly as completely with each alteration. With frequency reaching 10,000 cps, the magnetic fields no longer have time to collapse completely between alterations. Rather, they are pushed away (radiated) from the wire by fields produced by each succeeding alterations, thereby giving rise to electromagnetic energy, commonly known as *radio waves.*

[2]The term "cycle" means one complete set of events or phenomena that occurs periodically.

[3]Multiples of the hertz are indicated by prefixes: "kilo" for one thousand, "mega" for one million, and "giga" for one billion and also a new designation, "tera" (trillion) hertz (THz). Thus, a million hertz—a million cycles per second—is expressed as one megahertz (abbreviated "MHz").

[4]In free space (no atmosphere) the speed of light expressed in feet is 984,300,000 ft/s, or 300,000,000 m/s (i.e., 186,000 mi/s). The speed is constant—so the more cycles (see Fig. 3-2) that pass a given point in a given amount of time, the higher their frequency, the shorter their wavelength. The equation that determines wavelength can, therefore, be expressed as:

$$\text{Wavelength} = \frac{\text{Velocity}}{\text{Frequency}}$$

Using this formula, one can calculate the distance a wave travels in one cycle (the wavelength, see Fig. 3-3)

[5]Due to equipment limitations, the first radio sets operated at the low frequency (LF) and medium frequency (MF) end of the radio-frequency (RF) spectrum. These two frequency bands offered good voice and low-speed teletype communications, but their transmission distance was limited due to the rather low power outputs available at that time. In the 1890s, experiments began with the use of the higher frequencies.

[6]High-frequency communication was first made practical in the 1920s when Marconi set up one of the first transatlantic wireless stations in County Galway, on Ireland's western fringe. On June 15, 1919, with generators fueled by peat, the station notified London of the successful flight of two British aviators. From this point on, radio technology developed rapidly. World War II also had a major impact on the use of radio-frequency spectrum. Radar was developed in the early part of the war. Higher radar frequencies needed the development of components and equipment to operate higher frequency radio systems at very high frequency (VHF) and ultrahigh frequency (UHF) bands. Higher frequency systems, as discussed later in this chapter, have weaknesses associated with their method of radio-wave propagation. In view of military's need to maintain uninterrupted communications, many experiments with low frequency (LF), very low frequency (VLF) and extremely low frequency (ELF) communications systems have been performed to demonstrate that in case of a nuclear detonation, these systems will be able to provide the communication links. The LF, VLF, ELF communications systems have been undergoing development, tests, and evaluation since the early 1960s.

[7]Carrier refers to a radio wave that is used to transmit information. Information being sent is impressed onto the carrier for it to carry the signal to its destination. At the receiver the carrier is filtered out, thereby allowing the original messages to reemerge.

[8]This section is drawn from *"Oulu Space Physics Textbook,"* spaceweb@oulu.fi. The author wishes to thank Dr. Reijo Rasinkangas and Ilya Usoskin, Thomas Ulich, Reijo Manninen, Jouni Jussila, and Jyrki Manninen for their generous intellectual contributions and for granting permission to reproduce in full various texts and figures.

[9]Earth is one of the planets that has a strong internal magnetic field. The forcing by the solar wind is able to modify this field, creating a cavity called the magnetosphere. This cavity shelters the surface of the planet from the high energy particles of the solar wind. The outer boundary of the magnetosphere is called the magnetopause. In front of the dayside magnetopause, another boundary called the bow shock is formed because the solar wind is supersonic. The region between the bow shock and the magnetopause is called the magnetosheath. At low-altitude limit, magnetosphere ends at the ionosphere. The magnetosphere is filled with plasma that originates both from the ionosphere and the solar wind. (see Fig. 3-35)

[10]Refers to angular frequency of gyration, cyclotron frequency, or larmor frequency. The magnitude of the angular velocity of a charged particle gyrating around its guiding center is given by w = $|q|B/m$ (in radians/s). The smaller the particle mass, the larger its gyrofrequency, and the higher the magnetic field, the higher is the gyrofrequency.

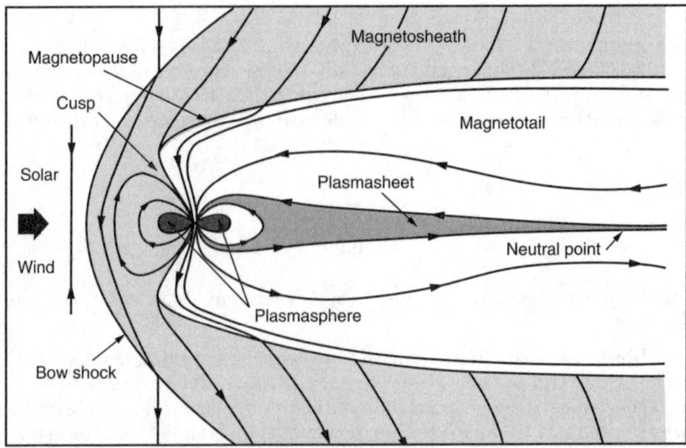

Figure 3-35 Effects of solar winds on Earth's magnetic field.

[11]Matter in the known universe is classified in terms of four states: solid, liquid, gaseous, and plasma. The basic distinction between solids, liquids and gases lies in the difference between the strength of the bonds that hold their constituent particles together. The equilibrium between particle thermal energy and the inter-particle binding forces determines the state. Heating of a solid or liquid substance leads to phase transition to a liquid or gaseous state, respectively. This takes place at a constant temperature for a given pressure and requires an amount of energy known as latent heat. On the other hand, the transition from a gas to an ionized gas, that is, plasma, is not a phase transition, for it occurs gradually with increasing temperature. During the process, a molecular gas dissociates first into an atomic gas which, with increasing temperature, is ionized as the collisions between atoms are able to free the outermost orbital electrons. Resulting plasma consists of a mixture of neutral particles, positive ions (atoms or molecules that have lost one or more electrons), and negative electrons. In an ionized plasma that is weak, the charge-neutral interactions are still important, while in ionized plasma that is strong, the multiple Coulomb interactions are dominant. On the Earth, plasmas usually do not occur naturally except in the form of lightning bolts, which consist of narrow paths of air molecules, of which approximately 20 percent are ionized, and in parts of flames. Most of the universe, however, consists of matter in the plasma state. The ionization is caused by high temperatures as described above (e.g., inside the sun and other stars), or by radiation, as in interstellar gases (e.g., solar wind) or, closer to the Earth, in the ionosphere defined in Fig. 3-36 and magnetosphere discussed earlier. Ionization appears at a number of atmospheric levels, producing layers or regions which may be identified by their interaction with radio waves. These layers are known as the D, E, and F layers, and their locations are shown in the Fig. 3-36 for both night and day conditions at mid-latitudes.

The first ionospheric layer found was the so called E layer or region at about 110 km altitude. It is used by radio operators as a surface from which signals can be reflected to distant stations. It is interesting to note that this works also the other way around and, for example, the auroral kilometric radiation, which refers to the burst of electromagnetic emission at about 100 to 500 kHz related to auroral arcs, created by the precipitating particles high above the ionosphere does not reach the ground because of the ionospheric E layer. Above the E layer, an F layer consisting of two parts can be found: F1 is at about 170 km, and F2 at about 250 km altitude. The F layer also reflects radio waves. The lowermost region of the ionosphere below 80 km altitude, D layer, however, principally absorbs radio waves. These will be discussed in greater detail later in this chapter.

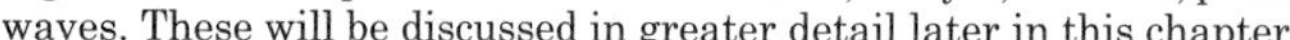

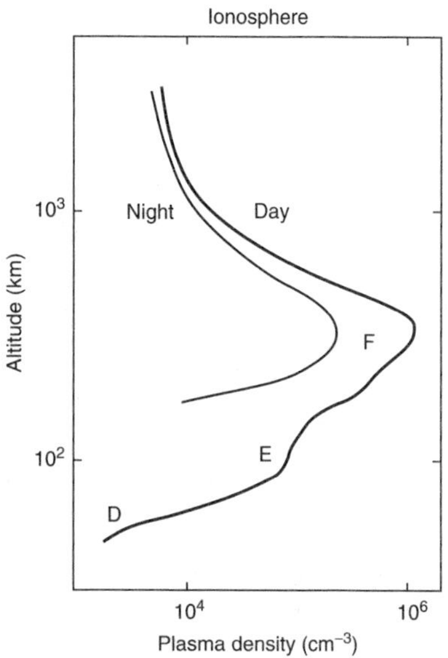

Figure 3-36 Ionization at various atmospheric layers.

[12]Stewart, B, On the great magnetic disturbance which extended from August 28 to September 7, 1859, as recorded by photography at the Kew Observatory, *Philos. Trans. R. Soc. London*, 423, 1861.

[13]Refer to Jacobs, J. A., Y. Kato, S. Matsushita, and V. A. Troitskaya, "Classification of geomagnetic micropulsations," *J. Geophys. Res., 69*, 180, 1964; Anderson, B. J., "An overview of spacecraft observations of 10 to 600 s period magnetic pulsations in the Earth's magnetosphere," in *Solar Wind Sources of Magnetospheric Ultra-Low-Frequency Waves*, M. J. Engebretson, K. Takahashi, and M. Scholer (eds.), *AGU Geophysical Monograph 81*, 25-43, 1994; Anderson, B. J., "Statistical studies of Pc 3-5 pulsations and their relevance for possible source mechanisms of ULF waves," *Ann. Geophysicae, 11*, 128-143, 1993.

[14]C.S. Clay, H. Medwin, "*Acoustical Oceanography: Principles and Applications*," Canada, Willey Interscience, 1977.

[15]Nicolas, B., J. Mars, J-L. Lacoume, and D. Fattaccioli, "*Are Ultra Low Frequency Waves Suitable For Detection*?" LABORATOIRE DES IMAGES ET DES SIGNAUX, Institut National Polytechnique de Grenoble, France, http://www.lis.inpg.fr/pages_perso/nicolas/recherche/oceans02.pdf

[16]Refer to the research articles regarding BioTENS—Ultra-Low-Frequency TENS: (1) Allgood JP. Transcutaneous electrical neural stimulation (TENS) in dental practice. Compend Contin Educ Dent 1986 October,7(9):640, 642-4; (2) Gomez CE, Christensen LV. Stimulus-response latencies of two instruments delivering transcutaneous electrical neuromuscular stimulation (TENS). *J Oral Rehabil.* 1991 January,18(1), 87-94.

[17]VLF transmissions are capable of higher data rates than ELF transmissions, but they are still limited. VLF broadcast systems use minimum shift keying (MSK) and operate low-speed, 50 Bd, secure teletypes. A very common mode of operation on VLF circuits is ICW. An anti-jam (AJ) capability does exist, but it reduces the data transmission rate dramatically—to about three characters every 12 s.

[18]Current shore-based LF communications systems use 50 to 100 kW transmitters and use frequency shift keying (FSK) for secure teletypewriter or International Morse Code (IMC) for communications operations. Using FSK and appropriate COMSEC equipment, LF can transmit in a secure teletype mode at 75 Bd (which equates to approximately 100 words per minute (wpm)).

[19]The 550 to 1600-kHz part of this frequency band is mainly used for AM broadcasting. A 10-kHz separation standard between stations results in 105 available audio channels. The MF band can support low-capacity multichannel circuits for both voice and teletype operations, with the latter limited to 75 Bd (100 wpm). Security is available through voice and data COMSEC devices.

[20]A wave striking a thin, very highly ionized layer, may be bent back so rapidly that it will appear to have been *reflected* instead of refracted back to Earth. To reflect a radio wave, the highly ionized layer must be approximately no thicker than one wavelength of the radio wave. Since the ionized layers are often several miles thick, ionospheric reflection is more likely to occur at long wavelengths (low frequencies).

[21]At any point in time, each ionospheric layer has a maximum frequency, known as the *Critical Frequency*, at which radio waves can be transmitted vertically and refracted back to Earth. Radio waves with frequencies higher than the critical frequency of a given layer will pass through the layer and be lost in space; but these same waves entering an upper layer with a higher critical frequency will be refracted back to Earth. Radio waves with frequencies lower than the critical frequency will also be refracted back to Earth unless they are absorbed or have been refracted from a lower layer. The lower the frequency of a radio wave, the more rapidly the wave is refracted by a given degree of ionization. Figure 3-14 illustrates these variations.

[22]The rate at which a wave of a given frequency is refracted by an ionized layer depends on the angle at which the wave enters the layer. For example, Figure 3-16 shows three radio waves of the same frequency entering a layer at different angles. Wave A strikes the

Figure 3-37 Simplified model of propagation in the ionospheric. (*Source: http://www.scs.ptc.com*)

layer at an angle that is too nearly vertical for the wave to be refracted to Earth. As a consequence, it is bent slightly but passes through the layer and is lost. If the wave is reduced to an angle that is less than vertical (refer to wave B), it strikes the layer and is refracted back to Earth. The angle made by wave B is called the *Critical Angle* for that particular frequency. Any wave that leaves the antenna at an angle *greater* than the critical angle will penetrate the ionospheric layer for that frequency and then be lost in space. Wave C strikes the ionosphere at the smallest angle at which the wave can be refracted and still return to Earth. At any smaller angle, the wave will be refracted but will not return to Earth. To understand the complex dynamics of the ionosphere, also refer to the material on Simplified Model of Propagation in the Ionosphere.

[23]Materials for this section heavily draw upon "*Ionospheric Physics of Radio Wave Propagation*," http://www.ecjones.org/physics.html. The author wishes to thank Dr. Edwin C. Jones for granting permission to reproduce in full various texts, charts, tables, and figures. A special word of thanks is directed to Dr. Jones for helping the author to improve the contextual coverage of this chapter.

[24]For PACTOR (PT) system design and operational details, refer to http://ecjones.org/pactor.html

[25]As of December 31, 2002, a total of 41,000 mi of driving have been logged in various HF propagation experiments.

[26]From statistical mechanics, the electrical conductivity of a free electron gas is given by

$$s = \frac{ne^2t}{m}$$

where e is the electron charge; n, the density of electrons; m, the electron mass; and t, the average time between electron collisions. The average time between electron collisions t can be calculated from

$$t = l_{\text{mfp}} \left(\frac{m}{3kT} \right)^{1/2}$$

where l_{mfp} is the mean free path between electron collisions; k, the Boltzmann constant; and T, the absolute temperature in Kelvins. The conductivity of the ionosphere also depends of the angle between the incidence of RF with the earth's magnetic field. Because this effect amounts to less than one order of magnitude difference between the perpendicular and parallel directions, the simplified model given above has ignored this effect.

[27]Inmarsat satellites have taken over a major portion of the maritime communications previously provided by HF systems. While flying over or near land, airplanes still use the HF channels when they are out of range of the VHF stations they communicate with; however, future aeronautical mobile-satellite service (AMSS) systems will likely replace those HF channels.

[28]Amplitude modulation of a carrier frequency (see Figure 3-5) results in two sidebands. The frequencies above the carrier frequency constitute what is referred to as the *upper sideband*, those below the carrier frequency, constitute the *lower sideband*. Transmission in which only one sideband is transmitted is called *single sideband* (SSB) transmission. Source: Federal Standard 1037C and Mil-Std-188.

[29]At recent World Radio Conferences, more and more HF spectrum from the fixed service has been reallocated to broadcasting, aeronautical mobile, and maritime mobile services. As this trend continues, lesser developed countries have become concerned about the amount of HF spectrum available in the fixed services for their domestic radiocommunications, such as for national broadcasting, mobile and fixed point-to-point radiocommunications.

[30]Single-channel VHF radios are portable, vehicular, or airframe mounted, and can usually be operated in motion. The larger multichannel systems are commonly mounted aboard ships and on 2 1/2 or 5-ton trucks in shelters, and require careful placing of directional antennas. Typical uses include short-range FM combat radio nets, radar, radio navigation, wideband line-of-sight (LOS) multichannel systems (repeatered or nonrepeatered) and television broadcasting. Many UHF systems are transportable by vehicle, aircraft, or ship. Some UHF satellite terminals are small enough and lightweight enough to be manpack portable. Common UHF applications are seen daily in local ambulance, fire, and police radio nets, with repeater operations being typical. On military installations, the nontactical intrabase radio (IBR) nets are usually VHF or UHF. UHF systems are capable of high-quality, reliable, and high-capacity transmissions with data rates of 2.4 Kbps and higher. UHF is used widely to provide secure/nonsecure voice, record, data, and facsimile service in both mobile and fixed configurations. Along with VHF, UHF is the band preferred for television.Military satellite terminals and troposcatter terminals have been designed for tactical operations, using SHF systems. These systems are transportable by 2 1/2 to 5-ton truck and have antenna dishes of varying size.

[31]Representatives of GPS equipment makers and users argue that "There is no question that UWB operations would increase the background noise in a given spectrum.... Increases in background noise in GPS frequency bands may reduce the ability of the GPS receiver to acquire a GPS signal or even to maintain tracking of a GPS signal, or cause errors in position or time accuracy. Any of these consequences is intolerable for a safety-of-life service such as GPS."

[32]For further details, refer to Bennett A. Kobb, "Wireless Spectrum Finder: Telecommunications, Government and Scientific Radio Frequency Allocations in the U.S., 30 MHz to 300 GHz," McGraw-Hill, New York 2001.

[33]This section draws heavily from the scholarly work done by Dedric Carter, Andrew Garcia, David Pearah (*Massachusetts Institute of Technology*), Stuart Buck, Donna

Dutcher, Devendra Kumar, Andres Rodriguez (*Harvard Law School*), "*Spread Spectrum: Regulation in Light of Changing Technologies*," Fall 1998, http://www-swiss.ai.mit.edu/6095/student-papers/fall98-papers/spectrum/whitepaper.html. The author is indebted to these scholars for their insight into the spread spectrum problem.

[34]The secure communications in and of itself is not sufficiently interesting as strong encryption and spoofing countermeasures can be added (perhaps at great cost) to existing narrow-band communications. The property of interest in spread spectrum transmission is the scheme's ability to provide point-to-point communications without explicit coordination of the speakers.

[35]Kobb, Ibid. 32.

[36]Reproduced with permission from Space Environment Laboratory, 325 Broadway, Boulder, CO 80303-3323, http://www.sel.noaa.gov/info/Radio.pdf

[37]This is reproduced with permission from Peter F. Guerrero, Director, Physical Infrastructure Issues, whose statement "Telecommunication: History and Current Issues Related to Radio Spectrum Management," was given before the Committee on Commerce, Science and Transportation, U.S. Senate and released for delivery, Tuesday, June 11, 2002. Also, refer to "United States National Spectrum Requirements: Projections and Trends," NTIA, U.S. Department of Commerce, March 1995.

Chapter

4

Applications

4.1 Transport of Information/Energy

The use of the radio frequency spectrum is evident in all the facilities, equipment, and many possibilities of today's society. For example, in the public and private sectors:

1. Telecommunication by radio, radio navigation, safety-of-life services and information exchange by radio are of fundamental importance.
2. Broadcasting serves our culture, education and entertainment, as depicted in Fig. 4-1.
3. Remote control tools add to our comfort.
4. Radio enables important research in many fields of science.

This is only a partial list, but the communications industries, regardless of the interests or the priorities, are now making life in our modern society function the way we have become accustomed to. Many industry observers discuss the several modern-day applications of radio frequencies,[1] including the:

1. Principal role that radio and television play in meeting information needs of more than two-thirds of the world's population, as illustrated in Fig. 4-2.
2. Applications of radio waves by millions of indispensable household microwave ovens.
3. Crucial use of radio for security and economy, nationally and worldwide.

Figure 4-1 Broadcasting from space. (*Source: Ryszard Struzak,* "Access to Spectrum/Orbit Resource & Principles of Spectrum Management," *School on Data and multimedia Commuinications, ICTP 7-25 February 2000 http://www.ictp.trieste.it/~radionet2000_school/lectures/struzak/AccesSpctrOrbICTP/tsld001.htm.*)

4. Creation of business. In spite of the present day economic fluctuations, the telecommunication sector alone has been one of the most profitable industries, after pharmaceuticals and diversified financials.[2] The industry analysts underscore that such a development leads to an increasing demand for radio frequencies.
5. Many industrial processes (e.g., electric power from orbit, as depicted in Fig. 4-3) and scientific experiments that radio waves have triggered or improved through their ingenious use.[3]

To better understand the various radio frequency applications, we ought to clearly identify and differentiate between two types of applications: "passive" radiocommunication service and "active" radio frequency applications.

Radio astronomy is a field of science that uses various kinds of radio telescopes to investigate radio waves generated by cosmic radio sources far into the depths of the universe. To conduct their research and investigations, radio astronomers commonly use "receive-only" facilities. A "receive-only" facility is called *passive*. When both a transmitter and a receiver are required for a particular scientific experiment, we define it as an *active* application. Thus, radio astronomy is exclusively a passive

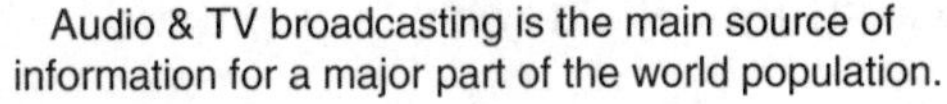

Audio & TV broadcasting is the main source of
information for a major part of the world population.

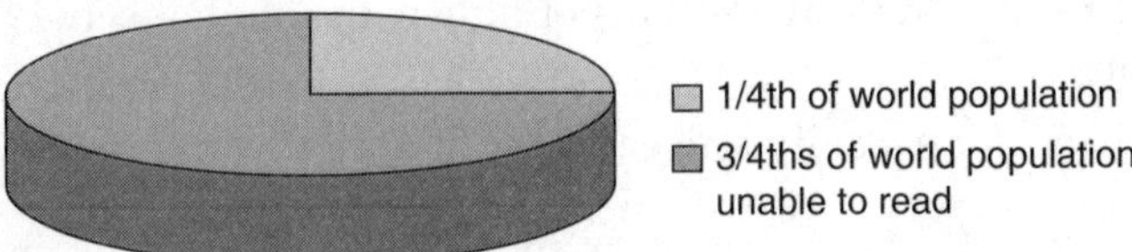

Figure 4-2 Role of radio and television.

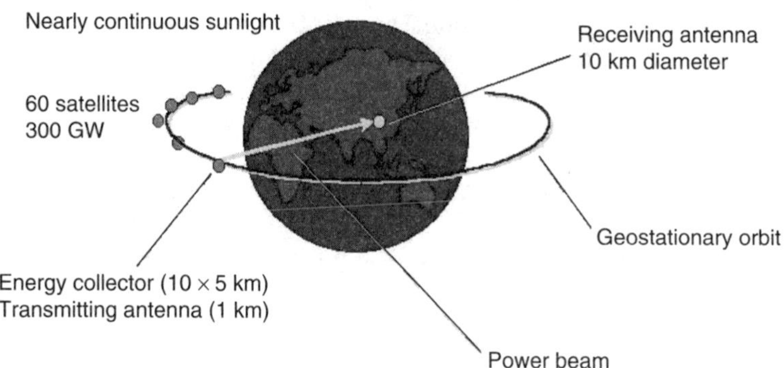

Figure 4-3 Electric power from orbit. (*Source: Struzak, Ibid.,* 1.)

radiocommunication service. Examples of active radio frequency applications include active sensing with a radiosonde for meteorological purposes and various radar-related experiments and facilities. Table 4-1 provides a more comprehensive listing of passive and active applications.[4]

While it is beyond the scope of this book to describe each application area in such great detail as would allow the reader to think that the subject has been fully discussed, the author wishes to elaborate on a number of common ones to state more clearly and precisely the way passive and active radio frequency applications are opening new windows on the universe and contributing to national defense, protection of the president and foreign officials, federal law enforcement, disaster relief, protection of national resources, the public safety of air and water transportation, the security of power generation and nuclear material, the

TABLE 4-1 Radio Frequency Applications

Passive applications	Active applications
Communications: voice, video, data	Defense, security
Exploration: universe and Earth	Disaster relief, emergency
Industrial processes	Weather prediction
Particle physics	Command, control
Medicine	Management
Veterinary	Location
Domestic applications	Navigation
Measurements	Education, training
Monitoring	Mass-media, culture
	Propaganda
	Amateur services

health and well-being of our military veterans, and the efficient operation of our postal services and several other societal needs.

Before entering into such discussions, a few introductory remarks about the factors that govern the choice of the radio frequencies of interest for particular applications may be relevant and useful. In radio astronomy, there are some constraints on frequency selection imposed by nature:

- In many instances, we cannot select at will the radio frequencies important for scientific research—nature tells us at which radio frequency, for example, an atom or molecule likes to transmit its emission.
- Celestial radio source never informs us beforehand when it likes to behave the way it does and at which frequency it prefers to do so.
  - Celestial source radiates radio waves that vary with time and frequency, with the physical conditions of the source determining the intensity and the polarization.
  - Propagation characteristics of the terrestrial troposphere and ionosphere provide guidance to science for the selection of its preferred radio frequencies.[5]
  - No human being will ever be able to have any control over the characteristics of cosmic radio sources.
- Specific information about a source is given by each part of the spectrum, thereby making it practically impossible to find two parts of the radio spectrum which would convey the same information and could be substituted for each other. Radio astronomy has to follow these constraints on frequency selection imposed by nature.
- Similarly, for atmospheric studies, or aeronomy, in cases of studies based on observation of atmospheric gases, the gases generate radio emission at one or more discrete frequencies that are called *spectral lines*.[6]

Having made a cursory review of the constraints and limitations on frequency selection, we should hasten to add that science has made steady progress, and it is in our best interest to accelerate and facilitate the scientific development, for in the absence of such development:

- Weather forecasting will become unreliable, causing alarm (e.g., public safety concern for air traffic) or an unwanted economic impact (e.g., unexpected flooding or storms).
- Measurement on deforestation or urbanization effects may suffer from inaccuracy, resulting in major environmental problems or erroneous major political decisions.
- Ionospheric data may suffer from insufficient quality, with negative impact on radiocommunications.[7]

- Radio frequencies used in various geodetic studies will feed the scientists with erroneous and misleading information, potentially causing dislocation for safety-of-life services.[8]

In industrial and commercial applications, the fundamental laws of physics do not constrain the frequency selection to the same extent as they do in the scientific use of the radio spectrum. Industrial and market planning and developments largely determine the spectrum-need of industrial and commercial applications. These developments can be actively influenced by the same industries, with well-steered technological development. When constraints develop, they usually tend to be of an industry-political nature or from a market point of view. Both depend on management knowledge, ability, and will.

Often the industrial-political problems arise because of the frequency-sharing issue between passive and active applications. The International Telecommunication Union (ITU) Regulations state that if a frequency band is allocated to more than one radiocommunication service, both of which have a primary status in that particular frequency band, *frequency sharing* must occur. If, however, a service has a secondary status in that band, it cannot interfere with the primary service nor claim protection. Instead, it must accept interference from a primary service unless some conditions are put on that primary service. In other words, a radiocommunication service with a secondary status in a frequency band in which other services have a primary status does not create a sharing situation.

Usually the requirements of the passive and active services are very different, thereby adding to the complicated issue of sharing that carries a potential threat of harmful interference. Table 4-2 summarizes the interference issues and the possible mutual relations between two radiocommunication services: the passive and active services.

If one radiocommunication service is adequately coordinated with another with which the frequency band is to be shared, harmful interference can be avoided. A National Regulatory Authority undertakes the coordination task. In regional coordination issues, it is the regional administrative[9] organization that plays a role. Chapter 5, Spectrum Management, will discuss the coordination process more thoroughly.

Against this backdrop of the constraints and limitations on frequency selection imposed by nature as well as the industry-political-market developments, the author will:

1. Provide sampling of a few passive and active applications to discuss the way radio frequencies are currently used.
2. Create a more elaborate table that lists particular unique applications supported by the various spectrum bands.

TABLE 4-2 Possible Mutual Relations between Radiocommunication Services

Relation between radiocommunication services	Sharing situation	Potential threat of harmful interference	Coordination required/recommended
1.—Frequency band allocated to more than one service, that is, these services have a primary allocation.	Yes	Yes	Yes—seek advice of Administration.
2.—Frequency band is allocated secondary service.	Yes	Yes—but secondary service has no right to complain.	Yes—support by Administration depends on goodwill of these Administrations only.
3.—Frequency band is not allocated but used by a radiocommunication service on the basis of an ITU-R footnote.	No	Yes—but secondary service has no right to complain.	No—support by Administration depends on good will of these Administrations with respect to the victim radiocommunication service.
4.—Adjacent band is used in which the service has no regulatory status.	No	No—but out-of-band or spurious emissions may occur when systems are not designed properly.	No—but in case of harmful interference from out-of-band or spurious emissions, the victim may submit a formal complaint to the national administration.
5.—A frequency band is used in which the service has no regulatory status.	No	Yes—the victim service has to accept this situation and try to find a solution.	No

SOURCE: *Science and Spectrum Management*, Committee on Radio Astronomy Frequencies (CRAF) Handbook for Frequency Management, February 2002, p. 110.

4.2 Active Applications

4.2.1 Defense, security

To perform congressionally-mandated functions dealing with national security and protection of federal property and military installations, effective and reliable radiocommunications are required for the U.S. Department of Defense (DoD).[10]

To support the above missions, the armed services (Army, Navy, Air Force), components, and agencies heavily use radio frequencies to support air-to-ground, air-to-air, and air-ground-air tactical communications.

For example, the 225–399.9 MHz band consisting of several band segments is used in support of the national security mission. Tactical fixed and mobile communications are performed in the 225–328.6 and 335–399.9 MHz portions of the band. The systems operating in these band segments support wartime functions; however, extensive peacetime training and alert exercises are conducted to maintain combat readiness. Military mobile satellite operations are conducted in the 235–322 and 335.4–399.9 MHz segments of the band. Military satellite communications are essential to link the activities of ground, air, surface, and subsurface mobile platforms. The communications functions performed in the 225–399.9 MHz band are critical to DoD operations, and this band is the single most critical spectrum resource of the tactical military forces.

The 399.9–401 MHz segment of the band is allocated to the radionavigation satellite and mobile satellite services.

The 420–450 MHz band is occupied by military land-based, airborne, and shipborne long-range search-and-surveillance radar systems used for national defense. The DoD has spectrum requirements for the radars that operate in this frequency band.

The 902–928 MHz band effectively supports a number of critical U.S. federal government requirements and a wide array of consumer and commercial applications. It is allocated for primary use by the federal government for radiolocation, (shipborne air search radar systems), and by users of industrial, scientific, and medical (ISM) devices. Use of the spectrum by federal government fixed and mobile and automatic vehicle monitoring (AVM) systems is secondary to both of these uses. The remaining users of the 902–928 MHz band, licensed amateur radio operators and unlicensed devices operating under Federal Communications Commission (FCC) Part 15 rules, operate on a secondary basis to all other uses, including AVM.

4.2.2 Emergency response

The 148–149.9 MHz band is used by many U.S. federal government agencies, including law enforcement, transportation, natural resource protection, emergency response, utility operation, and for tactical aviation communications. Satellite uplink operations in this band are authorized for the U.S. National Aeronautical and Space Administration (NASA), Department of Energy (DOE), National Science Foundation (NSF), and Department of State (DOS). This band is also essential to the support of the Civil Air Patrol and the United States Coast Guard (USCG) for search and rescue operations.

The 149.5–150.05 MHz band is one of the few internationally allocated bands to the mobile satellite service (MSS) (Earth-to-space).

The 406–406.1 MHz band is allocated to the mobile satellite service and is used for the Emergency Position Indicating Radio Beacon (EPIRB). The 406–406.1 MHz EPIRB signal frequency has been designated internationally for distress use only. The International Cospas-Sarsat Program announced it will terminate satellite processing of distress signals from 121.5 and 243 MHz emergency beacons. Mariners, aviators, and other persons will have to switch to emergency beacons operating at 406 MHz in order to be detected by satellites. The Emergency Locating Transmitters (ELT) for aircraft also operate in this band.

4.2.3 Navigation

Aviation communication and radionavigation. To ensure safe travel in the air, aeronautical communications and radionavigation services must have interference-free use of radio frequency spectrum. In the United States, the frequency bands 108–118, 118–137, 328.6–335.4, and 960–1215 MHz are allocated to the aeronautical radionavigation, aeronautical mobile (route), and radionavigation satellite services. These bands support a variety of civil, nonfederal, and DoD, provided aeronautical safety functions. These are shared government/nongovernment bands.

The Instrument Landing System (ILS), VHF Omnidirectional Range, and Local Area Augmentation Systems for Global Positioning System (GPS), which are used extensively by civil and federal aircraft, all use the 108–118 MHz band. The estimated number of ILS sites is projected at 1020 by the year 2005.

Air traffic control communications use the 118–137 MHz band. The United States Federal Aviation Administration (FAA) operates over 3000 radio stations in this band to serve all elements of air traffic services, weather alerting, landing and take-off, en route control, and emergency communications. In addition, FAA provides radio frequency support for the aeronautical communications needs of public safety agencies such as the U.S. Treasury, Customs, and Forestry for law enforcement and firefighting requirements on an *as required* basis.

A wide variety of aeronautical radionavigation systems, including:

- Distance measuring equipment
- Tactical air navigation system
- Air traffic control radar beacon system
- Universal access transceiver

use the 960–1215 MHz band. In addition, the band is authorized for use by DoD's Joint Tactical Information Distribution System on a non-harmful-interference basis. At the 2000 World Radio Conference, a new allocation was added to the 1164–1215 MHz portion of the band for a new GPS signal, GPS L5, to be used by international civil aviation. The FAA provides management of this band, as well as the other bands designated for the Aeronautical Assignment Group.

Maritime mobile service. The maritime mobile bands are used to communicate between and among fixed coast stations, ships, and other offshore vessels. National planning for the maritime mobile bands closely follows international use. In the United States, specific uses include the command, control, and communications of U.S. Navy and U.S. Coast Guard ships and vessels, distress and safety communications, treaty and law enforcement, drug interdiction, geological survey operations, and national marine fishery operations.

The maritime mobile service allocation includes 4808 kHz of spectrum in the 12 shared high frequency (HF) bands. This spectrum is used exclusively by the stations to the maritime mobile service. There are times, however, when it is used on a shared basis with other radio services. The federal agencies have approximately 13,000 assignments in the maritime mobile bands, while there are almost 10,000 assignments for nonfederal users.

4.2.4 Mass-media, culture, and propaganda

For decades, nations have made increasing use of the electromagnetic spectrum to conduct public diplomacy by broadcasting speech and music directly to receivers throughout the world. Broadcasting services that are transmitted across international borders are HF broadcast operations subject to the ITU Radio Regulations. In the United States, a total of 2930 kHz of HF spectrum is presently allocated for the broadcasting service within eight HF subbands. Their use includes: (1) operations by U.S.-based broadcasters who are trying to reach non-U.S. audiences; (2) operations by both U.S.-based and foreign-based broadcasters who are trying to reach U.S. audiences.

4.2.5 Amateur services

Amateur radio is a popular, technical hobby among its many users. In the United States, there are in excess of 650,000 amateur radio operators. It is estimated that there are over 2.4 million such operators worldwide. They provide a unique service to the public, for the amateur radio plays an important role in disaster-relief communications where amateurs provide radiocommunications independent of the telephone network or other radio services, particularly in the first few days before relief agencies are at the scene and have set up disaster telecommunications services.

In the HF band, the amateur operator has a choice of narrow frequency bands each with different propagation properties for long-distance communications. The amateur operator can follow the changing maximum usable frequency (MUF) as propagation conditions change in the amateur bands and still be able to communicate. Having a good selection of frequencies is critical to maintain reliable communications for both voice and data operations in the HF band.

Radio amateurs have made significant contributions to the field of radio propagation, HF single-sideband radio, HF data communications systems, packet radio protocols, and communications satellite design.

4.3 Passive Applications

4.3.1 Exploration: Universe and Earth

One of mankind's oldest and strongest fascinations and of great scientific, cultural, and practical value for many centuries is the exploration of the universe. For centuries, scientists and astronomers have been observing the electromagnetic spectrum at all wavelengths—both from the ground and space. These observations, especially the more recent ones, have led to the phenomenal progress in all areas of astronomy, including the exploration of the solar system, discoveries of the echo of the Big Bang, and the beginnings of the structure in the universe.[11]

Radio astronomy is a passive radio service based on the reception of radio waves of cosmic origin through the use of radio telescopes. Radio astronomy shares the radio frequency spectrum with other radio services. The radio astronomy community makes observations and conducts research in the radio frequency spectrum and other portions of the electromagnetic spectrum. This research has helped to evolve our knowledge of stellar physics, star formation, the interstellar medium, the evolution of the galaxy and the universe. The research has also made major contributions and led to innovation in telecommunications, medicine, industry, defense, and the environment, as shown in the Tables 4-3 through 4-9.[12]

The spectrum requirements for the radio astronomy service are different from those of most other radio services. Traffic requirements are not predictable and, in the case of spectral lines, involve very specific frequencies. The General Assembly of the International Astronomical Union (IAU), 1991, approved the lists of frequencies of the astrophysically most important spectral lines. Tables 4-7, 4-8, and 4-9 highlight the preferred frequency bands for radio astronomical measurements based on Recommendation 314-8, International Telecommunication Union (ITU)-R Recommendations, 1994 RA Series, Radioastronomy.

Man-made environmental problems of rapidly growing severity are jeopardizing continued scientific studies of the origin and evolution of the universe and mankind's place within it. For instance, the future of

TABLE 4-3 Medical Applications of Astronomical Techniques

Astronomical technique or device	Medical uses
3-D image reconstruction from one- and two-dimensional images	Imaging for CAT scans
	Magnetic resonance imaging
	Positron emission tomography
Microwave receivers	Scans for breast cancer
Image-processing software (IRAF and AIPS) developed by NRAO, NOAA, and NASA	Cardiac angiography
	Monitoring neutron activity in brain
Positive pressure clean rooms for assembly of space instruments	Cleaner hospital operating rooms
Detection of faint x-ray sources	Portable x-ray scanners (Lixiscope) for neonatology and third world clinics

TABLE 4-4 Industrial Applications of Astronomical Techniques

Astronomical technique or device	Industrial uses
Image-processing software (AIPS, IRAF)	General Motors Corp. study of automobile crashes; Boeing Co. tests of aircraft hardware
Holographic methods for testing figures of radio telescopes	Testing communications antennas
Development of low-noise receivers	Components for communications industry
FORTH computer language developed by NRAO for control of radio telescopes	20 vendors supply FORTH for applications including analysis of auto engines in 20,000 garages, quality control for films at Kodak, 50,000 hand-held computers used by express mail firm
Gold sensitization of photographic plates	Development of Tri-X and 400-ASA films by Kodak
Infrared-sensitive films for spectroscopy	Aerial reconnaissance and Earth resources mapping
X-ray detectors for NASA telescopes	Baggage scanners at airports
Gas Chromatographs to search for life on Mars	

TABLE 4-5 Defense Applications of Astronomical Techniques

Astronomical technique or device	Defense uses
Stellar observations and model atmospheres	Discrimination of celestial objects from rocket plumes, satellites, and warheads
Infrared all-sky survey by NASA's IRAS satellite	
Detectors for gamma-ray and x-ray astronomy	Vela satellite monitors for nuclear explosions
Positions of quasar and stars	Precision navigation for civil and military purposes

radio astronomy and the operation of scientific satellites are of major concern owing to the interference at radio frequencies from telecommunications satellites and their ever-increasing demand for frequency space.[13] In the same vein, space debris interferes with ground-based observations and presents a growing threat to scientific satellites.[14] Observational astronomy is also negatively impacted by several projects that launch bright objects into space for earth illumination, artistic, celebratory, or advertising purposes. Also, on the ground, man-made light pollution is making many large areas of the world unsuitable for astronomical observations.

Given these symptoms, most nations are cooperating at the national and regional level with industry and through the ITU to:

1. Preserve quiet frequency bands for radio astronomy[15] and find practicable technical solutions that reduce radio emissions and other undesirable side-effects from telecommunications satellites;

TABLE 4-6 Environmental Applications of Astronomical Techniques

Astronomical technique or device	Environmental uses
Millimeter wave spectroscopy	Study of ozone depletion
Models of planetary atmospheres	Global change modeling
Study of sunspots and solar flares in sun and stars	Short- and long-term predictions of terrestrial effects
Models of astrophysical shocks	Study of terrestrial storms
Precision measurement of quasars	Geodesy and study of tectonic drift
Composite materials for orbiting infrared telescope	Design of solar collectors
Theory of cosmic rays, solar flares, and stellar fusion	Design of fusion reactors

TABLE 4-7 Radio Frequency Lines of the Greatest Importance to Radio Astronomy at Frequencies below 275 GHz

Substance	Rest frequency	Suggested minimum band	Notes (1)
Deuterium (DI)	327.384 MHz	327.0–327.7 MHz	2,3
Hydrogen (HI)	1420.406 MHz	1370.0–1427.0 MHz	
Hydroxyl radical (OH)	1612.231 MHz	1606.8–1613.8 MHz	4
Hydroxyl radical (OH)	1665.402 MHz	1659.8–1667.1 MHz	4
Hydroxyl radical (OH)	1667.359 MHz	1661.8–1669.0 MHz	4
Hydroxyl radical (OH)	1720.530 MHz	1714.8–1722.2 MHz	3, 4
Methyladyne (CH)	3263.794 MHz	3252.9–3267.1 MHz	3, 4
Methyladyne (CH)	3335.481 MHz	3324.4–3338.8 MHz	3, 4
Methyladyne (CH)	3349.193 MHz	3338.0–3352.5 MHz	3, 4
Formaldehyde (H_2CO)	4829.660 MHz	4813.6–4834.5 MHz	3, 4
Methanol (CH_3OH)	6668.518 MHz	6661.8–6675.2 MHz	3, 6
Helium ($3He^+$)	8665.650 MHz	8657.0–8674.3 MHz	3, 6
Methanol (CH_3OH)	12.178 GHz	12.17–12.19 GHz	3, 6
Formaldehyde (H_2CO)	14.488 GHz	14.44–14.50 GHz	3, 4
Cyclopropenylidene (C_3H_2)	18.343 GHz	18.28–18.36 GHz	3, 4, 6
Water vapor (H_2O)	22.235 GHz	22.16–22.26 GHz	3, 4
Ammonia (NH_3)	23.694 GHz	23.61–23.71 GHz	4
Ammonia (NH_3)	23.723 GHz	23.64–23.74 GHz	4
Ammonia (NH_3)	23.870 GHz	23.79–23.89 GHz	4
Silicon monoxide (SiO)	42.821 GHz	42.77–42.86 GHz	
Silicon monoxide (SiO)	43.122 GHz	43.07–43.17 GHz	
Carbon monosulphide (CS)	48.991 GHz	48.94–49.04 GHz	
Deuterated formylium (DCO^+)	72.039 GHz	71.96–72.11 GHz	3
Silicon monoxide (SiO)	86.243 GHz	86.16–86.33 GHz	
Formylium ($H^{13}CO^+$)	86.754 GHz	86.66–86.84 GHz	
Silicon monoxide (SiO)	86.847 GHz	86.67–86.93 GHz	
Ethynyl radical (C_2H)	87.3 GHz	87.21–87.39 GHz	5
Hydrogen cyanide (HCN)	88.632 GHz	88.34–88.72 GHz	4
Formylium (HCO^+)	89.189 GHz	88.89–89.28 GHz	4
Hydrogen isocyanide (HNC)	90.664 GHz	90.57–90.76 GHz	
Diazenylium (N_2H^+)	93.174 GHz	93.07–93.27 GHz	
Carbon monosulphide (CS)	97.981 GHz	97.65–98.08 GHz	4
Carbon monoxide ($C^{18}O$)	109.782 GHz	109.67–109.89 GHz	
Carbon monoxide (^{13}CO)	110.201 GHz	109.83–110.31 GHz	4
Carbon monoxide ($C^{17}O$)	112.359 GHz	112.25–112.47 GHz	6
Carbon monoxide (CO)	115.271 GHz	114.88–115.39 GHz	4
Formaldehyde ($H_2{}^{13}CO$)	137.450 GHz	137.31–137.59 GHz	3, 6
Formaldehyde (H_2CO)	140.840 GHz	140.69–140.98 GHz	
Carbon monosulphide (CS)	146.969 GHz	146.82–147.12 GHz	
Water vapor (H_2O)	183.310 GHz	183.12–182.50 GHz	
Carbon monoxide ($C^{18}O$)	219.560 GHz	219.34–219.78 GHz	
Carbon monoxide (^{13}CO)	220.399 GHz	219.67–220.62 GHz	4
Carbon monoxide (CO)	230.538 GHz	229.77–230.77 GHz	4
Carbon monosulphide (CS)	244.953 GHz	244.72–245.20 GHz	6
Hydrogen cyanide (HCN)	265.886 GHz	265.62–266.15 GHz	

TABLE 4-7 Radio Frequency Lines of the Greatest Importance to Radio Astronomy at Frequencies below 275 GHz (*Continued*)

Substance	Rest frequency	Suggested minimum band	Notes (1)
Formylium (HCO^+)	267.557 GHz	267.29–267.83 GHz	
Hydrogen isocyanide (HNC)	271.981 GHz	271.71–272.25 GHz	

NOTES:

1. If Note 4 or Note 2 is not listed, the band limits are the Doppler-shifted frequencies corresponding to radial velocities of ± 300 km/s (consistent with line radiation occurring in our galaxy).
2. An extension to lower frequency of the allocation of 1400—1427 MHz is required to allow for the higher Doppler shifts for HI observed in distant galaxies.
3. The current international allocation is not primary and/or does not meet bandwidth requirements. See the Radio Regulation for more detailed information.
4. Because these line frequencies are also being used for observing other galaxies, the listed bandwidths include Doppler shifts corresponding to radical velocities of up to 1000 km/s. It should be noted that HF had been observed at frequencies redshifted to 500 MHz, while some lines of the most abundant molecules have been detected in galaxies with velocities up to 50,000 km/s, corresponding to a frequency reduction of up to 17 percent.
5. There are six closely spaced lines associated with this molecule at this frequency. The listed band is wide enough to permit observations of all six lines.
6. This line frequency is not mentioned in Article 8 of the Radio Regulations.

TABLE 4-8 Radio Frequency Lines of the Greatest Importance to Radio Astronomy at Frequencies between 275 and 811 GHz (Not Allocated in the Radio Regulations)

Substance	Rest frequency (GHz)	Suggested minimum band (GHz)
Diazenylium (N_2H^+)	279.511	279.23–279.79
Carbon monoxide ($C^{18}O$)	329.330	329.00–329.66
Carbon monoxide (^{13}CO)	330.587	330.25–330.92
Carbon monosulphide (CS)	342.883	342.54–343.23
Carbon monoxide (CO)	345.796	345.45–346.14
Hydrogen cyanide (HNC)	354.484	354.13–354.84
Formylium (HCO^+)	356.734	356.37–357.09
Diazenylium (N_2H^+)	372.672	372.30–373.05
Water vapor (H_2O)	380.197	379.81–380.58
Carbon monoxide ($C^{18}O$)	439.088	438.64–439.53
Carbon monoxide (^{13}CO)	440.765	440.32–441.21
Carbon monoxide (CO)	461.041	460.57–461.51
Heavy water (HDO)	464.925	464.46–465.39
Carbon (CI)	492.162	491.66–492.66
Water vapor ($H_2{}^{18}O$)	547.676	547.13–548.22
Water vapor (H_2O)	556.936	556.37–557.50
Ammonia ($^{15}NH_3$)	572.113	571.54–572.69
Ammonia (NH_3)	572.498	571.92–573.07
Carbon monoxide (CO)	691.473	690.78–692.17
Hydrogen isocyanide (HNC)	797.433	796.64–798.23
Formylium (HCO^+)	802.653	801.85–803.85
Carbon monoxide (CO)	806.652	805.85–807.46
Carbon (CI)	809.350	808.54–810.16

TABLE 4-9 Frequency Bands Allocated to the Radio Astronomy Service That Are Preferred for Continuum Observations (Secondary Allocations Are Contained within Brackets)

Frequency band (MHz)	Bandwidth (%)	Frequency band (GHz)	Bandwidth (%)
13.360–13.410	0.37	10.6–10.7	0.94
25.550–25.670	0.49	15.35–15.4	0.33
[37.5–38.25]	[1.98]	22.21–22.50	1.30
73–74.6*	2.17	23.6–24.0	1.68
150.05–153†	1.95	31.3–31.8	1.58
322–328.6	2.03	42.5–43.5	2.33
406.1–410	0.96	86–92	6.74
608–614‡	0.98	105–116	9.95
1400–1427	1.91	164–168	2.41
1660–1670	0.60	217–231	6.25
2690–2700 [2655–2690]	0.37 [1.31]	265–275	3.70
49905000 [48004990]	0.20 [3.88]		

NOTES:

For the allocation of frequencies the world has been divided into three Regions as shown in Fig. 4-4

*Allocation (primary) in Region 2, protection recommended in Regions 1 and 3.

†Allocation (primary) in Region 1, Australia, and India.

‡Allocation (primary) in Region 2, China, and India.

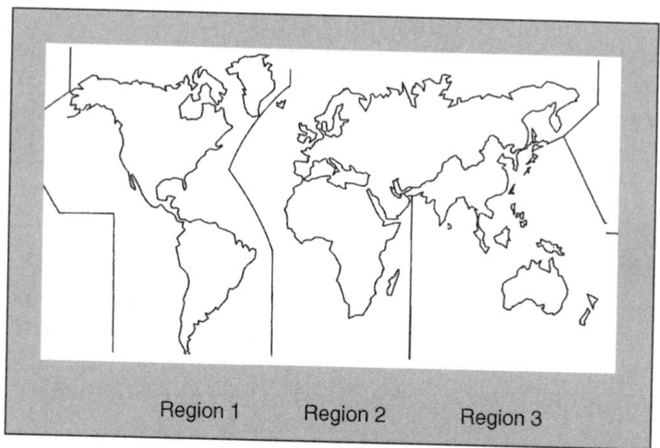

Figure 4-4 International Telecommunication Union regions of the world. (*Source:* U.S. Department of Commerce, National Telecommunications and Information Administration, *Tables of Frequency Allocations and Other Extracts* From: *Manual of Regulations and Procedures For Federal Radio Frequency Management*, September 1989 ed., p. 4–30, as referenced in *The 1992 World Administrative Radio Conference: Issues for U.S. International Spectrum Policy*, November 1991, OTA-BP-TCT-76, NTIS order #PB92-157601.)

2. Protect particular regions of the Earth and space from radio emissions (radio quiet zones) with innovative techniques that optimize the conditions for scientific and other space activities. Such optimization will result in the coexistence of these activities in radio frequency space and physical space in the future;

3. Protect "the near and outer space environment through further research in, and implementation of, measures to control and reduce the amounts of space debris and unwanted emissions at all wavelengths of the electromagnetic spectrum."[16]

4.3.2 Weather prediction

The 401–402, 402–403, and 403–406 MHz bands are allocated on a worldwide basis for the meteorological aids service. In addition, the 402–403 MHz band is also allocated to Earth Exploration Satellites and Meteorological Satellites. Equipment employed in these bands provides many of the observations and weather reports, issues forecasts, and warns of weather and flood conditions affecting national safety and the economy. The data collected by the equipment used in these bands are shared among academic research programs, private weather-forecasting firms, federal agencies, state, and local governments. Because of the national and agency specific meteorological functions and requirements, the federal government is the largest user of meteorological aids equipment in these bands. These bands are important for weather prediction as well as the services provided to the public by the systems operating in these bands.

The 608–614 MHz band is allocated for radio astronomy observations and land mobile (medical telemetry and telecommand operations only). These frequencies are important for radio astronomy and, accordingly, are allocated on an exclusive basis for radio astronomy observations and the extreme sensitivity of the receivers.

4.3.3 Domestic applications

In 1996, President Clinton (by Executive Order No. 13010) recognized the railroad, water, and energy industries as part of the nation's critical infrastructure. These entities provide commodities and services that are essential to daily life. Table 4-10 illustrates the three industries and the spectrum and applications currently used by each.

The energy, water, and railroad industries currently use spectrum between 20 MHz and 25 GHz for technologies and applications vital to their core operations. This spectrum range is congested and quickly approaching critical mass, which could lead to problems of interference. The anticipated frequency requirements for each industry are detailed in the following sections.

TABLE 4-10 Spectrum and Applications Currently Used as Indicated by the Energy, Water, and Railroad Industries

	Energy industry	Water industry	Railroad industry
40 MHz	48–50 MHz: voice dispatch, alarms from remote stations		
	150–175 MHz: alarms from remote substations, PLMRS		
	470–512 MHz: PLMRS		End of train devices
	Trunked PLMRS		
	806–821 MHz: PLMRS; 821–824 MHz: PLMRS		
	851–866 MHz: PLMRS; 866–896 MHz: PLMRS		
	896–901 MHz: PLMRS		896 MHz: ATCS/PTC
	902–928 MHz: SCADA		902–928 MHz: LMS
	928–929 MHz: POFS	928 MHz: MAS	928 MHz: MAS
	928/932/941 MHz: MAS; 952/956/959 MHz: MAS	952 MHz: MAS	936 MHz: ATCS/PTC
	928–952 MHz: SCADA; 929–930 MHz: PLMRS, 932–935 MHz, POFS	956 MHz: MAS	
	932–941 MHz, SCADA, 935–940 MHz, PLMRS, 941–944 MHz, POFS		
	952–960 MHz, POFS, 956 MHz, mobile meter reading		952 MHz, 956 MHz, MAS
	2.4 GHz band, point-to-point microwave	Water operations network	Point-to-point microwave
	5.8 GHz, 5.9–6.4 GHz, point-to-point microwave		
	6.5–6.8 GHz, point-to-point microwave	Water operations network	Point-to-point microwave
	6.525–6.875 GHz, POFS	Water operations network	Point-to-point microwave
			Point-to-point microwave

a. 2.11–2.2 GHz, 2.45–2.5 GHz and 2.65–2.69 GHz. 47 CFR§ 101.147(a).
b. *Id.*
c. 5.925–6.875 GHz. 47 CFR§ 101.147(a).
d. *Id.*
e. 10.7–12.2 GHz. 47 CFR§ 101.147(a).
f. *Id.*
g. 18–19 GHz. 47 CFR§ 101.147(a).
h. 23–23.6 GHz. 47 CFR§ 101.147(a).

SOURCE: Marshall W. Ross and Jeng F. Mao, NTIA Special Publication 01–49, *Current and Future Spectrum use by the Energy, Water, and Railroad Industries,* Response to Title II of the Departments of Commerce, Justice, and State, the Judiciary, and Related Agencies Appropriations Act, 2001, Public Law 106–553, U.S. Department of Commerce, January 2002.

The energy industry. The United Telecom Council recommends spectrum in the 450, 800, and 900 MHz bands be allocated exclusively for voice and data communications for utilities. DTE Energy recommends access to unused television channels (bands between 1 and 12 GHz for fixed narrow and medium-wide data channels as other preferred spectrum) should be allocated to utilities on a low-powered noninterfering basis for voice and data communications.

Likewise, Itron, Inc., suggests that the 1427–1432 MHz band should be licensed for utility telemetry services such as Automatic Meter Reader and Supervisory Control and Data Acquisition. The National Rural Telecommunications Council states that access to the 220 MHz band for Supervisory Control and Data Acquisition applications allows rural electric and telephone cooperatives to transmit telemetry data over wide distances at reduced costs compared to land line or high frequency wireless alternatives.

The water industry. The American Water Works Association argues that the 1998 Final Report of the Utilities Spectrum Assessment Taskforce (USAT) of the United Telecom Council underestimated spectrum requirements for the utilities industries based on industry trends and the pace of development of telecommunications technology. Table 4-11 summarizes the spectrum prediction of the USAT report, derived from projections of future wireless applications and growth.

Another company, Data Flow Systems, specifically identifies the 216–220 MHz band as important for water utility telemetry uses nationwide.

The railroad industry. The Association of American Railroads (ARA) suggests that the 700 MHz "guard band" should be a source for additional spectrum, divided into geographic sectors, each with a separate band manager. The 700 MHz band is currently occupied by broadcast television stations.

The ARA also suggests the 1.4 GHz band as a source for the proposed Land Mobile Communications Service for itself and other members of the Land Mobile Communications Council (LMCC). The ARA and other members of the LMCC have previously asked the FCC for spectrum in

TABLE 4-11 USAT Final Report Spectrum Requirements

Year	2000	2004	2010
Additional bandwidth required	1.0 MHz	1.9 MHz	6.3 MHz

SOURCE: Marshall W. Ross and Jeng F. Mao, NTIA Special Publication 01–49, *Current and Future Spectrum Use By the Energy, Water, and Railroad Industries*, Response to Title II of the Departments of Commerce, Justice, and State, the Judiciary, and Related Agencies Appropriations Act, 2001, Public Law 106–553, U.S. Department Of Commerce, January 2002.

TABLE 4-12 Summary of Frequency Bands That Could be Used as Indicated by the Energy, Water, and Railroad Industries

Energy industry	Water industry	Railroad industry
220 MHz band	216–220 MHz band	700 MHz band*
450 MHz band	6 GHz band	1.4 GHz band
800 MHz band	11 GHz band	
900 MHz band	23 GHz band	
1427–1432 MHz band		
1–12 GHz band		

*Although the AAR mentioned the 700 MHz Guard Band, this spectrum will also be available to the energy and water industries by leasing spectrum from the "Guard Band Managers." More information on the 700 MHz band can be found on page 6-3, Ross and Mao.

SOURCE: *Marshall W. Ross and Jeng F. Mao*, Ibid.

the 1.4 GHz band (specifically, the 1390–1395 MHz/1427–1429 MHz/1432–1435 MHz bands), and to limit auctions in the 1392–1395 MHz and 1432–1435 MHz bands to band managers.

The events of September 11, 2001, have underlined the importance of these industries and the role they play not only in our daily lives, but in times of disaster response and recovery. When the World Trade Center collapsed, utilities needed to respond. When airlines were grounded, people and commerce relied more on the railroad industry for transportation.

Observations in this section are based predominantly on comments from the industries and information from their oversight and regulatory federal agencies. The author is not able to validate specific requirements and issues such as exclusivity and congestion; however, Table 4-12 presents a summary of the spectrum bands identified by the energy, water, and railroad industries to meet U.S. domestic applications in critical infrastructure areas.

Section 4.4 provides the summary Appendix for a quick overview of the various spectrum uses. The summary is based purely on U.S. spectrum requirements and allocations. It draws heavily upon the PRIMARY and Secondary allocations listed in the U.S. national table and footnotes. It is not all-inclusive in its portrayal of passive and active applications that were discussed in the preceding sections; however, it serves its underlying purpose to emphasize, to the greatest extent possible, the unique applications supported by the various spectrum bands.

4.4 Appendix: Spectrum Use Summary 137 MHz to 10 GHz

Frequency (MHz)	Non-government allocation	Non-government use	Government allocation	Government use
137–138	Meteorological Satellite (space-to-Earth) Space Operation (space-to-Earth) Space Research (space-to-Earth) Mobile Satellite ((space-to-Earth) 137–137.025 and 137.175–137.825) Mobile Satellite ((space-to-Earth) 137.025–137.175 and 137.825–138)	The FCC has allocated this band for operations using nongeostationary nonvoice mobile satellite systems (Little LEOs).	Meteorological Satellite (space-to-Earth) Space Operation (space-to-Earth) Space Research (space-to-Earth) Mobile Satellite ((space-to-Earth) 137–137.025 and 137.175–137.825) Mobile Satellite ((space-to-Earth) 137.025–137.175 and 137.825–138)	Worldwide use of polar orbiting satellites for transmission of weather pictures occurs in this band via the TIROS system. The satellite also transmits tracking and telemetry information. NASA conducts satellite operations for the Advanced Technology Satellite (ATS) and High Energy Transient Experiment (HETE). Government use of the mobile-satellite service is limited by US319 to earth stations operating with nongovernment satellites.
138–144			Fixed Mobile	This band is primarily used for nontactical military land-mobile communications essential to maintain DoD infrastructure-related functions. It is also used throughout the U.S. for critical military air-traffic and tactical communications. Specific functions for tactical training include air-ground-air communications for combat weapons training carried out in the vicinity of all major bases and military training areas in the U.S. Also, this band is essential to the activities of the Air Force Auxiliary (Civil Air Patrol) and Coast Guard Auxiliary for support of search and rescue operations.

144–148	Amateur Amateur Satellite (144–146)	Weak signal modes (144–144.3), repeaters and other modes (144.3–147.99) Active use by amateur satellites worldwide (145.8–146)		
148–149.9	Mobile Satellite (Earth-to-space) Space Operations (Earth-to-space)(FN608)	The FCC has allocated this band for operations using nongeostationary nonvoice mobile satellite systems (Little LEOs).	Fixed Mobile Mobile-Satellite (Earth-to-space) Space Operations (Earth-to-space)(FN608)	This band is primarily used for nontactical military land-mobile communications essential to maintain DoD infrastructure-related functions. It is also used throughout the U.S. for critical military air-traffic and tactical communications. Specific functions for tactical training include air-ground-air communications for combat weapons training carried out in the vicinity of all major bases and military training areas in the U.S. A TIROS command link operates in the band in accordance with Footnote 608. NASA conducts satellite operations for the ATS. Also, this band is essential to the activities of the Air Force Auxiliary (Civil Air Patrol) and Coast Guard Auxiliary for support of search and rescue operations. Government use of the mobile-satellite service is limited by US319 to earth stations operating with nongovernment satellites.

(*Continued*)

Frequency (MHz)	Non-government allocation	Non-government use	Government allocation	Government use
149.9–150.05	Radionavigation satellite Mobile Satellite (Earth-to-space)	The FCC has allocated this band for operations using nongeostationary nonvoice mobile satellite systems (Little LEOs). Commercial shipping makes extensive use of TRANSIT-SAT signals for radionavigation.	Radionavigation Satellite Mobile-Satellite (Earth-to-space)	Government use of the mobile-satellite service is limited by US319 to earth stations operating with nongovernment satellites.
150.05–150.8			Fixed Mobile	This band is primarily used for nontactical military land-mobile communications essential to maintain DoD infrastructure-related functions. It is also used throughout the U.S. for critical military air-traffic and tactical communications. Specific functions for tactical training include air-ground-air communications for combat weapons training carried out in the vicinity of all major bases and military training areas in the U.S.
150.8–156.2475	Land Mobile	Land transportation (150.8–150.98, 152.255–152.465), public safety (150.98–151.4825, 154.6375–156.2475), industrial (151.4825–151.4975,		

		152.465–152.495, 152.855–153.7325, 154.4825–154.6375), industrial and public safety (151.4975–152, 153.7325–154.4825), domestic public (152–152.255, 152.495–152.855), Earth telecommand (154.2)		
156.2475–157.0375	Maritime Mobile	In accordance with international agreements, this band is used worldwide for maritime communications.		
157.0375–157.1875			Maritime Mobile	This band is critical to national VHF distress system communications associated with response to distress signals.
157.1875–157.45	Maritime Mobile	In accordance with international agreements, this band is used worldwide for maritime communications.		
157.45–161.575	Land Mobile	Land transportation (157.45–157.725, 159.48–161.575), public		

(Continued)

Frequency (MHz)	Non-government allocation	Non-government use	Government allocation	Government use
		safety (158.715–159.48), industrial (157.725–157.755, 158.115–158.475), domestic public (157.755–158.115)		
161.575–161.625	Maritime Mobile	In accordance with international agreements, this band is used worldwide for maritime communications.		
161.625–161.775	Land Mobile	Remote pickup broadcast		
161.775–162.0125	Maritime Mobile	In accordance with international agreements, this band is used worldwide for maritime communications.		
162.0125–174	Fixed (173.2–173.4) Land Mobile (173.2–173.4) Fixed (US13 hydrological and meteorological data-designated frequencies)	Industrial, public safety, police radio for stolen vehicle recovery systems (173.075).	Fixed (162.0125–173.2 and 173.4–174) Mobile (162.0125–173.2 and 173.4–174)	This band supports many federal nontactical fixed and land-mobile uses. These uses are critical to Departments of Agriculture and Interior fire fighting, FAA windshear reporting, NOAA weather radio, Department of Interior land and resource management, including flash flood warning, earthquake/volcano monitoring, wildlife telemetry and law enforcement activities

				throughout the federal government. law enforcement applications include land-based and maritime operations. federal agencies began shifting operations to narrowband technologies in 1995.
174–216	Broadcasting	This band is used for VHF TV channels 7–13. Also, wireless microphones and auxiliary broadcasting systems operate on a secondary basis.		
216–220	Maritime Mobile Fixed Land Mobile Aeronautical Mobile Amateur (219–220)	This band is used on inland waterways by Automated Maritime Telecommunications Systems. The FCC has set aside the 218–219 portion of this band for the interactive video data service (IVDS). Amateurs use this band for fixed point-to-point digital message forwarding systems.	Maritime Mobile Radiolocation Fixed Aeronautical Mobile Land Mobile	Though allocated secondary, there continue to be critical federal radiolocation requirements in this band. The U.S. Navy operates the SPASUR system in the band 216.88–217.08 MHz at several locations in the southern U.S. for the purpose of detecting Earth orbiting satellites. Assignments to the fixed and mobile service may be made on condition of no harmful interference to the SPASUR system (US229).
220–222	Land Mobile	Various trunked and conventional data users operate mobile systems. The band is broken into 200 5-kHz channel pairs.	Land Mobile Radiolocation	This band is shared by the federal government and private sector for narrowband technologies. However, the federal government will relinquish its co-primary status on 125 non-nationwide channels concurrent with the

(*Continued*)

Frequency (MHz)	Non-government allocation	Non-government use	Government allocation	Government use
				FCC's adoption of final and effective rules to license those channels pursuant to competitive bidding. Though allocated secondary, there continue to be critical federal radiolocation requirements in this band.
222–225	Amateur	Weak signal modes (222–222.15), repeaters, packet radio and other modes (222.15–225)	Radiolocation	Though allocated secondary, there continue to be critical federal radiolocation requirements in this band.
225–328.6			Fixed (G27 military only) Mobile (G27 military only) Mobile-Satellite (G100 235–322 and 335.4–399.9, military only)	These bands are heavily used worldwide for critical military air-traffic control and tactical training communications. Specific functions of tactical training include air-ground-air communications for combat weapons training carried out at and in the vicinity of all major air bases and military training areas worldwide. Tactical and strategic military satellite communications, essential to linking the activities of ground, air, surface, and subsurface mobile platforms, are conducted in this band under G100. Also, rocket test and test data telemetry operations are performed in this band.
328.6–335.4	Aeronautical Radionavigation	Commercial aircraft use the Instrument Landing Systems (ILS)	Aeronautical Radionavigation	This band is set aside on a worldwide basis for operation of aircraft ILS glideslope signal and serves as a critical

		glideslope for approach and landing.		part of the National Airspace System. ILS service to international carriers is required under agreements with the International Civil Aviation Organization (ICAO).
335.4–399.9			Fixed (G27 military only) Mobile (G27 military only) Mobile-Satellite (G100 235–322 and 335.4–399.9, military only)	These bands are heavily used worldwide for critical military air-traffic control and tactical training communications. Specific functions of tactical training include air-ground-air communications for combat weapons training carried out at and in the vicinity of all major air bases and military training areas worldwide. Tactical and strategic military satellite communications, essential to linking the activities of ground, air, surface, and subsurface mobile platforms, are conducted in this band under G100. Also, rocket test and test data telemetry operations are performed in this band.
399.9–400.05	Radionavigation Satellite Mobile Satellite (Earth-to-space)	Commercial shipping makes extensive use of TRANSIT-SAT signals for radionavigation.	Radionavigation Satellite Mobile-Satellite (Earth-to-space)	TRANSIT-SAT (polar orbiting satellite) downlink transmissions in this band support worldwide navigation. Government use of the mobile-satellite service is limited by US319 to earth stations operating with nongovernment satellites.
400.05–400.15	STD FREQ & TIME		STD FREQ & TIME	This band is set aside on a worldwide basis for distribution, via satellite, of standard time and frequency signals

(Continued)

Frequency (MHz)	Non-government allocation	Non-government use	Government allocation	Government use
				used for purposes such as industrial and scientific research. There is presently no use within the U.S.
400.15–406	Meteorological Aids (Radiosonde) Space Research ((space-to-Earth) 400.15–401) Space Operation (primary 401–402, secondary 400.15–401) Mobile Satellite ((space-to-Earth) 400.15–401) Meteorological Satellite (401–403) Earth Exploration Satellite ((Earth-to-space) 401–403)	Meteorological radiosondes and satellites The FCC has allocated the 400.15–401 MHz portion of this band for operations using nongeostationary nonvoice mobile satellite systems (Little LEOs).	Meteorological Aids (Radiosonde) Space Research ((space-to-Earth) 400.15–401) Meteorological-Satellite ((space-to-Earth) 400.15–401) Meteorological-Satellite ((Earth-to-Space) 401–403) Space Operation (primary 401–402, secondary 400.15–401) Mobile-Satellite ((space-to-Earth) 400.15–401) Earth Exploration Satellite ((Earth-to-space) 401–403) Fixed (G6 military 403–406) Mobile (G6 military 403–406)	This band is extensively used worldwide for gathering meteorological data for weather prediction, severe storm warning, public safety and research. The data are gathered by three technologies: satellite imagery, radiosondes, and wind profiler radars. The Department of Commerce operates the GOES and TIROS-N satellites used for weather tracking and prediction. This information is essential for severe storm notification and public safety, and is used daily in TV and radio broadcast weather reporting to the public. The DoD plans to implement Defense Meteorological Satellite Program (DMSP) downlinks to furnish weather data to light-weight, highly transportable DoD terminals intended for a variety of tactical missions. Radiosondes are operated nationwide by numerous federal agencies to gather local weather data. These small, inexpensive transmitters are attached to balloons and provide wind velocity, temperature, atmospheric pressure and humidity at various altitudes. Their availability is essential to aviation

				activities, as well as space launches. The data gathered by radiosondes are exchanged internationally for worldwide weather prediction and research. Government use of the mobile-satellite service is limited by US319 to earth stations operating with nongovernment satellites.
406–406.1	Mobile Satellite (Earth-to-space)	Emergency position beacons	Mobile Satellite (Earth-to-space)	Emergency position beacons are operated in this band on a worldwide basis, supported by the joint U.S. SARSAT/ Soviet COSPAS satellite network for worldwide air, sea, and land rescue.
406.1–420	Radio Astronomy (406.1–410)	Fixed (US13 hydrological and meteorological data-designated frequencies)	Fixed Mobile Radio Astronomy (406.1–410) Space Research ((space-to-space) 410–420)	This band is one of the principal bands supporting federal land-mobile communications. Important functions include law enforcement, protection of the president and other dignitaries, resource management, disaster and emergency response, security alarms, command destruct of launch vehicles to avoid loss of life and property, and support for public health and power generation activities. This band will accommodate future growth from the very congested 162–174 MHz band. Communications using trunking techniques are being implemented by many agencies to ensure efficient spectrum use. Federal agencies began shifting operations to narrowband technologies in 1995.

(Continued)

Frequency (MHz)	Non-government allocation	Non-government use	Government allocation	Government use
				Fixed links are used in this band for transmission of airport windshear data, flood warning and other environmental data, for law enforcement, for public dissemination of weather warning and disaster information, and for other critical activities. There are also radio astronomy observations at several sites across the U.S.
420–450	Amateur Amateur Satellite (FN664 435–438)	Amateur weak signal modes (432–433), television (420–432, 438–444), repeaters (442–450), auxiliary links (433–435). There is also some use of spread spectrum and other modes. Amateur satellite activities are conducted (435–438) under RR 664. Land mobile systems are operated along the Canadian border in accordance with US230.	Radiolocation	This band is used for long-range surveillance on land-based, ship, and airborne platforms. These uses are essential to the nation's early warning capability, law enforcement, and tracking objects in space. These systems operate with very high power and wide bandwidths. This band is becoming increasingly important for detection of low observable targets. This band is the only military radiolocation band currently available for this frequency sensitive function. Federal agencies operate wind profilers at 449 MHz for measurement of wind speed and direction at various altitudes. NASA and military use of telemetry and telecommand is also extensive.

450–470	Land Mobile Earth Exploration-Satellite (US201 460–470) Space Research and Space Operations (FN668 450 MHz)	Remote pickup broadcast (450–451, 455–456) Public safety, industrial, land transportation (451–454, 456–459, 460–462.5375, 462.7375–467.5375, 467.7375–470) Domestic public (454–455, 459–460) Personal (462.5375–462.7375, 467.5375–467.7375)	Meteorological Satellite (460–470) Earth Exploration-Satellite (US201 460–470) Space Research and Space Operations (FN668 450 MHz)	GOES satellite downlinks for integration of data collection platforms operate in this band. Veteran's medical programs depend on the use of biomedical telemetry and telecommunications in conjunction with nongovernment medical activities.
470–512	Broadcasting Land Mobile	TV channels 14–20, public safety, industrial, land transportation, domestic public		
512–608	Broadcasting	TV channels 21–36, Auxiliary broadcasting		
608–614	Radio Astronomy		Radio Astronomy	There are few federal assignments in this band for other than experimental use. However, radio astronomy usually involves passive operations that do not require an assignment. This band is used for international collaborations in Very Long Baseline Interferometry and will continue to be used for this purpose as the VLBA antennas come on line. The band is also used for observations by the Air Force Radio Solar Telescope Network.

(*Continued*)

Frequency (MHz)	Non-government allocation	Non-government use	Government allocation	Government use
614–806	Broadcasting	TV channels 38–69, Auxiliary broadcasting		
806–902	Land Mobile	Private land mobile (806–824, 851–869, 896–901) Domestic public land mobile (824–849, 869–894) Aeronautical public correspondence-airphone (849–851, 894–896) General purpose mobile (901–902)		This band is used for high-power U.S. Navy shipborne long-range search radars under footnotes US268 and G2. These radars serve a critical role in defense of the fleet.
902–928	Amateur	Amateur weak signal modes (902–904), digital communications, repeaters, spread spectrum and other modes (904–928). Automatic vehicle monitoring (902–912 and 918–928 as authorized by FN US218), ISM, and Part 15 spread spectrum devices. This band is also used for a variety of	Radiolocation Fixed (G11) Mobile (G11)	This band is used predominantly for military radiolocation systems. These include low-power devices, such as those for tactical and nontactical intrusion detection at military facilities, and high-power radars used for long-range search, many of which are employed on U.S. Navy ships and aircraft or at shore stations. These radars serve a critical role in defense of the fleet. Federal mobile communications applications include video surveillance for law enforcement missions, transmission of infrared scanner imagery during overflights of disaster areas, and use of high power packet radio systems.

		ISM applications, particularly industrial heating and food processing.	Fixed use includes point-to-point TV links for monitoring unmanned ports of entry along borders. Though most low capacity links will be moving to the 932–935 MHz and 941–944 MHz bands, this band will continue to be used for a variety of resource management, power administration and law enforcement purposes, as necessary.
928–929	Fixed	Private fixed microwave, domestic public land mobile, private land mobile. Systems in this band provide one-way and two-way interrogate/response data transmission services such as: remote control of electric power networks, burglar and fire alarm monitoring, and other telemetry applications. For two-way systems, the band is paired with 952–953 MHz band.	
929–932	Land Mobile	Domestic public land mobile, private land mobile	

(Continued)

Frequency (MHz)	Non-government allocation	Non-government use	Government allocation	Government use
932–935	Fixed	This band is paired with the 941–944 MHz band and channelized for point-to-point voice and data services. The 932–932.5 MHz end of the band is used for the single channel response from a remote location for point-to-multipoint multiple address services.	Fixed	The 932–935 and 941–944 MHz bands are shared by government and nongovernment fixed service users. It has recently been allocated for federal use. Use for low-capacity fixed systems is anticipated. Many federal agencies expect heavy government and nongovernment use for point-to-point and point-to-multipoint communications. Functions include support for aviation activities, remote meter reading for electric power marketing, and light route radio relay. The latter includes reaccommodation of light route systems from higher bands.
935–941	Land Mobile	Private land mobile trunked and conventional systems in 12.5 kHz channels paired with 896–901 MHz.		
941–944	Fixed	This band is paired with the 932–935 MHz band and channelized for point-to-point voice and data services. The 932–932.5 MHz end of the band is used for the single channel response from a remote location	Fixed	The 932–935 and 941–944 MHz bands are shared by government and nongovernment fixed service users. It has recently been allocated for federal use. Use for low-capacity fixed systems is anticipated. Many federal agencies expect heavy government and nongovernment use for point-to-point and point-to-multipoint

			for point-to-multipoint multiple address services.	communications. Functions include support for aviation activities, remote meter reading for electric power marketing, and light route radio relay. The latter includes reaccommodation of light route systems from higher bands.
944–960	Fixed		Auxiliary broadcasting, domestic public fixed, international fixed public, private fixed microwave. The 944–952 MHz is portion is used primarily for radio broadcast stations studio-to-transmitter links (STLs) and intercity relays. These carry frequency modulated stereophonic audio program material, plus ancillary carriers for remote control of transmitters and Subsidiary Communications Authorization (SCA) channels. The 952–953 MHz portion is used in combination with 928–929 MHz. The 953–960 MHz portion is	

(*Continued*)

Frequency (MHz)	Non-government allocation	Non-government use	Government allocation	Government use
		primarily used for fixed point-to-point communications. The band is segmented as 953.00–956.15 MHz for go and 956.55–959.75 MHz for return operation.		
960–1215	Aeronautical Radionavigation	This band is heavily used for safety-of-life services within the national and international airspace systems. Nearly all aspects of aircraft identification, tracking, control, navigation, collision avoidance, and landing guidance are carried out. Major aeronautical radionavigation systems in this band include the Distance Measuring Equipment (DME/P), Air Traffic Control Beacons (ATCRBS), Mode-S, and the Collision Avoidance System (T-CAS). All	Aeronautical Radionavigation	This band is heavily used for safety-of-life services within the national and international airspace systems. Nearly all aspects of aircraft identification, tracking, control, navigation, collision avoidance, and landing guidance are carried out. Major aeronautical radionavigation systems in this band include the Distance Measuring Equipment (DME/P), Air Traffic Control Beacons (ATCRBS), Mode-S, the military's tactical air navigation system (TACAN) and IFF/SIF systems, and the Traffic Alert and Collision Avoidance System (T-CAS). These aeronautical systems are not only essential to civil and military aircraft but also to special users such as the U.S. Space Shuttle Program. These systems are used throughout the world under International Civil Aviation Organization agreements.

		systems support civil and military aircraft. Systems in this band are developed internationally and agreed to by Civil Aviation Organization for standardization of air travel throughout the world.		Under US224, the military departments are using this band for integrated communications and navigation through the Joint Tactical Information Distribution System (JTIDS) on a non-interference basis. JTIDS is part of an updated NATO system that provides highly secure, jam resistant communications in a hostile environment.
1215–1240	Earth Exploration-Satellite and Space Research (FN713 using radiolocation)		Radionavigation Satellite (space-to-Earth) Radiolocation Earth Exploration-Satellite and Space Research (FN713 using radiolocation)	The frequency 1227.6 MHz is designated for the Global Positioning System (GPS) as part of the radionavigation satellite service. This is a multisatellite system (up to 24 are planned) with large numbers of U.S. and international users. This band is jointly used by the FAA and DoD for radiolocation performing long-range air surveillance and safety-of-flight en route air-traffic control under Joint surveillance System agreements. The military services make use of the band for high-power long-range surveillance radars on land and ships in support of national defense missions. The DoD and FAA are implementing a joint program to field a modernized Air-Route Surveillance Radar Model 4 (ARSR-4) in this band for air-defense, drug interdiction and air-traffic control. A recent radiolocation application, having high national priority, is the use of radar

(Continued)

Frequency (MHz)	Non-government allocation	Non-government use	Government allocation	Government use
				equipment in support of drug interdiction efforts. In this application, radar equipment is mounted on tethered balloons along the southern border of the U.S. to detect low-flying aircraft entering U.S. airspace. Data are relayed to ground, and appropriate action is taken. Space research and Earth-exploration satellite activities for microwave sensor measurements of ocean wave surface are performed by NASA.
1240–1300	Amateur Amateur Satellite (FN664 1260–1270) Earth Exploration-Satellite and Space Research (FN713 using radiolocation)	Amateur television (1240–1246, 1252–1258, 1276–1282), weak signal modes (1295.8–1297), other modes through the band. Active use of amateur satellite (Earth-to-space) in accordance with Footnote 664.	Radiolocation Aeronautical Radionavigation (FN714) Earth Exploration-Satellite and Space Research (FN713 using radiolocation)	This band is used heavily for radiolocation and radionavigation performing long-range air surveillance and en route air-traffic control functions. The FAA and aviation users depend on air-route surveillance radars (ARSRs) to obtain aircraft position information in support of en route air-traffic control. The military makes use of it for high-power long-range surveillance and air-traffic control in support of national defense missions. A recent radiolocation application, having high national priority is the use of radar equipment in support of drug interdiction efforts. In this application, radar equipment is mounted on tethered balloons along the southern border of the U.S. to detect low-flying aircraft

			entering U.S. airspace. Data are relayed to ground and appropriate action taken. NASA radiolocation activities in the 1240–1300 MHz band are for an experimental multi-spectral imaging radar using synthetic aperture (side-looking) techniques. NASA also uses this band for space research and Earth-exploration satellite in conjunction with microwave sensor measurements of ocean wave surface.
1300–1350	Aeronautical Radionavigation	Aeronautical Radionavigation Radiolocation	This band is used heavily for radiolocation and radionavigation performing long-range air surveillance and en route air-traffic control functions. The FAA and aviation users depend on air-route surveillance radars (ARSRs) to obtain aircraft position information in support of en route air-traffic control. The Air Force and Navy make use of it for high-power long-range surveillance radars and air-traffic control radars, in support of national defense missions. The DoD and FAA are implementing a joint program to field a modernized Air-Route Surveillance Radar Model 4 (ARSR-4) in this band for air-defense, drug interdiction and air-traffic control. A recent radiolocation application, having high national priority is the use of radar equipment in support of drug

(Continued)

Frequency (MHz)	Non-government allocation	Non-government use	Government allocation	Government use
				interdiction efforts. In this application, radar equipment is mounted on tethered balloons along the southern border of the U.S. to detect low-flying aircraft entering U.S. airspace. Data are relayed to ground and appropriate action is taken. Radio astronomy observations of highly redshifted hydrogen atoms occur in the 1330–1350 MHz band.
1350–1400			Fixed Mobile Radiolocation Aeronautical Radionavigation (FN714 1350–1370) Fixed Satellite ((space-to-Earth) G114 1381.5) Mobile Satellite ((space-to-Earth) G114 1381.5) Earth Exploration-Satellite and Space Research (FN720 1370–1400)	This band is heavily used for various military radiolocation applications for high-power long-range surveillance radars. The DoD and FAA are implementing a joint program to field a modernized Air-Route Surveillance Radar Model 4 (ARSR-4) in this band for air-defense, drug interdiction and air-traffic control. GPS operates on 1381.05 to relay data on nuclear bursts detected by orbiting satellites. GPS is a multisatellite system with large numbers of U.S. and international users; however, this specific requirement is limited to U.S. users. Radio astronomy observations of highly redshifted hydrogen atoms occur in this band. Knowledge of other galaxies and the early universe comes from these observations. NASA performs passive

				space research and earth-exploration satellite observations. This band is seeing increased use for fixed links and mobile links, since the federal fixed and mobile service allocations were upgraded to primary in 1989. The DoD uses this band for drone telecommand at military test ranges. National Telecommunications and Information Administration (NTIA) reallocated the 1390–1400 MHz portion of this band for nonfederal use after January 1999.
1400–1427	Radio Astronomy Earth Exploration Satellite (Passive) Space Research (Passive)		Radio Astronomy Earth Exploration Satellite (Passive) Space Research (Passive)	This band has been set aside nationally for passive operations, and no stations are authorized to transmit. There are no federal assignments in this band. Radio astronomy, including the spectral line observations of neutral atomic hydrogen, continuum observations and Radio Solar Telescope Network observations allow study of the structure of our galaxy as well as others. NASA performs passive space research and earth-exploration satellite observations. This band is extremely important for measurements of land moisture and salinity, and ocean surface characteristics.

(*Continued*)

Frequency (MHz)	Non-government allocation	Non-government use	Government allocation	Government use
1427–1435	Space Operation ((Earth-to-space) 1427–1429) Land Mobile Fixed	Private land mobile, Satellite communications	Fixed Mobile Space Operation ((Earth-to-space) 1427–1429)	This band is used to support a variety of military fixed and mobile applications. Functions include tactical/training operations, light route radio relay, telemetry and telecommand including command of missiles and RPVs, and automatic target scoring. There are also some fixed operations planned for use in federal resource management programs. NTIA will reallocate the 1427–1432 MHz portion of this band for nonfederal use after January 1999.
1435–1530	Mobile	This band is heavily used for aeronautical telemetry and telecommand. This is crucial to industry research, development, and testing of aircraft and missile systems.	Mobile	This band is heavily used for aeronautical telemetry and telecommand. This is crucial to NASA, and DoD research, development, and testing of aircraft and missile systems. Many of the assignments in the 1435–1525 MHz band are for missile test telemetry. The use of small devices with omnidirectional antennas require frequencies below 3 GHz. Aeronautical telemetry needs extensive spectrum, and minimal in-band and adjacent band interference. Equipment using this band have been built into many missile and aircraft platforms and have been tailored to those specific electromagnetic environments. This band is congested in many areas, and new systems are being moved to 2360–2390 MHz.

				Also, the DOE uses the band for telemetry in nuclear research and development efforts. Use of this band is dictated by the need for equipment mobility and small antennas.
1530–1544	Maritime Mobile Satellite (space-to-Earth) Mobile (aeronautical telemetry 1530–1535)	The major use in this band is for INMARSAT downlinks providing distress, safety, and general communications. This system is used currently by 17,000 ships throughout the world, including extensive operations within inland waterways for ship-to-shore communications. The number of users is expected to reach 40,000 within the next 10 years. Its use for distress and safety communications is part of the Global Maritime Distress and Safety System (GMDSS). This international application is tied to and required by international treaty	Maritime Mobile Satellite (space-to-Earth) Mobile (aeronautical telemetry 1530–1535)	Federal vessel operators participate in the use of the INMARSAT/GMDSS system. The Navy uses the INMARSAT system for international communications in ocean areas from its Military Sealift Command vessels operated by civilian crews. The Air Force and Navy use the 1525–1535 MHz portion of this band for aeronautical telemetry on a secondary basis when such operation does not conflict with the primary operation.

(*Continued*)

Frequency (MHz)	Non-government allocation	Non-government use	Government allocation	Government use
		resulting from the Safety of Life at Sea (SOLAS) Convention. INMARSAT also provides satellite supported aeronautical public correspondence and some land mobile satellite service downlinks.		
1544–1545	Mobile Satellite (space-to-Earth)	Solely used for distress and safety.	Mobile Satellite (space-to-Earth)	This band is used by SARSAT for a downlink to relay satellite EPIRB transmissions.
1545–1559	Aeronautical Mobile Satellite (R) (space-to-Earth) Mobile Satellite ((space-to-Earth) primary 1549.5–1558.5, secondary 1545–1549.5)	This band will be used for the downlink for the nationwide mobile satellite system authorized by the FCC. This system was operated by the American Mobile Satellite Corporation, a consortium of eight U.S. companies. Mobile satellite services are expected to grow rapidly. Included within this frequency range is the internationally	Aeronautical Mobile Satellite (R) (space-to-Earth) Mobile Satellite ((space-to-Earth) primary 1549.5–1558.5, secondary 1545–1549.5)	Federal agencies will make use of mobile satellite operations in this band.

		allocated 1545–1555 MHz AMS(R)S allocation (space-to-Earth) to support the worldwide interoperable AMS(R)S through a number of satellites. In accordance with US308, AMS(R)S has priority and real-time preemptive access in this band segment.		
1559–1610	Aeronautical Radionavigation Radionavigation Satellite (space-to-Earth) Aeronautical Mobile (US260)	Private sector use of GPS is extensive for land, sea, and air radionavigation. Other uses of GPS include surveying, aircraft landing aids, position location, traffic management, and scientific research.	Aeronautical Radionavigation Radionavigation Satellite (space-to-Earth) Aeronautical Mobile (US260)	The Global Positioning System operates on a center frequency of 1575.42 MHz in this band as part of the radionavigation satellite service. This is a multisatellite system with large numbers of U.S. and international users. ICAO has recognized the GPS and GLONASS as the two principal candidates for the Global Navigation Satellite System.
1610–1626.5	Aeronautical Radionavigation Aeronautical Radionavigation Satellite (FN732) Radio Determination Satellite (Earth-to-space)	There is one active private sector RDSS systems. However, the FCC has granted two applications for low-earth and geostationary orbit mobile satellite	Aeronautical Radionavigation Aeronautical Radionavigation-Satellite (FN732) Radio Determination-Satellite (Earth-to-space)	This band has been reserved on a worldwide basis for development and use of airborne electronic aids to air navigation. Recent changes were made to allocate this band to radiodetermination satellite service uplinks on a primary basis. Federal agencies have begun leasing

(*Continued*)

Frequency (MHz)	Non-government allocation	Non-government use	Government allocation	Government use
	Mobile Satellite (Earth-to-space) Radio Astronomy (1610.6–1613.8) Mobile Satellite ((space-to-Earth) 1613.8–1626.5) Aeronautical Mobile (US260)	systems to provide voice and high data rate communications (Big LEOs).	Mobile-Satellite (Earth-to-space) Radio Astronomy (1610.6–1613.8) Mobile-Satellite ((space-to-Earth) 1613.8–1626.5) Aeronautical Mobile (US260)	access to systems in this service, and increased use is expected. Radio Astronomy observations of the OH radical are carried out between 1610.6 and 1613.8 MHz. The OH line observations are crucial to understanding interstellar medium and star formation. Government use of mobile-satellite and radiodetermination-satellite services is limited to earth stations operating with nongovernment satellites.
1626.5–1645.5	Maritime Mobile Satellite (Earth-to-space) Mobile Satellite (Earth-to-space)	A major use of this band is for INMARSAT ship earth stations. These systems are used currently by 17,000 ships throughout the world, including extensive operations within inland waterways for ship-to-shore communications. The number of users is expected to reach 40,000 within the next 10 years. Its use for distress and safety communications is part of the Global Maritime	Maritime Mobile Satellite (Earth-to-space) Mobile Satellite (Earth-to-space)	Federal agencies make active use of the mobile-satellite operations for land, air, and maritime communications. The Navy uses the INMARSAT system for international communications in ocean areas.

		Distress and Safety System. This international application is tied to and required by international treaty resulting from the Safety of Life at Sea (SOLAS) Convention. INMARSAT also provides satellite supported aeronautical public correspondence and some land mobile satellite service downlinks.		
1645.5–1646.5	Mobile Satellite (Earth-to-space)	Solely used for distress and safety.	Mobile Satellite (Earth-to-space)	There are no operational federal assignments in this band; however, this band is used for distress and safety operations. Plans exist for satellite EPIRBS and relay of distress and safety signals between satellites.
1646.5–1660.5	Aeronautical Mobile Satellite (R) (Earth-to-space) Mobile Satellite ((Earth-to-space) primary 1651–1660, secondary 1646.5–1651) Radio Astronomy (1660–1660.5)	INMARSAT II operates 1646.5–1649.5 MHz This band is used for the uplink for the nationwide mobile satellite system operated by the American Mobile Satellite Corporation, a consortium of eight	Aeronautical Mobile Satellite (R) (Earth-to-space) Mobile Satellite ((Earth-to-space) primary 1651–1660, secondary 1646.5–1651) Radio Astronomy (1660–1660.5)	This band is used for the uplink for the nationwide mobile satellite system operated by the American Mobile Satellite Corporation, a consortium of eight U.S. companies. Mobile satellite services are expected to grow rapidly, including many federal users. Included within this frequency range is the internationally allocated 1646.5–1656.5 MHz AMS(R)S allocation

(Continued)

Frequency (MHz)	Non-government allocation	Non-government use	Government allocation	Government use
		U.S. companies. Mobile satellite services are expected to grow rapidly.		(Earth-to-space) to support the worldwide interoperable AMS(R)S through a number of satellites. In accordance with US308, AMS(R)S has priority and real-time preemptive access in this band segment. Passive radio astronomy observations of the redshifted spectral line of the OH radical, essential for understanding interstellar medium and star formation in other galaxies, are carried out in this band.
1660.5–1670	Radio Astronomy Space Research ((Passive) 1660.5–1668.4) Meteorological Aids ((Radiosonde) 1668.4–1670)		Radio Astronomy Space Research ((Passive) 1660.5–1668.4) Meteorological Aids ((Radiosonde) 1668.4–1670)	The 1660.5–1668.4 portion of this band has been set aside nationally for passive operations, and no stations are authorized to transmit. There are no federal assignments in this range. Passive radio astronomy observations are performed under the protection of US246 (transmissions prohibited). Observation of the two spectral lines of the OH radical (1665.402 and 1667.359 MHz), essential for understanding interstellar medium and star formation in other galaxies, are carried out in this band. The band is also used for continuum observations. Radiosondes are operated nationwide by numerous federal agencies to gather local weather data. These small inexpensive transmitters are attached to

				balloons and provide wind velocity, temperature, atmospheric pressure and humidity at various altitudes. The availability is essential to aviation activities, as well as space launches. The data gathered by radiosondes are exchanged internationally for worldwide weather prediction and research.
1670–1710	Meteorological Aids (Radiosonde) Meteorological Satellite (space-to-Earth) Fixed (1700–1710) Earth Exploration Satellite (FN671 1690–1710)		Meteorological Aids (Radiosonde) Meteorological Satellite (space-to-Earth) Fixed (1700–1710) Earth Exploration-Satellite (FN671 1690–1710)	This band is extensively used worldwide for gathering meteorological data for weather prediction, severe storm warning, public safety and research. These data are gathered by two technologies: radiosondes and satellite imagery. NTIA reallocated the 1670–1675 MHz portion of this band for nonfederal use after January 1999. Radiosondes are operated nationwide by numerous federal agencies to gather local weather data. These small inexpensive transmitters are attached to balloons and provide wind velocity, temperature, atmospheric pressure and humidity at various altitudes. The availability is essential to aviation activities, as well as space launches. The data gathered by radiosondes are exchanged internationally for worldwide weather prediction and research. Also, NASA uses this band for transmission of meteorological data from tethered balloons.

(*Continued*)

Frequency (MHz)	Non-government allocation	Non-government use	Government allocation	Government use
				The Department of Commerce operates the GOES and TIROS-N satellites used for weather tracking and prediction. This information is essential for severe storm notification and public safety and is used daily in TV and radio broadcast weather reporting to the public. Most of the meteorological satellite users are earth stations that receive the satellite data. The assignments in this band for earth terminals are primarily for fixed locations; however, over 40 are planned for shipboard use in the U.S. coastal waters. Some agencies have begun using the 1700–1710 MHz band for fixed line-of-sight data communications as the 1710–1850 MHz band gets crowded.
1710–1850	Radio Astronomy (US256 1718.8–1722.2)		Fixed Mobile Space Operation ((Earth-to-space) G42 (1761–1842) Radio Astronomy (US256 1718.8–1722.2)	1710–1850 MHz is the predominant federal medium-capacity line-of-sight fixed service band. Fixed links are operated by federal agencies for voice, data, and/or video communications where commercial service is unavailable, excessively expensive, or cannot meet required reliability. Applications include law enforcement networks and control links for various power, land, water, and electric-power management systems. Other specialized fixed links include video relay, data relay, and timing

distribution signals. Growth averages about 400 new assignments per year.

Specific agency applications of the fixed service include: FAA remote data transmission of critical flight safety data in support of essential aeronautical services, Army tactical radio relay systems, Department of Agriculture and Interior backbone links for control of land mobile radio systems necessary in fire fighting, law enforcement and disaster control within national forests and for provision of voice and data connections between sites where commercial service is not available, and Departments of Treasury and Justice microwave links related to law enforcement.

One example of a wide area fixed network is the Department of Energy's use of this band for supervision, control, and protection of power-administration-operated electrical power transmission systems and activities supporting nuclear weapons development. Power administration microwave must be capable of carrying hundreds of radio channels per system. The channels are used for high-speed relaying, supervisory control, load control, telemetering, data acquisition, land-mobile radio dispatching, operations and

(*Continued*)

Frequency (MHz)	Non-government allocation	Non-government use	Government allocation	Government use
				maintenance. The nuclear test facilities backbone microwave systems serve sites at greater than 10 miles and are more efficient in this band than in lower or higher bands. This band also allows for a greater range capability for robot control and video requirements. The present system connects all federal power marketing control facilities in the western half of the U.S. Common equipment exists with the nongovernment sector allowing interconnectivity for critical communications dealing with all aspects of generating and distributing power. This band is also used for a variety of mobile applications, including airborne telemetry, telecommand, video and data links automated target scoring, and air combat maneuvering instrumentation. Many military aeronautical mobile systems depend on frequencies in this band. Border surveillance through the use of aerostats is supported by narrowband uplink and downlink telemetry transmissions. The Air Force and Navy also use the band for space command and control. Uplink frequencies between 1761 and 1842 MHz are heavily used in certain locations in conjunction with a 2200–2290 MHz

			downlink. Use of these systems has national security implications. Telemetry, and telecommand and control of the NASA Space Shuttle is conducted on space-to-space links in this band. This band is also used by the USCG for vessel traffic safety systems, the VHF National Distress System and remote distress and safety communications and control networks. Radio astronomy observations are made of the 1720.530 MHz spectral line of the OH molecule. These observations are crucial to understanding the interstellar medium and star formation. NTIA reallocated the 1710–1755 MHz portion of this band for nonfederal use after January 2004 under conditions that some federal systems would be permitted to continue to operate.
1850–1990	Mobile Fixed	The FCC has reserved the 1850–1990 MHz band for personal communications services on a coprimary basis with fixed services. The band has been divided between frequencies used for metro trading areas, basic trading areas,	

(*Continued*)

Frequency (MHz)	Non-government allocation	Non-government use	Government allocation	Government use
		and nonlicensed use (1910–1930). Private fixed microwave. This band is used to provide fixed point-to-point voice, data, telemetry and control services for private (non-common-carrier) companies. Typical users include electric and gas utilities, police and fire departments, and local governments. Most of these uses are being moved to other frequency bands within the next few years to allow development of PCS.		
1990–2110	Fixed Mobile Space Research and Earth Exploration-Satellite (US90 2025–2110)	Auxiliary broadcasting, cable television, domestic public fixed. This band is heavily used by TV broadcasters for one way transmission services such as: portable van and helicopter mounted transmissions of video	Space Research and Earth Exploration-Satellite (US90 2025–2110)	NASA's global ground network and TDRSS operations from 2025–2110 MHz are essential to NASA Earth exploration, space operations, and space research activities. This use includes Earth-to-space and space-to-space transmissions. Over 50 U.S. space missions, and, consistent with international agreements, additional foreign missions will be supported by NASA in the next five years. There will be varying degrees

		from remote news events; studio-to-transmitter links; and, intercity relay of video programming.		of support from launch and orbital transfer to full-time data relay. These telecommunications links are made available to private sector expendable launch vehicle operations. Some 123 satellites from nine countries are planned for or are operational in the 2025–2110 MHz and 2200–2290 MHz bands. These missions comprise 341 planned or existing assignments, not including earth stations. This band is also used for uplinks for the GOES weather satellite, supporting weather prediction efforts.
2110–2200	Fixed Mobile	The 2110–2130 MHz portion, paired with 2160–2180 MHz, is used by common carriers for "light-haul" radio relay routes, for control and repeater links used with land-mobile base stations, and by cellular telephone companies for cell site-to-cell site links. The 2130–2150 MHz portion, paired with 2180–2200 MHz, is used by private companies (non-common-carriers) for	Space Research (US252 2110–2120)	NASA uses the 2110–2120 MHz portion of this band for Deep Space Network Earth-to-space command links. These activities support or will support Voyagers 1 and 2, GALILEO, ULYSSES, and other deep space missions.

(Continued)

Frequency (MHz)	Non-government allocation	Non-government use	Government allocation	Government use
		applications similar to those in the 2110–2130 MHz band. The 2150–2162 MHz portion is used for omnidirectional transmission of point to multipoint video signals. This band is congested in many of the urban areas. The FCC has reserved the 2110–2150 and 2160–2200 MHz band for future emerging technologies on a coprimary basis with fixed services.		
2200–2290	Space Research, Space Operations, and Earth Exploration Satellite (US303 2285–2290)		Fixed Mobile Space Research (space-to-Earth, space-to-space) Space Operations (space-to-Earth, space-to-space) Earth Exploration-Satellite (space-to-Earth, space-to-space)	This band is predominantly used for federal terrestrial and space telemetry systems. Space applications include the NASA Tracking Data Relay Satellite System (TDRSS) and the Air Force Space Ground Link Subsystem (SGLS). These two systems provide the telemetry, telecommand and control for all federal satellite systems and some activities with national security implications. Telemetry, tracking and control functions for a new satellite ALEXIS will be performed in this band

as part of U.S. treaty verification efforts. Terrestrial telemetry is predominantly air-to-ground links for various operational and experimental systems. Growth averages about 80 new assignments per year.

TDRSS operations from 2200–2290 MHz are essential to NASA Earth exploration, space operations, and space research activities. This use includes space-to-Earth and space-to-space transmissions. Over 50 U.S. space missions, and, consistent with international agreements, additional foreign missions will be supported by NASA in the next five years. There will be varying degrees of support from launch and orbital transfer to full-time data relay. These telecommunications links are made available to private sector expendable launch vehicle operations. Some 123 satellites from nine countries are planned for or are operational in the 2025–2120 and 2200–2290 MHz bands. The band also supports similar space-to-Earth and space-to-space telemetry, telecommand and control for military satellites through the Air Force SGLS system.

Terrestrial telemetry is heavily used in this band for such purposes as nuclear testing, airborne weapons testing, aircraft flight testing, and a wide variety of experimental and research projects.

(*Continued*)

Frequency (MHz)	Non-government allocation	Non-government use	Government allocation	Government use
				Most of this equipment was moved to this band during the 1970s, at significant expense to the federal government, to reaccommodate requirements in lower bands for other uses. Other mobile applications include narrowband uplinks and downlinks in conjunction with radar-laden tethered balloons. These balloons are used in law enforcement and drug interdiction missions. Fixed microwave systems are also in this band for control of land-mobile radio systems to provide voice and data connections between sites where commercial service is not available, and where the 1710–1850 MHz band is saturated.
2290–2300	Space Research ((space-to-Earth) deep space only)		Space Research ((space-to-Earth) deep space only) Fixed Mobile	NASA uses this band for Deep Space Network space-to-earth telemetry. These activities support or will support Voyagers 1 and 2, GALILEO, ULYSSES, Cassini (radio science experiment), and other deep space missions. Radio Astronomy observations are also conducted in this band.
2300–2310	Amateur	Amateur weak signal modes (2304), other modes throughout the band.		

2310–2360	Broadcasting Satellite Mobile	The FCC has allocated this band for Broadcast-Satellite for high-quality radio, and Wireless Communications Systems.	Radiolocation Mobile Fixed	
2360–2390	Mobile	This band is used for telemetry and telecommand for expendable and reusable launch vehicles.	Mobile Radiolocation Fixed	This band is used for telemetry and telecommand for expendable and reusable launch vehicles. The Air Force and Navy use this band for aeronautical telemetry. Aeronautical telemetry needs extensive spectrum, and minimal in-band and adjacent-band interference. The 1435–1525 MHz band is filled, and new systems are being moved into this band. The Air Force uses the band for high-power long-range surveillance radars and air-traffic control radars, while the Army and DOE use it for air/ground ranging system tracking. NASA uses this band for the Venus Radar Mapper (VRM) synthetic aperture radar and associated telemetry. The National Science Foundation and NASA use planetary radars in coordination with research universities. Observations at the National Astronomy and Ionospheric Center (Arecibo) occupy 20 MHz centered around 2380 MHz.

(*Continued*)

Frequency (MHz)	Non-government allocation	Non-government use	Government allocation	Government use
2390–2450	Amateur (primary 2390–2400, 2402–2417, secondary 2400–2402, 2417–2450) Amateur Satellite (FN664)	Amateur mixed modes (2390–2400, 2410–2450). Amateur satellite operation (space-to-Earth) occur in accordance with FN664 from 2400–2450 MHz. The band 2390–2400 is available for unlicensed PCS operations under Part 15 of the FCC's Rules. The band 2400–2450 MHz band is available for use by a wide-variety of unlicensed devices under Part 15. This band is also used for microwave ovens (approximate operating frequency 2450) and a variety of industrial processes.	Radiolocation (2417–2450)	The Air Force uses the band for high-power long-range surveillance radars and air-traffic control radars. However, because of the operation of tens of millions of microwave ovens and other industrial, scientific, and medical (ISM) equipment little use is made of this band, and little growth is expected. The Navy uses this band for scoring applications for missiles and projectiles. There is some packet radio development by the Army going on in this band.
2450–2483.5	Fixed Mobile Radiolocation	This band is used for fixed and portable transmission of video by TV broadcasters for remote news events. In addition, the band is used for private	Radiolocation (FN41)	The United States Customs Service (Treasury) uses this band on a secondary basis for law enforcement related radiolocation.

		company fixed service radio relay transmission of voice and data transmissions by private companies. This band is available for use by a wide-variety of unlicensed devices under Part 15 of the FCC's rules. This band is also used for microwave ovens (approximate operating frequency 2450) and a variety of industrial processes.	
2483.5–2500	Radio Determination Satellite (space-to-Earth) Mobile Satellite (space-to-Earth)	Though this is the downlink band for the Radiodetermination Satellite service, private company fixed stations and TV broadcaster portable stations that were in operation prior to 1985 may continue to operate on a primary basis. These are multichannel equipment having 10 channels.	Radio Determination Satellite (space-to-Earth) Radiolocation (FN41) Mobile Satellite (space-to-Earth)

(*Continued*)

Frequency (MHz)	Non-government allocation	Non-government use	Government allocation	Government use
2500–2655	Broadcasting Satellite Fixed Space Research and Earth Exploration-Satellite (FN720 2640–2655)	Auxiliary broadcasting. The 2500–2686 MHz portion of this band is used for omnidirectional transmission of point to multipoint (multipoint MDS) that can be contained within 6 MHz channel bandwidths. Portions of the band are allocated to be used for pay television distribution, transmission of educational lectures by school systems (ITSF), and private video teleconferences.	Space Research and Earth Exploration-Satellite (FN720 2640–2655)	NASA performs passive space research and earth-exploration satellite observations that allow measurement of soil moisture and of coastal ocean salinity.
2655–2690	Broadcasting Satellite Fixed Earth Exploration Satellite (Passive) Radio Astronomy Space Research (Passive)	Auxiliary broadcasting. Private fixed microwave (above 2680 MHz). There are also fixed multipoint MDS and instructional television operations in this band.	Earth Exploration Satellite (Passive) Radio Astronomy Space Research (Passive)	This band is used in the U.S. and other countries for radio astronomy continuum measurements. It is used in addition to the 2690–2700 MHz band, which is too narrow to conduct some measurements. NASA performs passive space research and earth-exploration satellite observations. These observations allow measurement of soil moisture and of coastal ocean salinity.

2690–2700	Radio Astronomy Earth Exploration Satellite (Passive) Space Research (Passive)		Radio Astronomy Earth Exploration Satellite (Passive) Space Research (Passive)	This band is used extensively in the U.S. and other countries for radio astronomy. It is an excellent band for continuum measurement, because the galactic background continuum radiation is low. Observations of galactic and extragalactic radio sources at these frequencies help to define their spectra, which gives information on the physical parameters of the radiating source. The band is also being used by the U.S. Naval Observatory interferometer at Green Bank, WV. This program is used for accurate position determinations by the Navy. The band is also used for solar observations by the Air Force Radio Solar Telescope Network. NASA performs passive space research and earth-exploration satellite observations protected under Footnote US246. These observations allow measurement of soil moisture and of coastal ocean salinity.
2700–2900			Aeronautical Radionavigation Meteorological Aids Radiolocation	This band is used predominantly for air-surveillance radars. It is a critical safety-of-flight band for airport surveillance radars (ASRs) to provide aircraft position information for air-traffic control in the vicinity of airports. Similar use is for military Ground Control Approach radars (GCAs). The Air Force and Navy use it for high-power long-range surveillance radars and

(*Continued*)

Frequency (MHz)	Non-government allocation	Non-government use	Government allocation	Government use
				air-traffic control radars. NEXRAD is also being used here when not in conflict with the ASRs. Weather radars are also operated in this band in support of flight safety. NASA uses the band for tracking for range safety purposes (radiolocation), and for atmospheric research (meteorological aids).
2900–3100	Maritime Radionavigation Radiolocation	This band is primarily used for maritime radars and radar beacons (racons). Radars of this type are required on cargo and passenger ships by international treaty (SOLAS) for safety purposes. Racons operate in conjunction with maritime radars to provide electronic markers to identify maritime obstructions and navigation points.	Maritime Radionavigation Radiolocation	Federal agencies use this band heavily for shipborne radionavigation radars, vessel traffic systems, and racons. The military uses this band for high-power 3-D long-range surveillance radars, precision approach radars, and air-traffic control radars. Also, NEXRAD operates from 2900–3000 MHz. NASA performs airborne measurements of rainfall rates over selected ocean areas.
3100–3600	Radiolocation Space Research and Earth Exploration-Satellite (FN713 3100–3300) Amateur (3300–3500)		Radiolocation Aeronautical Radionavigation ((Ground-based) 3500–3600) Space Research and	This band is primarily used for military radiolocation, including several multi-billion dollar defense radar systems. Use of this band for these systems is considered critical to national defense. The high-power mobile radars include

			Earth Exploration-Satellite (FN713 3100–3300)	airborne, land-based, and shipborne applications.
3600–3700	Fixed Satellite Radiolocation	INMARSAT and INTELSAT have limited use for fixed satellite service earth stations. Each site must be actively coordinated with the U.S. Government with supporting EMC analysis.	Aeronautical Radionavigation (Ground-based) Radiolocation	The principal federal use of this band is to support a Navy radar used for landing operations on aircraft carriers. This high-power radar is operated on Navy ships and at certain shore locations for training. NTIA reallocated the 3650–3700 MHz portion of this band for shared nonfederal use after January 1999.
3700–4200	Fixed Fixed Satellite	Domestic public fixed, Satellite communications		
4200–4400	Aeronautical Radionavigation	This band is heavily used for radar altimeters on board nongovernment fixed-wing and rotary aircraft.	Aeronautical Radionavigation	This band is heavily used for radar altimeters on board government fixed-wing and rotary aircraft, as well as spacecraft. Methods for reducing the bandwidth necessary to perform this function are being studied within the ITU-R; however, some altimeter functions may not be able to be provided in a reduced bandwidth. Also, significant capital has been invested in current equipment.
4400–4990	Fixed Satellite ((space-to-Earth) 4500–4800)	The band 4660–4685 MHz is available for the General Wireless	Fixed (4400–4660, 4685–4990) Mobile (4400–4660,	This band is heavily used by the military services for tactical communications, both line-of-sight and troposcatter.

(Continued)

Frequency (MHz)	Non-government allocation	Non-government use	Government allocation	Government use
	Fixed (4660–4685) Mobile (4660–4685) Radio Astronomy (US203 4825–4835, US257 4950–4990) Space Research and Earth Exploration-Satellite (FN720 4950–4990)	Communications Services (GWCS) under Part 26 of the FCC's Rules.	4685–4990) Radio Astronomy (US203 4825–4835, US257 4950–4990) Space Research and Earth Exploration-Satellite (FN720 4950–4990)	In addition to extensive transportable fixed service use, the DoD operates air-to-ground data links, drone command and control systems, air-defense, and many other systems in this band. The DoD anti-air warfare systems employ high power spread spectrum techniques in a distributed network among ships, aircraft, and land forces. Additional uses are for emergency incident response for the Nuclear Emergency Search Team and target scoring and control. Narrowband and wideband uplinks and downlinks operate in conjunction with aerostats used in law enforcement and drug interdiction missions. The National Science Foundation performs some continuum observations in the 4950–4990 MHz portion of the band when the 4990–5000 MHz band does not provide adequate bandwidth. NTIA reallocated the 4635–4660 MHz portion of this band for nonfederal use after January 1997.
4990–5000	Radio Astronomy Space Research (Passive)		Radio Astronomy Space Research (Passive)	This band is used extensively in the U.S. and other countries for radio astronomy. It is an excellent band for continuum measurement because the galactic background continuum radiation is low. Observations of galactic and

				extragalactic radio sources at these frequencies help to define their spectra, which gives information on the physical parameters of the radiating source.
5000–5250	Aeronautical Radionavigation Aeronautical Mobile (R) (FN733) Fixed Satellite and Inter-Satellite (when in conjunction with Aeronautical Radionavigation or Aeronautical Mobile (R) FN797)		Aeronautical Radionavigation Aeronautical Mobile (R) (FN733) Fixed Satellite and Inter-Satellite (when in conjunction with Aeronautical Radionavigation or Aeronautical Mobile (R) FN797)	The Microwave Landing System is being deployed in the 5000–5150 MHz portion of this band as one of two internationally recognized systems for precision landing of aircraft.
5250–5350	Radiolocation		Radiolocation (G59 nonmilitary secondary)	This band is used for high-power DoD radar systems. NASA is performing experiments with spaceborne radar systems in this band in accordance with FN713.
5350–5460	Aeronautical Radionavigation (FN799 airborne radars and associated radar beacons only) Radiolocation	The 5350–5470 MHz band is used for airborne weather radars for storm avoidance. Some ground-based weather radars operate in this band to provide weather information for state and local governments, universities, and broadcast stations.	Aeronautical Radionavigation (FN799 airborne radars and associated radar beacons only) Radiolocation (G56 nonmilitary secondary)	The Navy operates its primary surface search radar in this band. The 5350–5470 MHz band is used for airborne weather radars for storm avoidance.

(Continued)

Frequency (MHz)	Non-government allocation	Non-government use	Government allocation	Government use
5460–5470	Radionavigation (FN799 aeronautical radionavigation limited to airborne radars and associated radar beacons only, US65 maritime radionavigation limited to shipborne radars) Radiolocation	Ship radars operate in this band to provide coastal navigation information. Some ground-based weather radars operate in this band to provide weather information for state and local governments, universities, and broadcast stations. Airborne weather radars also operate in this band.	Radionavigation (FN799 aeronautical radionavigation limited to airborne radars and associated radar beacons only, US65 maritime radionavigation limited to shipborne radars) Radiolocation (G56 nonmilitary secondary)	Ship radars operate in this band to provide coastal navigation information. The Navy operates its primary surface search radar in this band.
5470–5600	Maritime Radionavigation (US65 limited to shipborne radars) Radiolocation	Ship radars operate in this band to provide coastal navigation information. Some ground-based weather radars operate in this band to provide weather information for state and local governments, universities, and broadcast stations.	Maritime Radionavigation (US65 limited to shipborne radars) Radiolocation (G56 nonmilitary secondary)	Ship radars operate in this band to provide coastal navigation information. The Navy operates its primary surface search radar in this band. Above 5500 MHz, this band is used heavily for test range instrumentation radars.
5600–5650	Maritime Radionavigation	Ship radars operate in this band to provide	Maritime Radionavigation (US65	Ship radars operate in this band to provide coastal navigation information.

	(US65 limited to shipborne radars) Meteorological Aids Radiolocation	coastal navigation information.	limited to shipborne radars) Meteorological Aids Radiolocation (G56 nonmilitary secondary to military)	The Navy operates its primary surface search radar in this band. Terminal doppler weather radars provide windshear information in support of air-traffic control activities. This band is used heavily for test range instrumentation radars.
5650–5850	Amateur Amateur Satellite (FN664 (Earth-to-space) 5650–5670, FN808 (space-to-Earth) 5830–5850)	The 5725–5850 MHz portion of the band is available for a variety of unlicensed uses under Part 15 of the FCC's Rules.	Radiolocation (G2 limited to military)	The Navy operates its primary surface search radar in this band. This band is used heavily for test range instrumentation radars.
5850–5925	Fixed Satellite (Earth-to-space) Amateur	International satellite systems such as INTELSAT use this band for uplinks in the fixed satellite service.	Radiolocation (limited to military G2)	This band is used heavily for test range instrumentation radars used to track missiles and other airborne targets and is used to safeguard range personnel and surrounding civilian communities. This band is also used to control airborne target systems that help maintain air-defense and combat readiness.
5925–7075	Fixed (5925–6425, 6525–6875) Fixed Satellite (Earth-to-space) Mobile (6425–6525, 6875–7075)	The 5925–6425 MHz portion of this band, is used for the uplink of the fixed satellite service corresponding to the downlink in the 3700–4200 MHz band. The common carrier point-to-point Microwave Service		

(*Continued*)

Frequency (MHz)	Non-government allocation	Non-government use	Government allocation	Government use
		used by local exchange and long distance telephone companies shares the 5925–6425 MHz frequency range. Fiber-optics are replacing many of these telephone links; however, this is one of the bands to which fixed microwave incumbents are migrating from the spectrum reallocated for personal communications services and other emerging technologies. These fixed users are also migrating to the 6525–6875 MHz band. The 5925–6425 MHz portion of this band is also used by cellular carriers for backbone networks. The main terrestrial use of the 6425–6525 MHz range is for broadcast auxiliary television remote pickup.		

		The 6875–7025 MHz band is used for broadcast auxiliary services including electronic newsgathering, intercity relay, and studio transmitter links.		
7075–7125	Fixed Mobile	This band is used for broadcast auxiliary services including electronic newsgathering, intercity relay, and studio transmitter links.		
7125–7250	Space Research ((Earth-to-space), US252 deep space only 7145–7190)		Fixed Space Research ((Earth-to-space) 7190–7250) and (US252 deep space only 7145–7190)	This band is used for fixed microwave links associated with control of power distribution and dam flood gates, remoting of weather data, remoting of vessel traffic information, remoting of air-traffic control radar data, and military test range communications. NASA uses this band for Deep Space Network earth-to-space telecommand links. These activities support or will support GALILEO, Mars Global Surveyor, Mars Pathfinder, Cassini Near Earth Asteroid Rendezvous (NEAR), and other deep space missions.

(Continued)

Frequency (MHz)	Non-government allocation	Non-government use	Government allocation	Government use
7250–7750			Fixed (primary 7300–7750, secondary 7250–7300) Fixed Satellite ((space-to Earth) G117 military only) Meteorological Satellite ((space-to-Earth) 7450–7550) Mobile Satellite ((space-to-Earth) primary 7250–7300, secondary 7300–7750, G117 military only)	This band is used for fixed microwave links associated with control of power distribution and dam flood gates, remoting of weather data, remoting of vessel traffic information, remoting of air-traffic control radar data, and military test range communications. The band is used for Defense Satellite Communication Systems (DSCS), and NATO SATCOM downlinks to provide secure voice and data communications to globally deployed military units, and for FLTSATCOM telemetry.
7750–7900			Fixed	This band is used for fixed microwave links associated with control of power distribution and dam flood gates, remoting of weather data, remoting of vessel traffic information, remoting of air-traffic control radar data, and military test range communications.
7900–8025			Fixed Satellite ((Earth-to-space) G117 military only) Mobile Satellite ((Earth-to-space) G117 military only) Fixed	This band is used for fixed microwave links on a secondary basis. The band is also used for Defense Satellite Communication Systems (DSCS) uplinks that provide secure voice and data communications to globally deployed military units including mobile earth terminals, and for FLTSATCOM uplinks for fleet broadcasts.

8025–8400	Earth Exploration Satellite (space-to-Earth US258)	This band is used for land remote-sensing operations within the Earth exploration-satellite service (space-to-Earth).	Earth Exploration Satellite (space-to-Earth) Fixed Fixed Satellite ((Earth-to-space) G117 military only) Meteorological Satellite ((Earth-to-space) 8175–8215) Mobile Satellite ((Earth-to-space) no airborne transmissions, G117 military only)	This band is used for fixed microwave links associated with control of power distribution and dam flood gates, remoting of weather data, remoting of vessel traffic information, remoting of air-traffic control radar data, and military test range communications. The band is used for Defense Satellite Communication Systems (DSCS) uplinks that provide secure voice and data communications to globally deployed military units. This band is used for land remote-sensing operations within the Earth exploration-satellite service (space-to-Earth).
8400–8500	Space Research ((space-to-Earth) 8450–8500)		Fixed Space Research (space-to-Earth, deep space only, 8400–8450) Space Research (space-to-Earth, 8450–8500)	NASA uses this band for Deep Space Network space-to-earth telemetry. These activities support or will support GALILEO, Mars Global Surveyor, Mars Pathfinder, Cassini (radio science experiment), Near Earth Asteroid Rendezvous (NEAR), and other deep space missions.
8500–9000	Radiolocation	The 8750–8850 MHz band is used for airborne weather radars for storm avoidance.	Radiolocation (G59 nonmilitary secondary to military)	The 8750–8850 MHz band is used for airborne weather radars for storm avoidance. There is also increasing use for ground-based missile defense. NASA operates its Goldstone Solar System Radar at 8530 MHz.

(*Continued*)

Frequency (MHz)	Non-government allocation	Non-government use	Government allocation	Government use
9000–9200	Aeronautical Radionavigation (FN717 ground-based radars and associated transponders only) Radiolocation		Aeronautical Radionavigation (FN717 ground-based radars and associated transponders only) Radiolocation (G19 military only)	This band is used extensively for military precision approach radars. There is also increasing use for ground-based missile defense systems.
9200–9300	Maritime Radionavigation (FN823 limits use of 9200–9225 to shore-based radars) Radiolocation	This band is used by maritime radars for general surface use, navigation, and collision avoidance.	Maritime Radionavigation (FN823 limits use of 9200–9225 to shore-based radars) Radiolocation (G59 nonmilitary secondary to military)	There is increasing use for ground-based missile defense systems in this band.
9300–9500	Radionavigation (US66 aeronautical radionavigation—airborne radars and associated airborne beacons only, ground-based permitted 9300–9320) Meteorological Aids (US67 ground-based radars only) Radiolocation	This band is used by maritime radars for general surface use, navigation, and collision avoidance. These radars employ wide bandwidths for high resolution. Also, radar transponder beacons (RACONs) identify maritime hazards search, and search and rescue transponders (SARTs) identify people in distress at sea.	Radionavigation (US66 aeronautical radionavigation—airborne radars and associated airborne beacons only, ground-based permitted 9300–9320) Meteorological Aids (US67 ground-based radars only) Radiolocation	Ship radars operate in this band to provide coastal navigation information. These radars employ wide bandwidths for high resolution. The Coast Guard operates vessel traffic system radars for controlling ship movement around harbors and coastal areas with high ship traffic. Also, radar transponder beacons (RACONs) identify maritime hazards, and search and rescue transponders (SARTs) identify people in distress at sea. This band is used for airborne weather radars for storm avoidance. There is also increasing use for ground-based missile defense systems.

		This band is used by civil aircraft for airborne weather radars for storm avoidance.		
9500–10000	Radiolocation Meteorological Satellite (FN828 9975–10025 for weather radars)	This band is used for weather radars operated by users such as state an local governments, broadcasters, university researchers, and commercial weather forecasters.	Radiolocation Meteorological Satellite (FN828 9975–10025 for weather radars)	This band is used extensively for military airborne radars. There is also increasing use for missile defense systems.

SOURCE: Compiled by the author with the generous support of the US National Telecommunications and Information Administration, March 20, 2003.

4.5 Endnotes

[1] Struzak, Ryszard, *Ibid.*

[2] ITU, 1998a.

[3] Maitra, Amit, "*Benefits From Out of This World*," Paper IAA-88-565 presented at the *39th Congress of the International Astronautical Federation*, Bangalore, India, 8-15 October 1988, *Acta Astronautica* Vol. 19. No. 9, pp.763–769, 1989.

[4] Struzak, *Ibid*. 1.

[5] Meteorology uses radio frequencies to study structures of clouds and precipitation in space and time. We may think of the frequently launched radiosondes to measure temperature, air pressure, and humidity as a function of altitude. Meteorology also makes use of passive measurements of radio spectral line data, active applications of radio, and facilities based on radar technology. The basis for a weather forecast is the sum total of the results of all these measurements, with the quality of the measurements impacting the weather forecast. Meteorology wants to use the best possible means of extracting the pertinent information from atmospheric radio emission.

With the use of various techniques, radar technology makes it possible to study the upper atmosphere, particularly the ionosphere. To determine the frequencies at which the experiments have to be done (usually below 30 MHz), one looks at the characteristics of the ionosphere, specifically its degree of ionization, electron density, and the vertical distribution of the electrons. Ionospheric research also employs passive uses of radio. A case in point is the study of ionospheric refraction by accurate direction measurements towards satellites and celestial radio sources (4, Spoelstra, 1997).

The physics of the radiating medium does not exclusively determine the radio frequency selection. The propagation characteristics of the terrestrial atmosphere constitute another important parameter. In 1895, Marconi's experiments showed that the atmosphere at frequencies below ~ 30 MHz is not transparent for the radio waves that enable long distance radiocommunication by reflecting radio waves off the ionosphere.

As we approach frequencies above ~10 GHz, the troposphere becomes increasingly important. The tropospheric propagation characteristics are greatly impacted by the density of the different gases and the water vapor content. Some frequency intervals become completely opaque while in other regions of the radio spectrum the atmosphere remains transparent to various degrees. This fact generates different complementary interests: terrestrial radio astronomy prefers a transparent atmosphere while remote sensing scientists may be interested in measuring the transparency even into the frequency intervals where the troposphere is opaque.

[6] The laws of nature are the same from our immediate environment to the farthest depths of the universe, thereby making these spectral lines equally interesting for radio astronomy. Additionally, various scientific projects perform measurements of broadband or *continuum* emission. For these measurements, one needs well-chosen radio frequencies (often every octave) to investigate the variation of the intensity as a function of frequency, that is, the radio spectrum of the transmitting source.

[7] Radiocommunication industry prefers the use of radio frequencies above ~30 MHz, however, even at frequencies as high as 30 GHz, scientists have observed that the ionosphere has effects on radiocommunication [Mawira, A., 1990, as referenced in "Science and Spectrum Management," European Science Foundation (ESF) Committee on Radio Astronomy Frequencies (CRAF) 2002].

[8] Geodesy studies among other things the shape of the Earth and continental drift.

[9] Administration refers to any government department or service responsible for discharging the obligations undertaken in the Constitution of the International Telecommunication Union, in the Convention of the International Telecommunication Union and in the Administrative Regulations (ITU) Constitution—Annex 1002).

[10] Department of Defense (DoD) uses wireless platforms, such as, land mobile radio (LMR), HF, satellite, paging, cellular communications for clear and encrypted voice

communications, audio and video monitoring, alarm systems, electronic tags and tracers, and limited data collection and transfer, both nationally and internationally over diverse geographic conditions. DoD missions often require subscriber unit interoperability and the ability to communicate on a priority basis 24 h/day, 7 days/week.

[11] Draft Report of UNISPACE III, IAU-COSPAR-UN Special Environmental Symposium: "Preserving Astronomical Sky" (IAU Symposium No. 196), held in Vienna July 12-16, 1999, paragraphs 1,2,6,28.

[12] http://www.ntia.doc.gov/osmhome/reports

[13] *Ibid.*, 10, paragraph 158.

[14] *Ibid.*, paragraph 70.

[15] *Ibid*, paragraph 162.

[16] Article III(b), Vienna Declaration, as referenced in CRAF (Committee on Radio Astronomy Frequencies) Newsletter 1999/3, July 1999.

Chapter

5

Spectrum Management[1]

5.1 Demand

The number of radio systems in operation worldwide is huge and increasing rapidly. Liberalization and deregulation are introducing new services and new technologies, thereby generating unprecedented demand for radio frequencies. The International Telecommunication Union (ITU) records indicate more frequency assignments in the last few years—a number that exceeds all assignments recorded during the entire history of radio. Professor Dr. Struzak of the ITU Regulation Board observes that most of suitable frequencies have already been occupied and, within the existing arrangements, the demand exceeds what can further be assigned.[2] According to him, no place exists for new radio stations in certain frequency bands and geographical regions. Other industry experts also point to spectrum scarcity in VHF/UHF frequency bands if the population density exceeds 200 people per square kilometer and GNP10,000 USD per capita per annum.[3] Likewise, there may be no place for new satellites in some areas if the congestion of the geostationary satellite orbit continues. That scarcity impacts further development of telecommunications.[4]

There are several questions that need to be addressed here: First, is the spectrum congestion real? Second, if it is real, what can be done to solve it? Third, does the law of nature influence the spectrum/orbit scarcity[5] or does our mismanagement create the problem? The scarcity issue has serious implications for the future of services and applications and warrants critical review and analysis. The scarcity of radio spectrum is not a new problem. Even as far back as 1925, Herbert Hoover, U.S. Secretary of Commerce, declared, "There is no more spectrum available."[6] Since then, there have been many new inventions that have incorporated and successfully implemented any number of applications of radio waves. With every new invention, however, the problem of the shortage of radio frequencies has become more acute. This problem has been periodically raised at international conferences and at other occasions, thereby indicating that the spectrum shortage has a periodic or chaotic character. Professor Dr. Struzak observes that the periodicity is intrinsic to the development mechanism involving competition and cooperation and is strongly dependent on the progress in science and technology.[7]

Every year hundreds of thousands of experts arrange and attend numerous conferences and symposia to address and solve the spectrum scarcity and spectrum congestion problems. The international organizations that are involved in these meetings and conferences include specialized UN agencies such as the ITU, International Civil Aviation Organization (ICAO), International Maritime Organization (IMO), World Meteorological Organization (WMO), World Health Organization (WHO), World Trade Organization (WTO), and the World Bank. In Europe, the European Commission (EC), Conference of European Posts and Telecommunications Administrations (CEPT), European Radiocommunications Committee (ERC), and European Radiocommunications Office (ERO), among others, gather to address their regional concerns. The involvement of the great number and caliber of these international organizations[8] underscore the unprecedented demand for spectrum and the seriousness of the scarcity and congestion issues.

5.2 Objectives

Spectrum management at the international level,[9] with all the institutional arrangements, originated in the early twentieth century, initially focusing on the regulations and procedures for dealing with technical issues, operating and licensing, and administration.[10] Today, spectrum management includes activities relevant to regulations, planning, allocation, assignment, use, and control of the radio frequency spectrum and the satellite orbits. For any spectrum management system to be useful and effective in today's world, it should take account of sound spectrum engineering,[11] monitoring, and enforcement.

The underlying objectives of any spectrum management system should be to:

1. Convey policy goals,
2. Apportion scarcity,
3. Avoid conflicts.

These objectives should be met, paying close attention to social, political, economic, ecological, and other concerns. The society has several groups, each trying to advance its particular interests, goals and views. Because of the spectrum scarcity, conflicts arise between those who have access to the spectrum resource and those who do not. Likewise, conflicts arise between the proponents of competing uses of the spectrum and also between those who must manage the spectrum and those who use it. Underlying all these conflicts are the commercial, political, physical interference, and other concerns, as discussed in Chap. 4 under domestic applications.

For those whose needs are satisfied, spectrum management should assure the continuation of the status quo. Any modification would jeopardize the benefits they attained. Now, for the newcomers that have no access to the spectrum they need, the principal aim of spectrum management should be to change the way the spectrum is assigned and to eliminate obstacles that prevent them from entering the competition. These policy or administrative arguments and counter arguments reflect the problem at hand: any decision that is the best for one group may not necessarily be good for the other. Therefore, the underlying objective of spectrum management rules and regulations has always been to reflect the relative balance of powers of the competing interest groups.

5.3 Tasks

5.3.1 Policy-making

The Table of Frequency Allocations[12] of the Radio Regulations[13] provides the frequency allocation principles on the basis of which the spectrum/orbit resources are used. It is important to understand the various references to the distribution of a frequency band: *Allocation* refers to the distribution of a frequency band to a wireless service, *allotment* to a country or area, and *assignment* to an individual radio station. The coverage area for allocations could be worldwide or regional to ensure uniformity throughout a particular region. As regards to an assignment, it could be to an individual station or to a group of stations, as needed by the country. This refers to the ad hoc frequency distribution method. Another option is a priori frequency distribution. It is important to identify and differentiate these two services:

1. Under ad hoc managed services—a system frequently described as first-come first-served—the priority of registration dates receives the protection.
2. Under a priori planning, an assignment receives protection from any other assignment.

Competent radio conferences coordinate international a priori frequency plans for specific applications, geographic regions, and frequency bands. A frequency plan refers to a table, or more generally, to a function that assigns appropriate characteristics to each radio station at hand. All the details necessary to operate the station are included in the design and operational frequency plans.[14] International plans provide minimal number of details and are general in nature.

Under a priori frequency plans, the expected or declared needs of the parties interested form the basis for the distribution of the spectrum resource. Reservations for specific frequency bands and associated service areas are made for particular application well in advance of their actual use. Case in point was the establishment of the plan for the Broadcasting-Satellite Service in the frequency bands 11.7 to 12.2 GHz in Region 3 and 11.7 to 12.5 GHz in Region 1 and plan for feeder links for the Broadcasting-Satellite Service in the Fixed Satellite Service in frequency bands 14.5 to 14.8 GHz and 17.3 to 18.1 GHz in Regions 1 and 3 by the World Radiocommunication Conference, Geneva, 1997. Both plans have been annexed to the Radio Regulations.

Those who support the a priori approach point out the unfairness of the ad hoc method, for it puts the burden on the latecomers to accommodate their requirements to the existing users. Opponents argue that a priori planning leads to "warehousing" the resources, thereby immobilizing the technological progress. The reality is that all usable frequency bands have been allocated to services, except that only a small part of them is the subject of international a priori planning. The underlying fear of many countries that currently lack the necessary financial resources is that they will lose access to unplanned frequency bands or positions on the geostationary satellite orbit. The countries with limited financial resources are apprehensive that these bands and positions will most likely be already occupied when they are ready to use them.

Essentially, it is the time horizon considered that differentiates the a priori approach from that of the ad hoc approach. The critics of a priori planning argue that future requirements cannot be predicted with any degree of accuracy or certainty, and as such, plans based on out of reach requirements have no practical value—they simply block frequencies, and stop the progress of development. Further, they add that the pace of technological development is so rapid that plans become outdated even before they are implemented.

The point is that the ITU Convention calls for minimizing the use of spectrum resources, but "...each country has an incentive to overstate its requirements, and there are few accepted or objective criteria for evaluating each country's stated need. In fact, the individual country itself may have only the dimmest perception of its needs over the time period for which the plan is to be constructed. ...Under these circumstances, it is easy to make a case that allotment plans are not only difficult to construct but when constructed will lead to a waste of resources as frequencies, and orbit positions are 'warehoused' to meet future, indeterminate needs…." These remarks do not apply to frequency planning at the design stage of wireless systems, for all requirements at that juncture are targeted, real, and immediate.

5.3.2 Trends

In a world that is changing rapidly and bringing about changes in the role of governments, the current spectrum management policies and practices, inherited from the days when radio was mainly under the state monopoly and the access to the spectrum resources was free, need further reviews and changes. Current trends indicate that:

1. Many countries are abandoning the state monopoly, while allowing the private sector and non-governmental international corporations to increase their share;
2. A single market encompassing a competitive worldwide market economy is emerging;
3. New technology developments are announced every day, including the accelerated pace of their introduction in the marketplace;
4. Digital signal processing with great potential for new integration of services is available;
5. New satellite and stratospheric station technologies, not yet fully exploited, are being deployed.

The framework of the present regulations is inadequate to handle all these new developments. Many industry experts believe it is time for redistribution and better use of radio waves.[15]

The basic complaint of every user is that the Radio Regulations are too complicated and excessively rigid. At the conclusion of every radio conference, the participants display unhappiness over the results achieved. While it sounds strange, this level of dissatisfaction of all parties involved points to the best compromise possible. If that were not the case, some participants would be more satisfied than the others. Over the years, the fundamental rules have remained unchanged even though various improvements have been proposed, resulting in few implementations.

Spectrum management: the shift toward competitive market economy mechanism. Industry experts cite the disparity among the member countries, their needs and their interests, as one of the reasons for slow adaptation of the ITU process to the changing environment. For instance, there are large differences between the governments of China, representing a billion people, and Tonga, representing a hundred thousand people. Nonetheless, the ITU constitution allows a single vote to each of them, as it does to any other member country. The ITU constitution also invokes two other most sacred principles—the national sovereignty and consensus-based decision process. The implication of these principles is that common decisions could be arrived at only if acceptable to the weakest and most conservative members. The third reason for the slowness is the separation of the decision-making process from economic mechanisms. Each member's financial contribution to the common budget is voluntary and does not have any correlation to the number of radio stations or satellites it uses.[16]

Economic incentives, such as "spectrum occupation fees," could be used as an instrument to rationalize the use of scarce resources. If introduced internationally, these fees could limit excessive demand and open "warehoused" frequencies and orbital positions. With the income, telecommunication infrastructure could be developed where needed. Proposals along these lines were presented[17] to the World Radiocommunication Conference, Geneva, 1997; however, the majority of ITU member countries preferred to continue with the administrative "due diligence" approach.

Another idea of spectrum management involves the replacement of the regulatory and fee system by a competitive market economy mechanism. This idea has been put forward in few countries, with initial action limited to selected frequency bands only. Its proponents argue that market forces automatically match the demand to the available resource capacity, and that the market-based management is inexpensive.[18]

The U.S. Federal Communications Commission (FCC)[19] recently conducted a series of spectrum auctions, thereby breaking the tradition. In previous occasions, operators planning to offer wireline communication services were awarded almost for free, based on lottery, comparative hearings ("beauty contests"), or "first-come first-served." That practice is changing. The FCC is now granting the licenses to the highest bidders. In 1994, the first auction held in the United States concluded in assigning three 1-MHz bands around 900 MHz for a total of about 650 million USD. In 1995, two pairs of 15-MHz bands around 1900 MHz for personal communication services were assigned for a total of 7.74 billion USD. Above and beyond this, the successful bidders are to pay the expenses for relocating microwave transmission facilities already using that portion of the spectrum. Depending on the demand and supply situations in various places, the above prices will fluctuate. In the final analysis, however, the consumers will have to bear the costs.

Once the national spectrum markets are introduced, the creation of an international spectrum market is the next logical step; however, there is no evidence that selling the spectrum on global market is going to solve the scarcity problem in a way satisfactory to all parties involved. Sovereignty is still an indisputable principle in the ITU. When the market approach is combined with that principle, spectrum management may be further fragmented. That fragmentation leads to differences in domestic regulations and legal provisions, manifesting some paradoxical situations.[20]

5.3.3 Laws and regulations[21]

Each sovereign state has its own administration[22] with the mandate to use all means possible to facilitate and regulate radiocommunication in that country. The possible structure of such a regulatory body within the country's administration is reviewed in Table 5-1. Except in dual structures, the regulatory authorities regulate and coordinate the radiocommunication interests of both public and private entities.[23]

The mandate and terms of reference of a regulatory authority are usually regulated by national telecommunication law. This national law also includes a national frequency allocation table, which is the national articulation of the ITU Radio Regulations.

However, the process of privatizing public facilities causes differences in a way this privatization is realized in different countries. Usually radio frequency management and regulation is retained as a public interest under the responsibility of a public administration. But in some countries this task is delegated to private organizations with immediate radio frequency interests. When this happens care must be taken that the interests of entities with different interests and requirements are properly managed.

TABLE 5-1 Location of a Radiocommunication Regulatory Body in a National Administration

Location in administration	Description	Country example
Department under ministry (single organization)	Radiocommunication authority	European countries
Department under ministry (dual structure)	1. Addressing private radiocommunication issues 2. Addressing public radiocommunication issues	United States
State commission (same level as a ministry)	"State Commission for Regulatory Affairs"	China

Global regulations. On a global scale the key role in spectrum management lies with the ITU as explained above. Other regulatory organizations must follow the resolutions, recommendations, and other guidance from the ITU. However, it should be noted that the WTO's role in this respect is developing strongly, but this does not replace the ITU position. The interest of the WTO in radio frequency issues is related to the commercial relevance of radio frequencies since the WTO is the international organization dealing with global rules of trade between nations. Its main function is to ensure that trade flows as smoothly, predictably and freely as possible, with the goal to improve the welfare of the people of the member countries.

Regional regulations. At a regional level, organizations such as the U.S. National Telecommunications and Information Administration (NTIA),[24] an agency of the U.S. Department of Commerce, and the FCC,[25] an independent agency, regulate spectrum for federal[26] and nonfederal[27] users,[28] and CEPT in Europe play a key role in spectrum management. The section on Telecom and Broadcasting by Regions, Chap. 6, refers to the major effort in frequency management of all regional spectrum management organizations. These organizations provide their respective administrations with a multitude of management elements in a framework reflecting from the ITU Radio Regulations, and these national administrations can tailor the Tables of Allocations to their national requirements.

In the United States, the expanding commercial and government demand for spectrum prompted the U.S. General Accounting Office (GAO) to examine whether future spectrum needs can be met with the current regulatory framework. The GAO Report[29] reviewed potential regulatory structure options in the United States, including possible policies to consider, and studied the details on spectrum management in foreign countries. For the readers interested in the details of the U.S. regulatory environment, the full report is available from www.gao.gov/cgi-bin/getrpt?GAO-03-277.

In Europe,[30] the CEPT *recommendations* and *decisions* in the legal sense have more or less the same status. The CEPT makes political agreements, decisions and recommendations. In the CEPT/ECC the national representatives have a status as delegated by the minister of the respective state.

Nevertheless, this status does not strengthen the legal status of the CEPT decisions, because the CEPT community of countries is not bound by a treaty regulating the legal status of decisions and directives, such as in the case of the European Union.

If the CEPT makes a decision or a recommendation, there is a chance that different National Regulatory Authorities (NRAs) will apply it in different ways. In a case of dispute where an action is before a national court against licensing conditions that are imposed, the outcome will most likely be that different conditions will apply in different CEPT member states.

The role of the European Commission, as indicated in Chap. 6, is different.[31] For European telecommunication regulation, CEPT decisions are binding only for those Administrations that committed to them. EU directives are stronger and are legally binding for all EU member states (because of the EU treaty). If national legislation is not in harmony with EU law, this incompatibility has to be removed in due course. On the other hand, CEPT decisions and recommendations also must not be incompatible with EU law. EU directives and CEPT decisions must be seen as instruments serving the interests of the community.

As far as the European Economic Area (EEA) and EC members are concerned, a CEPT decision or recommendation would have to be implemented and applied in accordance with EU law, including the Licensing Directive, 97/13/EC, and Council and Parliament Decision 710/97/EC, as a yardstick. In addition, other documents such as the EC EMC directive, which deals with harm caused by "unwanted emissions," are applicable. The same must be noted for the EU Directive 99/5/EC of the European Parliament[32] that states in paragraph 2 of Article 7, which deals with "Putting into service and right to connect," that "Notwithstanding paragraph 1, and without prejudice to conditions attached to authorizations for the provisions of the service concerned in conformity with Community law, Member States may restrict the putting into service of radio equipment only for reasons related to the effective and appropriate use of the radio spectrum, avoidance of harmful interference or matters relating to public health." This clause indicates on what basis an administration can still ask for a license for the use of radio. Such a respect for national sovereignty follows also from Articles 30 to 36 of the EU Treaty and articles of the treaty on free traffic of services.

These legal instruments allow national regulatory authorities to impose licensing conditions if they are linked to the efficient use of radio frequencies. Any condition must be justifiable and is subject to the principle of proportionality. Regulators always have to use the least restrictive regulatory means to achieve the required conditions.

Given the different mandates of the CEPT and the EC, the views on spectrum management and policy of these organizations are different.

To improve the strategic profile of the EC, it published a Green Paper[33] on European spectrum policy in 1999, which was meant to serve as a discussion document and to improve the influence of the Commission on European spectrum policy issues and to bring attention to the relevance of a debate on the current European spectrum policy. Unfortunately, several important issues, especially a coherent strategic view on public and scientific usage of radio frequencies, are not well developed. Also, the Green Paper does not include an analysis of strong versus weak aspects of the current practice. It is also observed that the EC overlooked relevant CEPT positions and publications in the preparation of the document, although the Council of Europe invited the Commission in its Resolution 92/C318/01 of November 19, 1992, "to give full consideration in future to the mechanism of ERC decisions as the primary method of ensuring the provisions of the necessary frequencies for Europe-wide radio services."

The proposed EC spectrum policy is expressed primarily in terms of interests and requirements for the active radiocommunication services. A strategic view on the specific interests and requirements of the passive (i.e., receive-only) services[34] and applications compared with those of the active services[35] is, however, lacking. Such one-sidedness, in the opinion of the European Science Foundation (ESF) Committee on Radio Astronomy Frequencies (CRAF), reduces the balance of the spectrum policy, since the requirements and characteristics of active and passive services are significantly different. An explanation for this is easily found in the priority the EC gives to commercial and industrial interest.

The EC policy is clear in its views on spectrum availability for each of the radiocommunication services, that is, on telecommunications, broadcasting, transport and also on research and development (research is understood in a generic way and applies to industrial as well as scientific research). However, availability in a regulatory sense is different from availability at a quality level sufficient for radiocommunication's requirements. Spectrum impurity degrades spectrum efficiency and reduces spectrum availability significantly, especially when inadequate spread-spectrum modulation techniques are used. Technological cooperation between the users in these different radiocommunication services may well lead to a significant alleviation of the problem. The results of such cooperation can be considered as beneficial for all users of the radio spectrum. The cooperation between active and passive users of the radio spectrum needs special attention because of the different requirements for the active and passive services.

For the benefit of the active and passive services, industry, operators and users of the provided facilities, cooperation between active and passive services could be improved, and the European Commission could play an important and stimulating role in this respect.

The need for such cooperation could be more clearly articulated as a strategic goal of the European Commission.

Another observation is that the spectrum policy of the EC is developed primarily from the perspective of terrestrial[36] use of radio. Coordination and regulatory procedures for terrestrial radiocommunication applications already exist to a great extent. This holds also for defective systems that have a harmful impact on other radiocommunication services. But a clear opinion of the EC on space[37] issues is lacking in this respect. At the present time, about half a dozen defective space systems can already be identified. In many respects the ITU Radio Regulations are not adequate to give guidance to the administrations in managing such situations. These systems defects may originate from: (1) malfunction of a system; (2) ignorance in the design and construction phase of the system; or (3) plain bad system design. These defects are known because they generate harmful interference or produce some other negative impact on other systems. But nonetheless, at present, operators of such systems meet no constraints.

A strategic position of the European Commission on this issue is urgent and needs to be developed in close cooperation with the CEPT. Such a strategy would be of great help to administrations, especially when it is known that there are reasons to expect that the extent and impact of this problem on other applications of radio (space based or terrestrial) is rapidly increasing. A strategy on defective space systems is complicated by the fact that some operators of space applications do not pay adequate attention to system quality if, in their view, this is not commercially justified.

Local regulation at national level. National telecommunication legislation[38] is developing rapidly in some countries while it hardly exists in other countries.[39] In the U.S. and Europe, national telecommunication legislation finds a framework in legislative instruments such as the bills introduced in the 107th Congress[40] and EU directives which are addressed to and binding upon the EU member states and requires implementation by national law. In other regions, the legislative role of regional organizations is weaker or even absent.[41]

Management and enforcement.[42] Radio spectrum management, that is, the efficient operation[43] of radiocommunication equipment and services without causing harmful interference, is the combination of administrative, scientific and technical procedures. Monitoring radio frequency usage and its characteristics is an ancillary function to administrative frequency management and regulatory effort.

The commercial-economic[44] impact of (radio)-telecommunication is steadily increasing.[45] This development has major consequences for the scope and work of NRAs.[46]

Operators are paying large amounts of money for access to parts of the radio spectrum with the intention of providing services to the public. As mentioned earlier under the policy approaches section, auctions, beauty contests or other means are often used by administrations to further this purpose. In most cases, the financial revenues from these actions will flow to the State budget and can be seen as an extra income for the government.

Because spectrum use is highly priced, it is obvious that telecom-operators demand a clean and usable spectrum. This means that the National Regulatory Authority being the body responsible for proper administrative management of the radio spectrum must be ready and able to perform this task. Today, National Regulatory Authorities or Radio Communications Agencies are made liable for taking care of the spectrum.

To achieve high-quality spectrum management, theoretical frequency planning is not sufficient; adequate information about the actual use of radio spectrum is needed to achieve successful spectrum management. This information can be obtained only by specific monitoring.[47]

It is obvious that besides administrations, various other users of the radio spectrum also deploy monitoring activities. A special case relates to passive[48] radio frequency use: the characteristic of all its measurement activity is a special kind of monitoring. Whenever monitoring is mentioned in this section, we refer to an activity performed by a National Regulatory Authority or Administration.

Role of monitoring in the past. The ITU-R *Handbook on National Spectrum Management*[49] describes monitoring as a tool to:

- Ensure compliance with national spectrum management rules and regulations through the verification of proper technical and operational characteristics of transmitted signals, and the detection and identification of unauthorized transmitters.
- Locate and resolve interference problems.
- Determine channel and band usage, including assessment of channel availability and verification of the efficacy of the frequency assignment process and spectrum analytical methods.

From the *Handbook on Spectrum Monitoring* (ITU, 1995c)[50] "monitoring" serves as the eyes and ears of the spectrum management process. Until today, in many countries hardly any work was done to assist the frequency management department of the national Administration and almost all monitoring activities were related to enforcement purposes. Handling interference issues was one of the main activities carried out in the daily monitoring work. If activities were carried out at the request of the frequency management department, it was mostly done on an ad hoc basis.

Role of monitoring in the future. The changing radiocommunication requirements and usage of radio frequencies, the technological development, changing user demands and regulatory developments imply that radio frequency monitoring must adequately cope with these developments.

We consider that monitoring has a wide variety of aspects and should not be limited to criteria described by the ITU-R *Handbook on National Spectrum Management*[51] (see above).

Because of the sometimes very high license-fees for telecom-operators, the regulatory authority wants to be informed whether license-holders are working in conformity with the license-conditions.

In the past, such verification could often be done by listening and simple measurements. Today sometimes more than one service is emitted within one transmission. For this reason it is necessary to look into the spectrum of such a transmission. In this respect specific transmissions such as in Digital Audio Broadcasting, in Digital Video Broadcasting, by cellular networks or by satellite systems can be identified.

The monitoring technique to enable such a task is signal-analysis, which will be one of the most important monitoring tools for the future development of the monitoring function.

Owing to the increasing use of (mobile) satellite systems, sharing of satellite services with terrestrial services and compatibility issues between space and terrestrial systems, the monitoring function becomes more and more important.

Administrative relationship between frequency management and monitoring facilities. Purely theoretical frequency planning is not sufficient for spectrum management. A well-structured relationship between the frequency management and monitoring departments is needed to accomplish high-quality spectrum management by the National Regulatory Authority. The Regulatory Authority's responsibility in controlling the radio spectrum via both the frequency management and monitoring departments is changing and growing in a structured relationship to each other. While in the past, 70 percent of activities were carried out by the monitoring service, and 30 percent were enforcement activities, nowadays the proportions are reversed.

The relationship between the monitoring and frequency management departments can be simply illustrated by Fig. 5-1.

This figure illustrates that:

- The spectrum is used for all kinds of radio transmissions. Frequency management is of overriding importance for the efficient and effective use of the radio spectrum. International and national authorities are setting the rules for the use of the radio spectrum via assignments, license parameters and so on.
- The Monitoring Service observes the radio spectrum, and the monitoring operators have the duty to compare whether the use of the

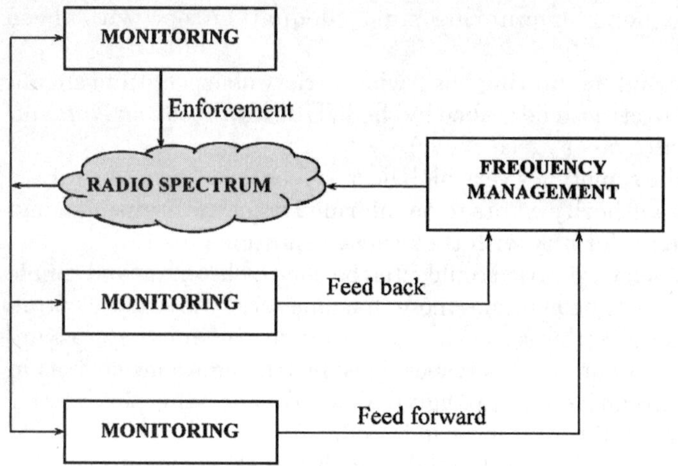

Figure 5-1 Spectrum management cycle.

radio spectrum matches the policy of frequency management. Via this loop, the monitoring service provides *feedback* to frequency management.

- In observing the radio spectrum, the monitoring service can also provide information to frequency management on hitherto unknown or new use of the radio spectrum. When Spectrum Management sets up an experiment for new services before a policy concerning that new service has been developed, the monitoring service can observe the experiment and advise on it. That is why a *feed-forward loop* is appropriate.
- The monitoring service can also address radio spectrum users directly in case of interference. The monitoring operators can give guidelines to the users to avoid interference. This is called *Enforcement*. In this case, no direct relation to Spectrum Management is needed.

Global and regional activities on monitoring issues. At a global level, the ITU radiocommunication sector addresses the issue of monitoring in its Working Party 1C (refer to Chap. 6, Table 6-3). Within this working party, representatives of administrations and manufacturers of monitoring equipment collaborate intensively. WP1C develops measuring protocols, working procedures between monitoring services, and so on, which are laid down in ITU-R Recommendations.

At a regional level, project teams from various administrations work on monitoring matters in support of the respective administrations'

Working Group Frequency Management to solve enforcement problems, to carry out monitoring campaigns to serve the administrations and the ITU, and to develop recommendations in the area of monitoring. For details on U.S. federal agency monitoring functions, refer to Chap. 6, Table 6A-1.

Monitoring as a key element in spectrum management complying with the ITU Radio Regulations. The ITU Radio Regulations contain several footnotes and tables giving power flux density levels, which must not be exceeded for a specified amount of time and in certain directions or area. Administration radio frequency monitoring is one of the instruments in spectrum management complying with these regulations, otherwise administrations choose to work on a revision of the ITU Radio Regulations in WRCs, while acknowledging at the same time that given these regulations, it is debatable whether the requirements can be met. This makes the radiocommunication services that are in victim position regulatory outlaws by definition and puts a consequential burden on administrations. This cannot be in the mandate of an administration.

Other organizations with related interests. Besides the public management organizations discussed so far, various other international and national organizations deploy spectrum management activities within the regulatory framework provided by the ITU Radio Regulations and regional and national regulations. Among these are organizations such as the North Atlantic Treaty Organization (NATO), the European Broadcasting Union (EBU), and the International Civil Aviation Organization (ICAO). Space organizations such as ESA (Europe), NASA (USA) and NASDA (Japan), and intergovernmental operators of space systems such as INTELSAT and EUTELSAT have their own frequency management facilities for managing the frequencies assigned to these organizations. Scientific organizations such as the *Scientific Committee on the Allocation of Frequencies for Radio Astronomy and Space Science* of UNESCO (IUCAF) as well as the Committee on Radio Astronomy Frequencies (CRAF) of the European Science Foundation (ESF), and the International Union of Radio Amateurs (IARU), work on the preservation of the frequencies allocated to the radiocommunication services they represent. Each of these organizations has internally a specific spectrum management responsibility resulting from the characteristics of this organization. In addition, by means of intensive communication and coordination with public spectrum management organizations and their participation in ITU-R and public regional activities[52], they can exploit the possibilities of making their case clear in the international and national spectrum management process.

5.3.4 Administration

Spectrum management process involves several tasks:[53]

- Administrative and legal support (e.g., record keeping, frequency assignment, license generation, and so on)
- Engineering (e.g., analyses of radiocommunication systems, including performance within the systems and the interference between the systems, EMC computation, band allocation)
- Licensing (license status, financial records, complaints received, resolved, and rejected)
- Coordination and notification.

Administrative and legal support functions are common to all spectrum management organizations and it is not necessary to discuss these in great detail. As Fig. 5-2 illustrates, with the help of software packages that are available one can perform comprehensive administrative

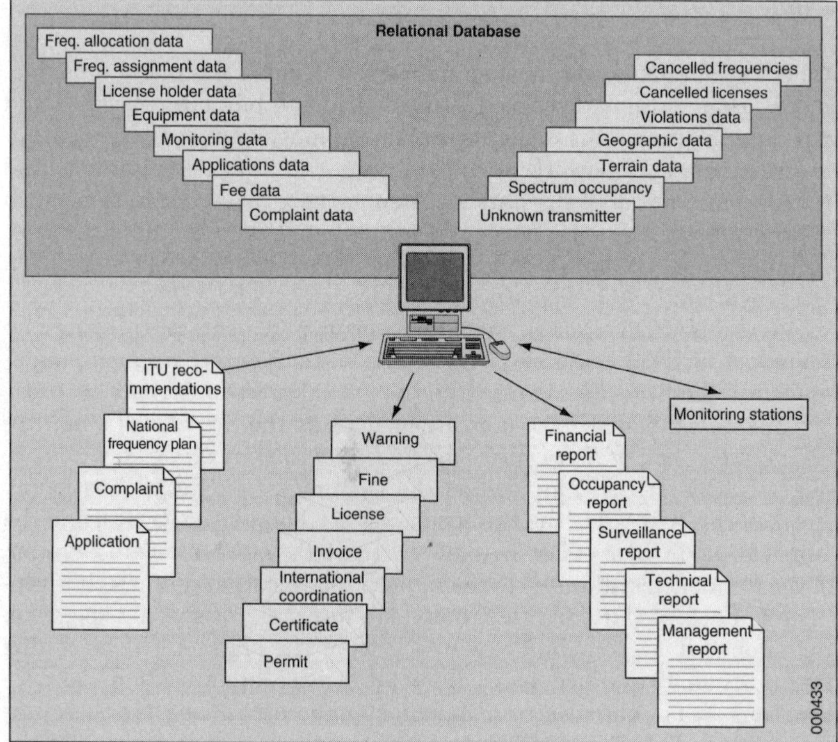

Figure 5-2 Spectrum management via software. (Any workstation can process applications, issue reports, notices, and other things.) (*Source: TCI Spectrum Management Software, 710 Model, http://www.tcibr.com/PDFs/710webs.pdf*)

functions, such as frequency assignment, licensing applications, issuing notices and invoices, paying fees and fines and processing reports much more easily and efficiently.

5.3.5 Engineering

Spectrum management calls for decisions about a field of technology, which require technical analyses of the information, capabilities, and choices concerning:

- RF spectrum
- Bandwidth
- Emission designators
- Power
- Receiver sensitivity
- Transmission lines
- Distance and azimuth calculations
- Radio wave propagation
- Antenna basics
- Coordination
  - Inter-system
  - Noninterference, nonlicensed bands
- Frequency planning
- Optimization
- Electro-magnetic compatibility (EMC)
- Electro-magnetic interference (EMI)
- Harmonics
- Intermodulation
- Radiation hazards (RADHAZ)
- Electronic warfare (EW)

Technical analyses in the above areas are critical to facilitate the resolution of many spectrum allocation issues involving spectrum use and future requirements, and also to prevent or resolve the interference issues. Through these analyses, equipment specifications and standards necessary to establish compatibility between systems are determined, using models or methods developed through engineering support. The models and methods also help assign frequencies.

The analytical tasks are very complex, with many steps and considerations needed for efficient radio spectrum use and interference-free services.[54] Given the high economic and political value of radio frequencies, these tasks are better handled by efficient computer-based integrated spectrum management systems.

At a national level, an integrated spectrum management system, where multiple users can access information, perform technical analyses, perform administrative processes, share work, pass on work tasks and store results and documents, is essential. The configuration of such an integrated system should include a central database that can store results from engineering analyses, calculations, and so on as well as documents generated by the spectrum management process. The system should also include a role-based security so that the right to change data in the database is properly controlled, while authorizing read-only access to all users. Data availability and integrity are always major concerns and an integrated system should address those concerns by scheduled automatic back-ups, including uninterruptible power supply.

Much of the work with radiocommunication systems is better performed against a set of geographical databases, including:

- Map images, typically derived from the scanning of paper maps (or by using digital sources directly, if available)
- Elevation data
- Terrain classification ("clutter" codes)
- Aerial photographs and/or satellite images.

For example, the methods for propagation calculations require detailed terrain features in order that we can accurately calculate the field strengths and signal levels. The ITU Digitized World Map (IDWM) has the coastal contours and national borders. This data provides geographical orientation for the user and is useful for calculations that require information about land/sea and rain climate zones. The propagation model ITU-R P.370 and the ITU Radio Regulations Appendix S7 (Earth station coordination contours) are good fit for IDWM.

An integrated system should also handle various station databases, such as the International Frequency List (now superseded by the ITU BR IFIC—International Frequency Information Circular) available from the ITU. The licensing authority in a given country maintains national station data comprising all radio emitters, except information on military frequency use. Military has specific allocations, and the civilian sector is not allowed to access the detailed frequency assignments and allotments. Now, the point to underscore is that if an integrated spectrum management system handles all three sources of spectrum data: international, national civil, and national military data, it may prove to be a very useful tool.

Software packages that are currently available in the marketplace cover complex engineering calculations, including sanity check, link performance analysis, frequency assignment, interference check, and so forth. AerotechTelub's WRAP, a fully integrated system for spectrum management, is an example of one such package. Appendix 5B provides details on WRAP, while Figs. 5-3 to 5-9 illustrate its various features and capabilities. WRAP and many similar products support nearly all spectrum management activities, including record keeping, forecasting and financial management related to licensing, as depicted in Fig. 5-8.

The benefits of implementing an integrated, computerized spectrum management system are many. For instance, the system is convenient in that the handling of the complete process of technical analysis, licensing and coordination becomes relatively easy, with a common station database holding all relevant data that all personnel working with various parts of the spectrum management process can access.

To be useful, the system should incorporate suitable functions that allow technical analysis of a wide variety of radiocommunication systems that are the responsibility of a spectrum management authority.

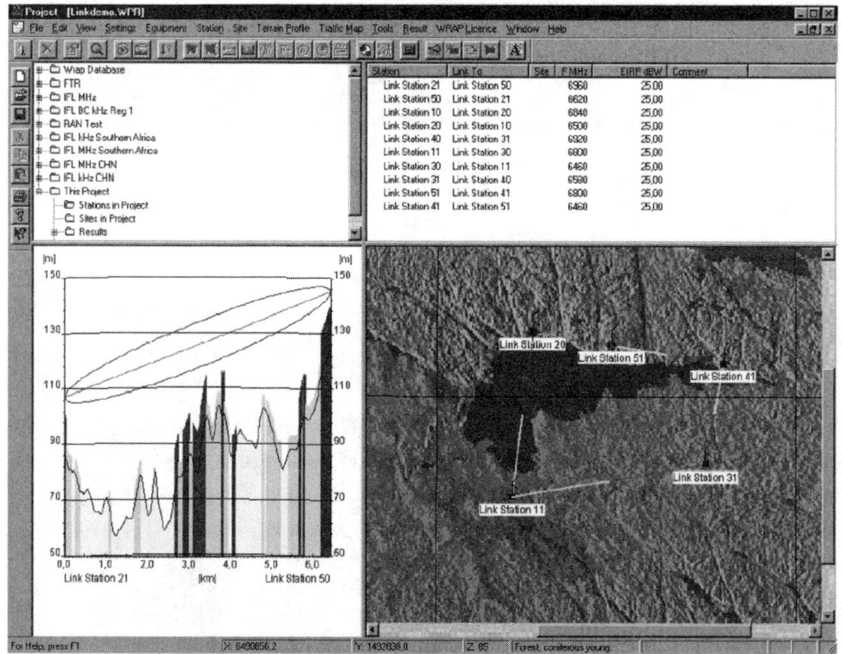

Figure 5-3 Sanity check performance. (*Source: Olov Carlsson, AertechTelub, Ibid*)

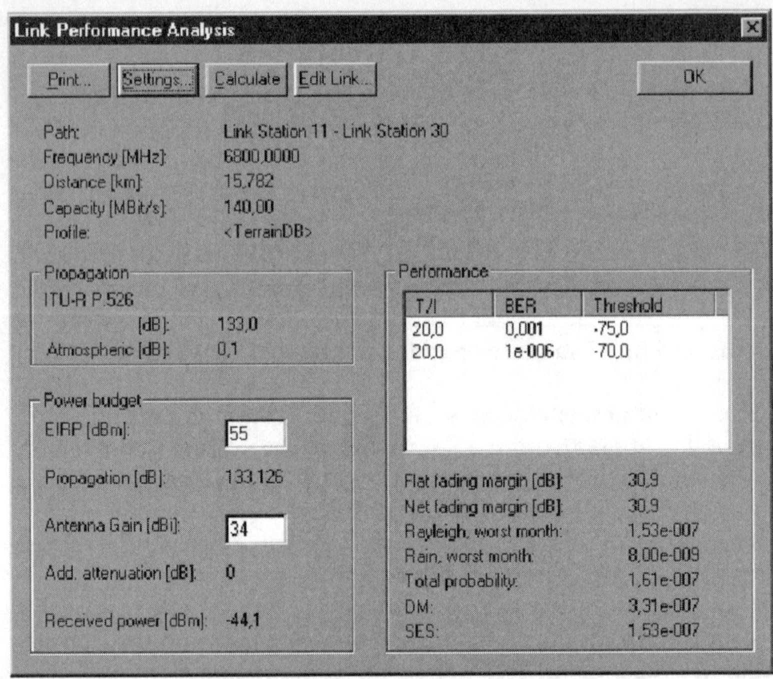

Figure 5-4 Link performance analysis. (*Source: AerotechTelub, Ibid.*)

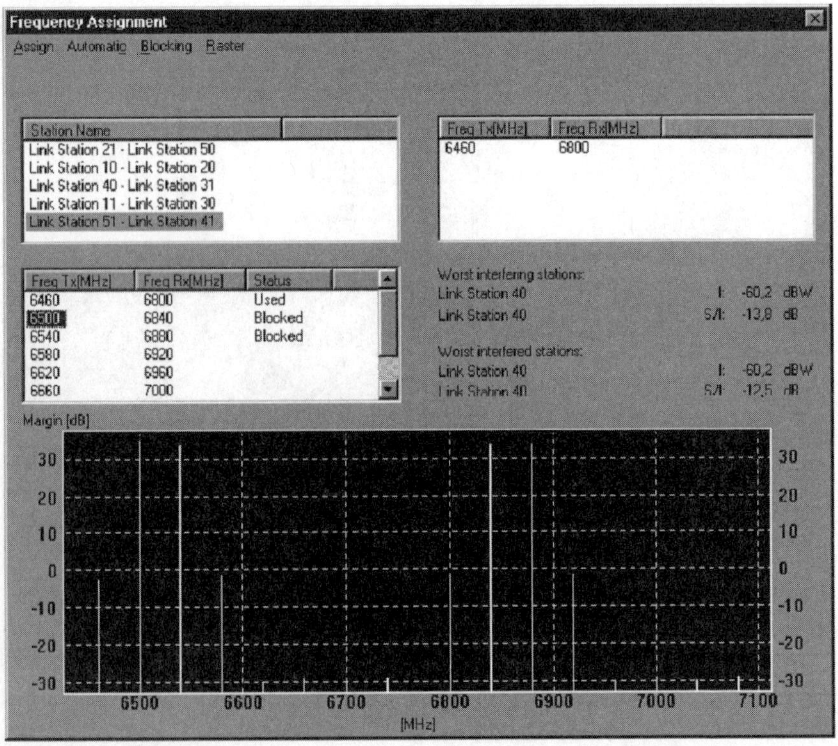

Figure 5-5 Frequency assignment performance. (*Source: AerotechTelub, Ibid.*)

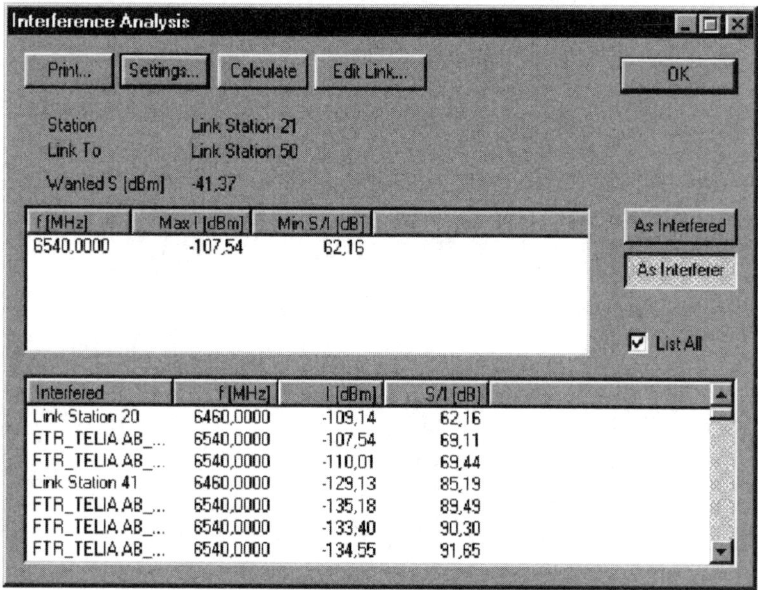

Figure 5-6 Interference analysis. (*Source: AerotechTelub, Ibid.*)

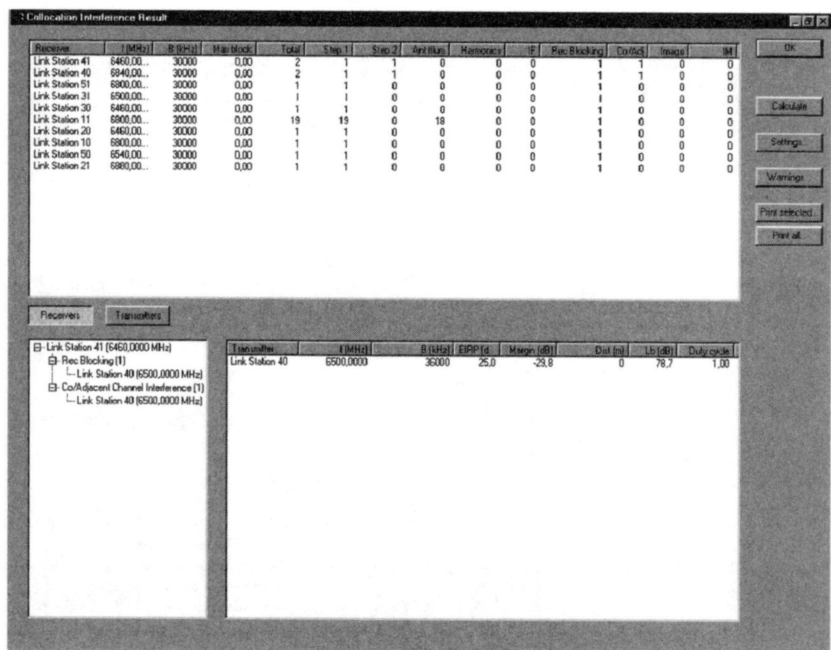

Figure 5-7 Collocation interference check. (*Source: AerotechTelub, Ibid.*)

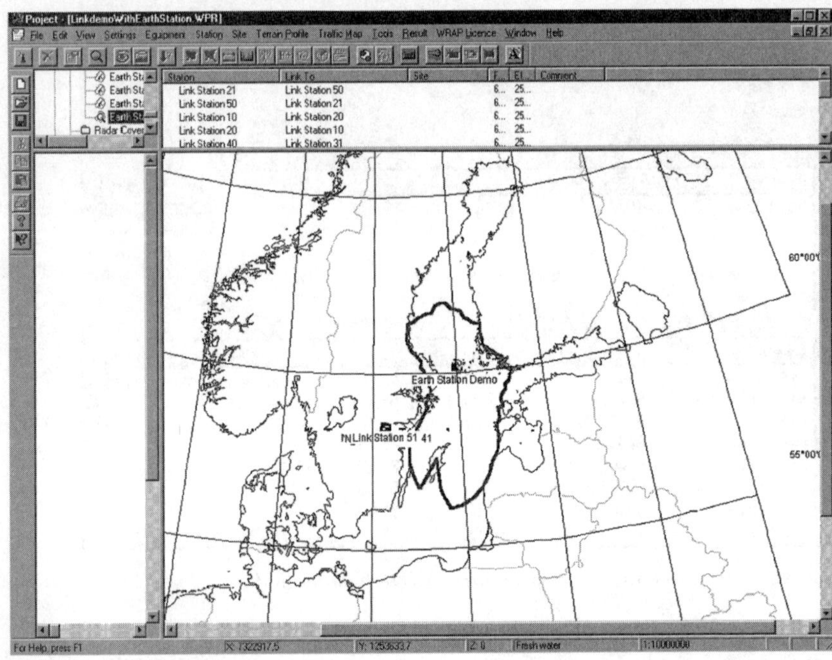

Figure 5-8 Coordination of interference check with neighbors. (*Source: AerotechTelub, Ibid.*)

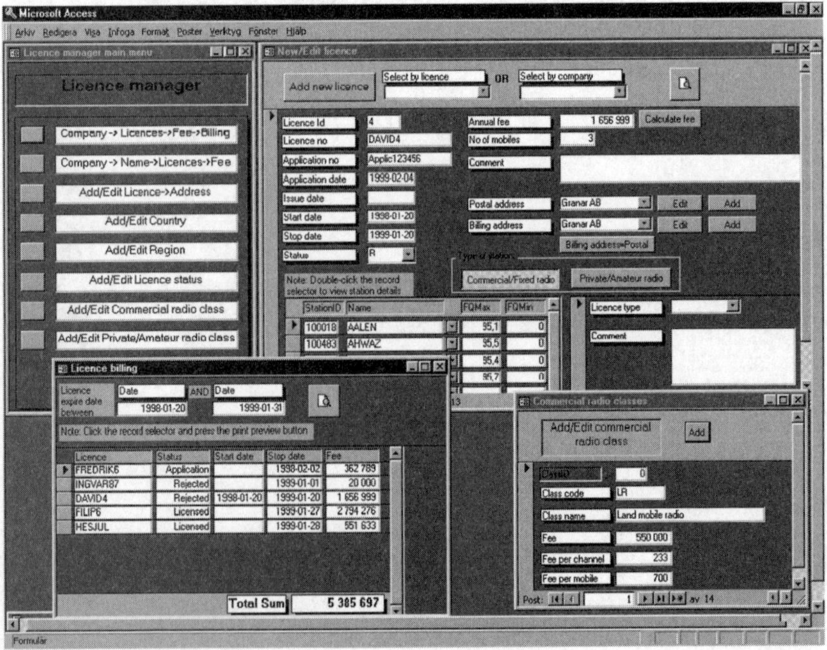

Figure 5-9 Performance of all spectrum management tasks via software. (*Source: AerotechTelub, Ibid.*)

This often requires the implementation of a number of radio wave propagation models as well as analytical methods, with general applicability. Some situations warrant service-specific methods in order to be consistent with international agreements and recommendations or to be responsive to systems that have unique features that need to be accounted for in detail.

5.4 Appendix A: Extract From: Report ITU-R SM.2012 "Economic Aspects of Spectrum Management"

5.4.1 Need for spectrum economic approach

The increasing use of new technologies has produced tremendous opportunities for improving the communications infrastructure of a country and the country's economy. Further, the ongoing technological developments have opened the door to a variety of new spectrum applications. These developments, though often making spectrum use more efficient, have spurred greater interest and demand for the limited spectrum resource. Thus, the efficient and effective management of the spectrum, while crucial to making the most of the opportunities that the spectrum resource represents, grows more complex. Improved data handling capabilities and engineering analysis methods are key to accommodating the number and variety of users seeking access to the spectrum resource. If the spectrum resource is to be used efficiently and effectively, the sharing of the available spectrum has to be coordinated among users in accordance with national regulations within national boundaries and in accordance with the Radio Regulations of the ITU for international use. The ability of each nation to take full advantage of the spectrum resource depends heavily on spectrum managers facilitating the implementation of radio systems and ensuring their compatible operation. Therefore, every available means including economic means is needed to improve national spectrum management.

This report has been developed to assist administrations in the development of strategies on economic approaches to national spectrum management and their financing. In addition, the report presents a discussion of the benefits of spectrum planning and strategic development and the methods of technical support for national spectrum management. These approaches not only promote economic efficiency, but can also promote technical and administrative efficiency.

Before the economic approaches can be discussed, it is first necessary to consider what is an effective spectrum management system and what areas of spectrum management can be appropriately supported by other means.

5.4.2 Requirements for national spectrum management

Effective management of the spectrum resource depends on a number of fundamental elements. Although no two administrations are likely to manage the spectrum in exactly the same manner, and the relative importance of these fundamental elements may be dependent on an administration's use of the spectrum, they are essential to all approaches. For further information on spectrum management functions, see the ITU Handbook on *National Spectrum Management*.

5.4.3 Goals and objectives

In general, the goals and objectives of the spectrum management system are to facilitate the use of the radio spectrum within the ITU Radio Regulations and in the national interest. The spectrum management system must ensure that adequate spectrum is provided over both the short and long term for public service organizations to fulfill their missions for public correspondence, for private sector business communications, and for broadcasting information to the public. Many administrations also place high priorities on spectrum for research and amateur activities.

In order to accomplish these goals, the spectrum management system must provide an orderly method for allocating frequency bands, authorizing and recording frequency use, establishing regulations and standards to govern spectrum use, resolving spectrum conflicts, and representing national interests in international fora.

Radiocommunications law. The use and regulation of radiocommunications must be covered within each nation's laws. In areas where radiocommunications use is not extensive, and where the need for management of the spectrum may not yet be crucial, national governments must still anticipate the increase of radio use and ensure that an adequate legal structure is in place.

National allocation tables. A national table of frequency allocations provides a foundation for an effective spectrum management process. It provides a general plan for spectrum use and the basic structure to ensure efficient use of the spectrum and the prevention of RF interference between services nationally and internationally.

5.4.4 Structure and coordination

Spectrum management activities may be performed by a government body or by a combination of government bodies and private sector organizations. The government bodies or organizations that are given the authority to manage the spectrum will, however, depend on the structure of the national government itself and will vary from country to country.

5.4.5 Decision-making process

The processes developed to allocate spectrum, assign frequencies to specific licensees, and monitor compliance with license terms are essential tools for implementing national goals and objectives. Administrative bodies responsible for developing rules and regulations governing the spectrum should develop an organized decision-making process to ensure an orderly and timely spectrum management process. The process should be set up to allow decisions that serve the public interest while reflecting national policies and plans relating to the spectrum, developments in technology, and economic realities. Often such processes will depend on the use of consultative bodies to make appropriate decisions.

5.4.6 Functional responsibilities

The spectrum management structure is naturally formed around the functions that it must perform. The basic functions are:

- Spectrum management policy and planning/allocation of spectrum
- Frequency assignment and licensing
- Standards, specifications, and equipment authorization
- Spectrum control (enforcement and monitoring)
- International cooperation
- Liaison and consultation
- Spectrum engineering support
- Computer support
- Administrative and legal support

Administrative and legal support functions will necessarily be a part of the spectrum management organization, but they are common to all organizations and thus it is not necessary to discuss these in relation to spectrum management.

Spectrum management policy and planning/allocation of spectrum. The national spectrum management organization should develop and implement policies and plans relating to the use of the radio spectrum, taking into account advances in technology as well as social, economic and political realities. National radiocommunications policy is commonly associated with regulation development because the regulations generally follow the establishment of policies and plans. Accordingly, it is often a primary function of the policy and planning unit to conduct studies to determine existing and future radiocommunications needs of the country and to develop policies to ensure the best combination of radio and wireline communications systems employed in meeting the identified needs.

The primary result of the planning and policy-making effort is the allocation of frequency bands to the various radio services. The designation of frequency bands for specific uses serves as the first step to promoting spectrum use. From allocation decisions follow further considerations such as standards, sharing criteria, channeling plans, and others.

Frequency assignment and licensing. Providing or assigning frequencies represents the heart of the daily operation of the spectrum management organization. The frequency assignment unit performs, or coordinates the performance of, whatever analysis is required to select the most appropriate frequencies for radiocommunications systems. It also coordinates all proposed assignments with regard to existing assignments.

Standards specification and equipment authorization. Standards provide the basis for equipments to work together and limit the impact of radio use to that which is intended. In many cases, such as aircraft navigation and communications systems, equipment must be capable of operating in conjunction with equipment operated by other users and often other countries. Standards can be used to require design characteristics that will ensure that such operation is possible. The second aspect of standards is their use to ensure electromagnetic compatibility (EMC) of a system with its environment and generally involves limiting transmitted signals to a specified bandwidth or maintaining a specified level of stability in order to prevent interference to other systems. In some cases an administration may choose to set standards for receivers, requiring a certain level of immunity to undesired signals. The establishment of an adequate program of national standards forms a basis for preventing harmful interference and, in some cases, for ensuring desired communications system performance.

Spectrum control (enforcement inspections and monitoring). Effective management of the spectrum depends on the spectrum manager's ability to control use of the spectrum through enforcement of spectrum regulations. This control is built primarily on enforcement inspections and monitoring. See the ITU Handbook on *Spectrum Monitoring.*

Enforcement inspections. Spectrum managers must be granted the authority to enforce regulation of spectrum use and set appropriate penalties. For instance, spectrum managers may be granted the authority to identify a source of interference and to require that it be turned off or to confiscate the equipment under appropriate legal mechanisms. However, the limits of that authority must also be specified.

Monitoring. Monitoring is closely associated with inspection and compliance in that it enables the identification and measurement of interference sources, the verification of proper technical and operational characteristics of radiated signals, and detection and identification of illegal transmitters. Monitoring further supports the overall spectrum management effort by providing general measurement of channel usage and band usage, including channel availability statistics and the effectiveness of spectrum management procedures. It obtains statistical information of a technical and operational nature on spectrum occupancy. Monitoring is also useful for planning, in that it can assist spectrum managers in understanding the level of spectrum use as compared to the assignments that are recorded on paper or in data files. Some administrations have chosen to use monitoring in place of license records.

International cooperation. Radiocommunications have a significance that goes beyond the borders of each nation. Navigation equipment is standardized to allow movement throughout the world. Satellite system transmissions facilitate worldwide communications. Radio wave propagation is unhindered by political boundaries. Communications system manufacturers produce equipment for many markets, and the more the markets encourage commonality, the simpler and less expensive the production process will be. For each of these reasons, the national spectrum manager's ability to participate in international fora becomes significant. International activities include those within the ITU, those within other international bodies, and bilateral discussions between neighboring countries concerned with ITU Radio Regulations.

Liaison and consultation. In order to be effective, the spectrum management organization must communicate with and consult its constituents, that is, the radio users composed of businesses, the communications industry, government users and the general public. This includes dissemination of information on the policies, rules and practices of the administration and provides mechanisms for feedback to evaluate the results of these policies, rules and practices.

Spectrum engineering support. Since spectrum management involves decisions pertaining to a field of technology, engineering support is required to adequately evaluate the information, capabilities and choices involved. Engineering support can assist the spectrum manager in many ways. For example, interference situations can often be prevented or resolved through technical analysis. The equipment specifications and standards necessary to ensure compatibility between systems can be determined. Frequencies can be assigned using models or methods developed through engineering support. Also, the resolution of many spectrum allocation issues can be facilitated by analysis of spectrum use and future requirements.

Computer support. The extent to which computer support facilities are available to be used and are used by the spectrum management authority depends on the resources, priorities, and particular requirements of the country concerned. Computer support may cover licensing records to complex engineering calculations and may include the development, provision, and maintenance of support facilities for nearly all spectrum management activities, including record keeping, forecasting and financial management related to licensing.

5.4.7 Performance of spectrum management functions

The previously described spectrum management functions need to be established in order to have an effective spectrum management system; however, not every aspect of each function needs to be performed by the national spectrum management organization. The policy or overall management authority must, however, remain with the national spectrum management organizations. The following chapters discuss the means by which spectrum management may be funded and the means by which economic approaches may improve the efficiency of spectrum use, methods of assessing the benefits of spectrum use and the use of other organizations to support and/or provide part, or all, of specific spectrum management functions.

5.4.8 Studies on economic aspects of spectrum management

The terms of reference for this economic study are given in questions ITU-R 206/1, ITU-R 207/1, and ITU-R 208/1. The Radiocommunication Assembly in 1995 approved these questions and recommended that Study Group 1 study these questions on an urgent basis. Study Group 1, through Working Party 1B, created a Rapporteurs' Group to address these questions and accelerate the development of this report. The decides from these questions are reproduced in the following to describe the subjects requiring study.

Question ITU-R 206/1: strategies for economic approaches to national spectrum management and their financing. *Decides* that the following questions should be studied:

1. What are the underlying principles that have been taken into consideration by various administrations in their approaches to financing the maintenance and development of national spectrum management?
2. What economic approaches have been, or are intended to be, used to promote efficient national spectrum management in different frequency bands?
3. What are the advantages and disadvantages of these various economic approaches to national spectrum management?
4. What are the factors (e.g., geographical, topographical, infrastructural, social, legal) that could affect these approaches and how would they vary with the use of radio in a country and the level of that country's development?

Question ITU-R 207/1: assessment, for spectrum planning and strategic development purposes, of the benefits arising from the use the radio spectrum. *Decides* that the following question should be studied:

1. What are the benefits that accrue to an administration from the use of radio within its country and how can they be quantified, allowing them to be represented in an economic form so as to enable a comparison of the benefits and costs of particular spectrum management options (e.g., in terms of employment or Gross Domestic Product)?
2. What models can be used to represent these benefits in an economic form and how can they be validated?
3. What factors could affect the benefits accruing to an administration from the use of the radio-frequency spectrum, including by national safety services?
4. How would the factors in (3) vary from country to country?

Question ITU-R 208/1: alternative methods of national spectrum management. *Decides* that the following question should be studied:

1. What are alternative spectrum management approaches including the use of nonprofit making user groups and private sector spectrum management organizations?

2. How can these approaches be categorized?
3. Which of these alternative spectrum management approaches would be responsive to the needs of the developing countries as well as for the least developed ones?
4. What measures, of a technical, operational and regulatory nature, would it be necessary for an administration to consider implementing when adopting one or more of these spectrum management approaches in the context of:
   a. The country's infrastructure;
   b. National spectrum management;
   c. Regional and international aspects (e.g., notification, coordination, monitoring)?

Note. The focus of this report in answering question ITU-R 206/1, decides 1, is placed on approaches to financing national spectrum management rather than underlying principles. Also, no models of the type discussed in question ITU-R 207/1 and decides 2, are presented as no consensus exists on how such models would be represented.

5.5 Appendix B: Example of an Integrated Spectrum Management System

5.5.1 WRAP—computerized spectrum management from AerotechTelub, Sweden

WRAP—AerotechTelub's system for spectrum management—is fully integrated as described before. Within a single main screen interface it provides the following functions:

The map viewer constitutes the main presentation tool and performs all presentations using geographical data, either as a profile viewer, a common 2-D map or as a full 3-D map. The map viewer works with any combination of geographical data, results and scanned map images.

The calculation tools are capable of performing a large number of different tasks, supporting different radio services:

1. Coverage: To calculate signal levels, field strengths, radio-optical visibility, signal-to-interference ratio, required antenna height, clearance, terrain clearance angle, best server and number of servers for single or multiple transmitters/receivers. This is used for the analysis of area coverage.
2. Interference: To calculate interference between stations located some distance apart.

3. Radio link performance: To calculate the performance of radio links, fading margins and unavailable time.
4. Spectrum viewer: To provide a visual overview of the spectrum occupancy at a selected geographical point, based on the contents of the station databases.
5. Collocation interference: To calculate the interference situation in confined locations where many stations may interfere within a small area.
6. Frequency assignment: To perform automatic and manual frequency assignment.
7. Radar coverage: To calculate the coverage for radar stations, based on radio-optical visibility or the two-way radar propagation path.
8. Coverage comparison: To perform comparisons and statistical analysis between measured coverage data and predicted coverage. To perform comparisons between coverage data that has been predicted with different propagation models or with different parameters.
9. Earth station coordination: To calculate the coordination contours around earth stations operating in bands shared with terrestrial services.
10. Traffic capacity: To calculate and design the traffic capacity of cellular networks.
11. Broadcast: For interference analysis and frequency assignment for sound and television broadcast networks, implementing the specific recommendations, protection ratios, and so on in common use for these services.
12. Interference between satellite networks: To calculate the interference between satellite networks. Can also be used to calculate the interference between terrestrial stations and satellites.

Many of the calculations involve the prediction of radio wave propagation, and there are several selectable models available to fit the desired applications.

Models based on detailed terrain information:

- ITU-R P.526: Propagation by diffraction.
- ITU-R P.452: Prediction procedure for the evaluation of microwave interference between stations on the surface of the earth at frequencies above about 0.7 GHz.
- DETVAG 90/FOA (The Swedish Defense Research Establishment), implementing several different diffraction methods, spherical earth methods and weighting/selection methods between these.

Models based on coarse terrain and land cover information:

- DETVAG 90/FOA for low-frequency calculations of ground wave propagation, based on the GRWAVE model with Millington and other extensions.
- ITU-R P.370: VHF and UHF propagation curves for the frequency range from 30 to 1000 MHz (uses the ITU Digitized World Map for land/sea determination).

Statistical models:

- Longley-Rice: Prediction of tropospheric radio transmission loss over irregular terrain.
- COST-231
  - Hata
- COST-231
  - Walfish-Ikegami

In addition to all these models the attenuation owing to atmospheric gases can be added in accordance with the ITU Recommendation P.676 attenuation of atmospheric gases. Rain attenuation and hydrometeor scatter are also handled. For earth-to-space paths the ITU Radio Regulations Appendix S8 "method of calculation for determining if coordination is required between geostationary-satellite networks sharing the same frequency bands" is implemented, extended with the ability to select a more accurate propagation model as defined in ITU-R Recommendation P.619-1, items "basic transmission loss," "Tropospheric scintillation" (from ITU-R Recommendation P.618-6), "Ray bending" and "Precipitation scatter" (from ITU-R Recommendation P.452-9).

The administrative tools consist of two different parts, both operating on the same station database as the technical system:

1. Integrated administrative functions. These functions form an integrated part of the WRAP system and provide access for reading and writing of data in the station database.
2. Extended administrative functions.
   a. License Manager, to handle licensing. The generation and handling of notification forms such as the ITU BR T-series forms are included in this function.
   b. Other specific administrative functions can be developed, operating on the open common station database.

The work procedure in an integrated spectrum management system. An example will be used to illustrate the benefits with an integrated system, based on a license application for a microwave radio link network. This is based on an anticipated work-flow starting with a license application and ending with issuing the license, and if required, the notification to the ITU BR of the new stations.

Application data. The applicant should have listed basic facts about the stations for which a license is sought. This typically includes geographical locations, antenna height above ground, transmitter power, antenna gains, required received signal level, required signal-to-interference ratio, and so on. The applicant may already have a preferred frequency band of operation and even suggestions of frequency assignment. Each licensing authority may have different requirements for the submission of data in the application, some requiring also a specification of which equipment will be used to check for compliance with type approval standards. For this example it is sufficient with the parameters that constitute important data for the analysis of link performance and interference.

Make project for calculations. Data for the stations are entered into the spectrum management system in a temporary workspace. The purpose for this is to have all relevant data under control to support the subsequent stages in the process. But this is still an on-going process, so the data should not yet be stored in the national station database.

Perform "sanity check," calculate link performance. The applicant should have performed basic link budget calculations and designed the links for the appropriate quality and availability. The licensing authority should nevertheless check the link performance for suitable design, for instance to make sure that the radiated power is reasonable so as not to create larger interference regions than necessary.

Search the national station database (and other databases, if available). To provide basis for the continued process for frequency assignment and interference analysis there is a need to examine the frequency use in the geographical area and frequency band of interest. This is done by searching the available station databases with geographically- and frequency-limited criteria. Stations that are found which are suspected to be subjected to interference from the new stations, or that may interfere with the new stations, should be entered into the calculation project for further analysis. This is done automatically (if selected) in WRAP. Refer to Fig. 5-10.

Perform frequency assignment. The frequency assignment function is used to assign frequencies to the new stations. Account is taken automatically to the interference from the new stations to the existing stations and the interference from the existing stations to the new stations. The frequency raster should naturally be compliant with the frequency

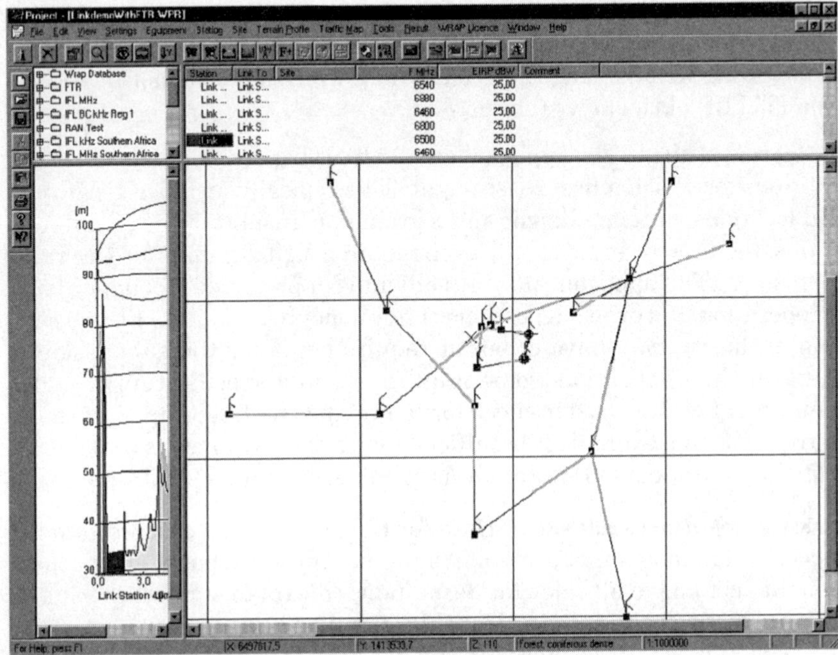

Figure 5-10 Search the national station database (and other databases).

plan for the band of operation. An appropriate propagation model should be selected to provide protection for the desired percentage of time.

Perform interference check. The automatic assignment process accounts properly for the interference contributions from new and existing stations, but there may be an interest to analyze and quantify the resulting interference margins by a specific interference check. In this process it is easy to find the most contributing interferer and to judge whether its contribution is acceptable.

Perform collocation interference check (may not be responsibility of the authority). The specific locations for the new stations would normally have been selected by the applicant to be suitable from a linking point of view. The applicant may or may not have the capabilities to judge whether the local interference situation at the intended site for installation may pose a problem. The licensing authority probably has a better knowledge about at least the already licensed stations at the sites and should have the capability to examine the frequency assignment for local interference such as intermodulation, harmonics, transmitter spectrum mask and receiver selectivity. For really detailed and accurate collocation interference analysis there is a need for quite detailed performance data of receivers and transmitters, as well as very detailed information on the exact locations of antennas at the site. This level of detail normally goes beyond the responsibility of the licensing authority, but it may anyway perform some calculations and provide guidance to the applicant or express concern that more detailed investigations should be performed.

Perform checks for potential international interference: BR IFIC, earth station coordination for shared frequency bands. The BR International Frequency Information Circular should be consulted for potential interference to or from stations in countries within a typical coordination distance. Bilateral or multilateral coordination agreements may already have been established incorporating coordination zones, in which case the search can be limited to within the zone along the border. Otherwise interference calculations should be performed, with station characteristics determined from the BR IFIC and complemented with appropriate additional data if needed.

Earth stations in frequency bands shared with the terrestrial services can be found in the BR IFIC/Space Radiocommunications Stations database. Data for these stations can be extracted and entered as characteristic earth station data for generation of the coordination contours using the Earth Station Coordination tool. Terrestrial stations falling within the coordination contour, with over-lapping frequency bands, should be examined further for interference potential by using the Interference tool.

Coordinate with neighbors. The investigations of potential interference or location of the new stations within pre-determined coordination areas as agreed with neighboring countries may necessitate coordination with authorities of neighboring countries. This should be performed in accordance with coordination agreements (if they exist). The generation of the appropriate ITU T-series notification form can be done to support the coordination.

Issue license. Pending the outcome of the coordination (if needed), a temporary license may be issued. In many cases it will be obvious for the licensing authority whether the coordination process will have a positive outcome, so that the license can be issued before the coordination is completed.

Perform notification to the ITU. The appropriate notification form can be completed and sent to the ITU BR (if needed). In this example it would be the T11 form. The system provides tracking of the notification process and allows for the suppression and modification of the notifications.

To sum up, WRAP offers many possibilities, which are better explained by a select few exercises[55] which follow.

5.5.2 Exercise A1: Planning a microwave link network

Input Data. The objective in this exercise is to plan and design a small 16 Mbit/s microwave link network in the 4 GHz band. Standard stations and equipment data available in the demo version of WRAPs station database should be used. Assume the following:

- Maximum height above ground for antenna towers is 60 m
- Maximum single-hop distance is 40 km
- The network shall link the following towns in the county of Östergötland in Sweden:
  - Motala
  - Mjölby
  - Linköping
  - Norrköping
  - Finspång
  - and back to Motala.
- Towers can be placed at the highest points within or near the urban areas of the towns
- The fixed service band 3810.0 to 4182.5 MHz, channel separation 14.5 MHz, duplex distance 213.0 MHz shall be used.
- The nonavailability requirement is an SESR (Severely Errored Seconds Ratio) in accordance with ITU-R Recommendation F.697-2, which is less than 1.5×10^{-4}.

Entry of stations. The **Samples** folder contains a project file that sets the map view to suit this exercise. Perform the following after starting WRAP as described in Endnote 54.

> Open the project **OstergotlandMap.WPR**. This is done by first selecting **File – Open** in the upper text menu bar. Then find the ...**\WRAP \Samples** folder, which contains this project file and other files used for exercises in the User's Manual. When you have opened the file, save it again (change the name), using **Save as**... to allow saving under a new name. Saving now and then is wise if errors are made and a saved, correct version of the project needs to be retrieved.
>
> Right-click in the map, select **Scale – 1:500 000** (or user-defined 1:300 000). Select **Properties**.... The first tab of the window that opens shows the available geographical vector themes. Check the theme **Text, town 10000 – 99999**. Accept and close the window by selecting **OK**.

The map now displays the names of the towns to be linked. The center position of the map can be adjusted by right-clicking at the desired new center position and selecting **Center position** in the menu.

Right-click in the left geographical viewer and select **Profile**. This will later conveniently allow the display of the terrain profiles along the path. Take the opportunity now to set the earth radius factor to 0.6 instead of the nominal 1.33. This is done by right-clicking, selecting **Properties** and entry of 0.6.

Note. An earth radius factor of 0.6 is commonly used when planning radio links for high time availability.

Right-click in the map and select **Snap to highest**. This will cause the selected station positions to automatically be placed at the highest point within a certain area around the cursor (scale-dependent size of the area).

Position the cursor within the urban (red) area of Motala. Right-click without moving the cursor and select **Tx Position**. Notice that a cross and circle appears near the cursor position.

Then position the cursor within the urban area of Mjölby. Right-click and select **Rx Position**. Notice that the Profile Viewer now displays the terrain profile between the two selected points.

The end points of the link have now been set. The next step is to create stations at these points. This is performed as follows:

Select **Station – New**... in the top text menu bar. A window with a number of selectable folders appears. Choose **Fixed service, digital links**. Select **Wrap Fixed station 1 to 19 GHz/16 Mbit/s**.

The **Edit Station** window opens. Give a name to the first station in the link hop (Motala). To allow easy identification of the individual links you may use the following naming convention: **Motala-to-Mjolby**.

Select the **Frequencies** tab. Enter **4110** in the **TX Fq (MHz)** box and **3897** in the **RX Fq (MHz)** box. The TX and RX equipment have been selected automatically as defined by the template for this type of station. Press **OK**, which closes the window and automatically opens the **Edit Station** window again, this time for entry of data for the opposite station in Mjölby.

Note. Observe that the Edit Station window will open twice, once for each end of the link, and you have to enter station names for both stations (and frequencies for the first station) and close each window with OK. Otherwise the link will not be created properly.

Enter the station name, for example as **Mjolby-to-Motala**. The frequencies have been set automatically to match those that were assigned to the Motala station, i.e., TX and RX frequencies are reversed. Press **OK** to accept the entries.

A feature that allows automatic entry of existing station coordinates to a new station should be used for the continued definition of the link network. This is done by right-clicking on an existing station in the upper right list view and selecting **Set As Tx**. Then the cursor is placed at the appropriate location in the map for the selection of **Rx Position**. This is a convenient way of building a chain of link hops.

Continue as follows:

> Right-click on station **Mjolby-to-Motala** in the list view. Select **Set As Tx**.
> Position the cursor within the urban area of Linköping. Right-click and select **Rx Position**. You may need to try this a few times to find a suitable high spot.

Hint. The map display can be changed in many respects. One way to allow quick identification of high spots is to right-click in the map and select **Properties – Raster Data – Type – Terrain heights, Floating colors**. This sets the map to display a grey-scale with the lighter shades being higher in elevation. Also make sure that the **Info** boxes are checked to display height and terrain codes in the status bar at the bottom of the main window. The z-value gives the height above sea level.

> After selection of the new Tx position and Rx position: Again perform the **Station – New** commands and enter station names with the same naming convention. **Observe that frequencies should be reversed**!

It is common practice to make sure that each link point transmits in the same frequency band. Otherwise interference may become severe from a transmitter operating in the same band as a receiver. So ensure that frequencies are reversed. This can be checked by showing the links on the map. Correct frequency distribution is indicated when each node shows the same color (green or red) in each of its two directions.

Note. If you are using the demo version you will only be able to enter in total six stations. Each link hop has two stations, so the maximum number of link hops you can enter is 3.

Users of the demo version: Save your project. Open the OstergotlandLinks.WPR project in the \wrap\samples folder and show the stations on the map. This is a complete network that you can use for the continuation of the exercise.

Continue in this manner until only the final link between Finspång and Motala remains. Perform the following (this has already been done in the **OstergotlandLinks.WPR** project):

> Mark **Finspang-to-Norrkoping, Set As Tx.** Mark **Motala-to-Mjolby, Set As Rx.** Continue to assign frequencies as before.

Now the complete network can be shown on the map, and each link can be analyzed for the proper clearance above the terrain. Perform the following to do this:

> Right-click in the bottom left viewer, select **Profile**. Set the earth radius factor to 0.6 by right-clicking and selecting **Properties**....

Mark all stations in the list view, select **Show In Map**. Check the frequency information (red/green) for consistency. Depending on the number of link stations there may be mismatch in a mesh network as this (even number OK, odd number there will be one mismatch). Mismatches may need to be taken care of later by proper frequency assignment or other means.

Mark one station at the time. Click on the icon to display the terrain profile. Attempt to achieve full clearance of the first Fresnel zone by editing the antenna heights. This is done simply by double-clicking in the Profile Viewer near the end points of the path. This opens the **Edit Station** window. Select the **Tx Equipment** and **Rx Equipment** tabs and edit the antenna heights (up to a maximum of 60 m). The profile view is updated when the window is closed with **OK**.

This procedure should work fine for all links except for the one between Finspång and Motala. A much higher antenna height would be required here. Even with 60-m-high antennas at both ends you will probably not achieve full clearance.

Here a relay site is required. A convenient way of finding a suitable location is by using an area-coverage calculation called **Minimum Required Antenna Height**. This will show the areas that have coverage to both Motala and Finspång with a certain antenna height above ground at the relay site. Perform the following:

Mark the stations **Finspang-to-Motala** and **Motala-to-Finspang** in the list view (any number of stations can be selected by standard Windows procedure: Press **Ctrl** and click on the station to select).

First an area of interest, within which an appropriate site would be located, should be marked in the map. This is done by right-clicking in the map and selecting **Calculation/Search Area – Circle**. Place the cursor at the center of the circular area to be drawn. Press the left mouse button and move the cursor until a circle of suitable size has been drawn. The circle will be defined as a purple color when releasing the mouse button.

Click in the icon **(Coverage)**. Make the following settings:

- **Calculation: Minimum Req Antenna Height**.
- Check **Above Ground Level**. *Note: If you are using version 3.3.1 or earlier there is no such selection. Instead, make sure that the "Above sea level" is not selected.*
- **Clearance: 100 percent**
- **Calculation Area: Circle, Defined in Map Viewer**
- **Resolution: High**
- **Effective Earth Radius Factor: 0.6**
- **Description:** A suitable name for the calculation.

Click on **OK** to start the calculation and close the window.

The calculation may take up to a few minutes if you are using a slow computer. The progress of the calculation can be seen in the upper left view. Perform the following to see this and set the subsequent map presentation:

> Double-click on the **Calculations** folder. The progress in percent can be seen. The result can be seen in the **Results – Coverage Results Area** folder when the calculation has finished. Open this and double-click on the result name that now has appeared.
>
> The **Edit Result** window appears. The **Information** tab contains some statistics on the result, such as maximum and minimum values (that is the minimum antenna height at the relay site to have 100 percent first Fresnel zone clearance to both Motala and Finspång). The **Presentation** tab defines the number of levels to present, their colors, patterns and line style. Press the **+** sign to display for instance three levels. Mark each level and set a suitable value. A selection of 30, 45, and 60 m may be appropriate. Press **OK** to close the window.
>
> Now right-click on the result name and select **Show Cursor Value** and **Visible**. The map will now display the areas within which a relay site should be placed. Move the cursor within these areas and read the result value in the status line. Select a suitable point, right-click and select **Tx Position**.

Now two link hops need to be created to divide the original Motala-Finspång link. This can be performed conveniently by first using the **Duplicate** function to create a complete copy of the original link. Then the respective end points can be moved to the new site defined by the **Tx Position**. The station names should obviously be changed to reflect the new locations, and the frequencies need to be checked for the proper frequency arrangement at each location.

Perform the following:

> Mark the link **Motala-to-Finspang/Finspang-to-Motala**. Right-click and select **Duplicate**. This results in a new link hop named **Copy of**... etc.
>
> Mark the link station **Copy of Motala-to-Finspang**. Right-click and select **Move To Tx Position.**
>
> Mark the link station **Copy of Finspang-to-Motala.** Right-click and select **Move to Tx Position.**
>
> Mark all stations in the list view and select **Show in Map**. Check the frequency indicators such that all link sites only have either red or green colors in both of their link directions. Double-click on station names that have the wrong frequency complement and edit (change Tx frequency and Rx frequency). This is done by marking the frequency, clicking on the right arrow to move the frequencies into the edit boxes and editing the frequencies there. Don't forget to click on the left arrow to move the changed frequencies back.
>
> Also take the opportunity to edit the station names to for instance:
>
> - **Relay-to-Motala**
> - **Motala-to-Relay**
> - **Relay-to-Finspang**
> - **Finspang-to-Relay**

Mark the **Motala-to-Finspang** link, right-click and select **Delete**.

As a final measure you should check the antenna heights for the new stations. They can probably be reduced significantly from the 60 m maximum height.

Calculate link performance. Now all links should have a reasonably free line-of-sight. The next step is to calculate the link performance. Do as follows:

Mark a link to be analysed for link budget and availability. Select the icon **Radio Link Performance**. The initial calculation is always performed with the Free-space propagation model, complemented with attenuation because of the atmosphere. Since all links now should have a free line-of-sight, this is a valid propagation model. This opens a window as follows:

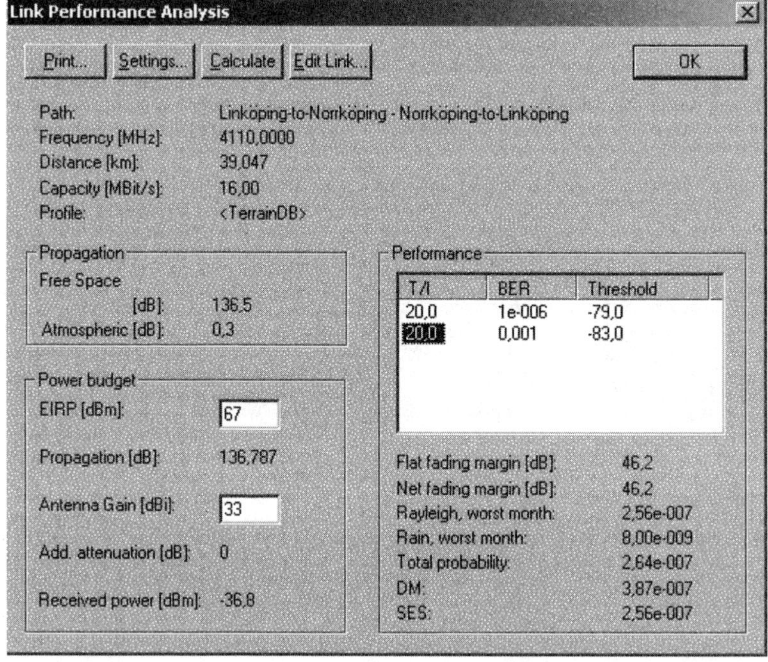

Note that the line for **T/I 20.0 – BER 0.001 – Threshold – 83.0** has been marked. This is necessary to show the result for **DM – Degraded Minutes** and **SES – Severely Errored Seconds**. The SES value constitutes the most important result, and in this case the probability of unavailability owing to too high bit error rate is only 2.56×10^{-7}, to be compared with the design objective of 1.5×10^{-4}. This means that the link budget can be reduced by, for instance, selecting antennas with lower gain or reducing the transmitter power. In this particular case it seems to be sufficient with about 40 dBm Effective Isotropically Radiated Power (EIRP). All links can be checked for availability.

Under **Settings**... you can find selections of various methods, propagation models and atmospheric conditions, and you may experiment with other settings.

Interference analysis

Interference within the network. The new network should be checked for potential interference within the network and against other use of the assigned frequencies. First you may check within the network by using the **Interference** tool. Perform the following:

> Mark the first station to be checked for interference. Select the **Interference** tool.
>
> Select **Settings... Propagation Model – Wanted – ITU-R P.526** and ...**Interferer – ITU-R P.452**. Use the default settings for other options. Press **OK** to start the calculation and close the window. A result screen similar to the following appears:

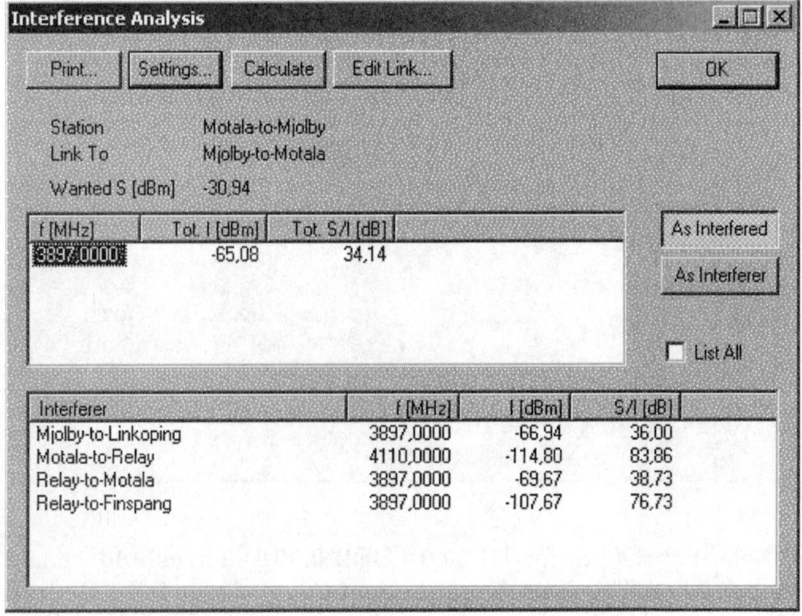

Note that the frequency entry in the top list view has been marked to display the individual interference contributions as listed in the lower box.

The total interference as received by the Motala-to-Mjolby station is –65 dBm, and the signal-to-interference ratio is 34 dB. These values are quite acceptable. Selecting the **As Interferer** function will show the interference that the transmitter of Motala-to-Mjolby causes in the other receivers in the network.

You may continue and check the remaining stations for interference.

Interference from other stations. Interference to and from other users of the assigned frequencies should be checked. This requires knowledge of the overall frequency use, such as a national master station database. WRAP supports this, but the demonstration version is limited in functionality. To demonstrate some of this, it is possible to search the IFL (ITU's International Frequency List, to be replaced with the BR IFIC—International Frequency Information Circular). This requires that the IFL in WRAP format is installed, which will show up as an additional station database named IFL in the upper left database folder view.

In this particular case it is of little interest, since Sweden does not have any entries in the IFL for this frequency. Norway, Denmark, and Germany have entries at both 3897 and 4110 MHz, but the distance is so large that there is no potential of interference.

Anyway, the procedure is to perform a geographically limited search on the frequencies of interest. Any matching station entries in the database can then be added to the WRAP project, and the interference contribution can be calculated.

Checking for collocation interference. There may be a risk for interference if the new stations are located near existing radio installations. This may give rise to interference such as intermodulation, harmonics, and so on. Analysis of this can be performed with the **Collocation Interference** tool. The normal procedure would be to search for other stations in the vicinity of each of the new installations and add these to the project.

Run the **Collocation Interference tool** just to illustrate the procedure. Perform the following:

> Mark all stations in the project. Start the **Collocation Interference** tool.
>
> The main window opens. Press **Calculate** (you may naturally study the settings).
>
> Maximize the window after the calculation has been performed. Mark the second receiver in the top list box. Then expand the view in the bottom left list box by clicking on the + sign. Mark the line as shown below. The "Margin" value indicates the sensitivity degradation, in this case owing to interference from the other transmitter at the same location. Since the wanted signal level is high, this degradation is negligible.

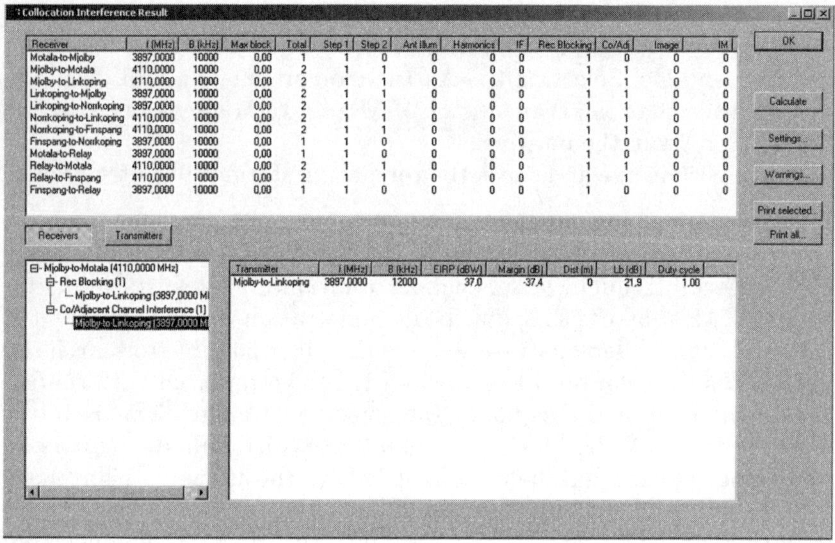

Receiver	f (MHz)	B (kHz)	Max block	Total	Step 1	Step 2	Ant Illum	Harmonics	IF	Rec Blocking	Co/Adj	Image	IM
Motala-to-Mjolby	3897,0000	10000	0,00	1	1	0	0	0	0	1	0	0	0
Mjolby-to-Motala	4110,0000	10000	0,00	2	1	1	0	0	0	1	1	0	0
Mjolby-to-Linkoping	4110,0000	10000	0,00	2	1	1	0	0	0	1	1	0	0
Linkoping-to-Mjolby	3897,0000	10000	0,00	2	1	1	0	0	0	1	1	0	0
Linkoping-to-Norrkoping	3897,0000	10000	0,00	2	1	1	0	0	0	1	1	0	0
Norrkoping-to-Linkoping	4110,0000	10000	0,00	2	1	1	0	0	0	1	1	0	0
Norrkoping-to-Finspang	4110,0000	10000	0,00	2	1	1	0	0	0	1	1	0	0
Finspang-to-Norrkoping	3897,0000	10000	0,00	1	1	0	0	0	0	1	0	0	0
Motala-to-Relay	3897,0000	10000	0,00	1	1	0	0	0	0	1	0	0	0
Relay-to-Motala	4110,0000	10000	0,00	2	1	1	0	0	0	1	1	0	0
Relay-to-Finspang	4110,0000	10000	0,00	2	1	1	0	0	0	1	1	0	0
Finspang-to-Relay	3897,0000	10000	0,00	1	1	0	0	0	0	1	0	0	0

Transmitter	f (MHz)	B (kHz)	EIRP (dBW)	Margin (dB)	Dist (m)	Lb (dB)	Duty cycle
Mjolby-to-Linkoping	3897,0000	12000	37,0	-37,4	0	21,9	1,00

Frequency assignment. The interference check showed that every link in the network operates satisfactorily with the first attempt to assign frequencies. This was such a simple case, requiring little or no support in the selection of frequencies. An automated method of frequency assignment should be used when it is difficult to find suitable frequencies. This is done with the F+ **Frequency Assignment** tool. This tool also gives a convenient graphic display of the interference situation for each station to support manual frequency assignment. You can try this out in the following way:

Mark all stations that should have transmitter frequency in the low band (for automatic assignment). To study the interference situation, just mark all stations. Start the F+ **Frequency Assignment** tool.

First an appropriate channeling plan must be selected. Click on **Raster – Select Raster**. Choose the **4 GHz, 14,5 MHz, 213 MHz** raster and click on **OK**.

Now select **Blocking – Settings**. Select the **ITU-R P.526** propagation model. Click on **OK** and then **OK** in the Blocking Settings window.

Select **Blocking – Calculate**. The list views become white when the calculation is ready. Mark the **Station Name** for each station in the upper left list view. This will display the margin diagram that gives information on the assignment margins.

The assignment can be changed manually by clicking on the vertical lines. The colors indicate the following:

- Blue: Already in use by the marked station
- Yellow: Assignable

- Red: Not assignable (owing to non-fulfilment of signal-to-interference criteria).

Frequencies can be removed by right-clicking and selecting whether to delete just this frequency or all frequencies. Before performing a complete automatic assignment it is common practice to delete all frequencies.

Automatic assignment can be performed by selecting **Automatic – Start**. There are several settings to control the assignment algorithm in the **Automatic – Settings** menu.

Finally: Save the stations to the station database by marking all stations in the project list view, right-clicking and selecting Save.

5.5.3 Exercise A2: earth station coordination

Many frequency bands in the microwave range are shared between fixed services and various satellite and space services. The ITU prescribes that a specific coordination procedure shall be performed to secure the compatibility between space and fixed services. The procedure is illustrated in the following.

The ITU publishes the Space Radiocommunication Stations on CD-ROM and regular updates through one of the BR IFIC CD-ROMs. There is associated software to search for earth stations and satellites to allow analysis of interference and to initiate, if needed, a coordination procedure.

A search was made to find a receiving earth station with which there may be a need to coordinate since its frequency use is overlapping with the 3897 MHz frequency of the link network that we just planned. The earth station NIT 5 is located in Norway, receiving the satellite Intelsat 5 Indo C1 at 63 degrees east in the geostationary orbit.

After establishing that a frequency overlap exists, the particular data for this earth station was entered in WRAP and stored in a project. The satellite was created and stored under the name Intelsat 5 Indo C1. Now perform the following to calculate the coordination area around earth station NIT 5:

Open the project **NIT5.WPR** in the ...**\WRAP\Samples** folder. (You should close the previous link project first).

This project uses another geographical database, the GTOPO30. This is a worldwide database having height information in a raster with 30 s resolution (about 900 m in north-south direction, 900 m or less in east-west direction) and a terrain classification of "land" and "sea". The link network that was created in the previous exercise has been included in this project.

Show the stations in the project in the list view. Show NIT 5 on the map. Mark NIT 5 and select the **Earth Station Coordination** tool.

Make the following settings:

- Exceed time RX: 0.01 percent
- Radiation pattern: 6
- Rx Ant Gain: 54 dBi (data from the SRS CD-ROM)
- Climate zone: Use IDWM (ITU Digitized World Map)
- Geometry: Calculate (means that the terrain mask will be derived from the GTOPO30 database).
- Effective earth radius factor: 1.33
- Maximum terrain height: 1000 m
- Number of auxiliary contours: 0
- Description: A suitable name for the calculation.

Click **OK** to start the calculation and close the window.

The result will appear in the folder **This Project\Results\Earth Station Coordination Results**. Double-click on the result name to open the **Edit result** window.

Select the **Results** tab. Check all three contours for display. Select thicker lines and nice colors to display well on the map. Press **OK**.

Right-click on the result name and select **Visible**. The map now shows three contours.

- Circle: Propagation mode 2 (rain scatter)
- Irregular: Propagation mode 1 (tropospheric)
- Circle + Irregular: The maximum distance of the two above.

The coordination contour is given by the maximum distance from the earth station, so the separate displays of Mode 1 and Mode 2 can be removed.

Show all the stations on the map and notice that all the links are within the contour, thus requiring the coordination procedure to be performed. The next step is to calculate the interference contribution in the earth station from the link transmitters that operate within the frequency band of the earth station. This is done with the Interference tool.

Perform the following:

Mark the **NIT 5** station. Start the **Interference** tool.

Make the following settings:

- Propagation model, Interferer: ITU-R P.452
- Propagation model, Wanted: Not applicable. Use Free-space
- Interference level: I [dBm] Above –170 dBm
- Interference level: S/I Below 50 dB
- ITU-R P.452 settings: Clear air, 0.01 percent, Worst month. Default for other settings.

Close the windows with **OK** to start the calculation. Notice that the interference level is very low (about –156 dBm). You may also try with the hydrometor-scatter model.

The resulting interference level is very low, far below the sensitivity of the earth station. This should mean that the links would pass the coordination procedure with Norway without problems.

5.5.4 Exercise A3: licensing

Licenses are handled in a separate application, the WRAP License Manager. This application operates on the same station database as the technical system, meaning that it can retrieve required technical data from the database and add various administrative data associated with the station(s) to be licensed.

Perform the following to start the License Manager:

> Select (Windows start menu) **Start – Programs – WRAP License Manager – WRAPLicMan**. This opens a main menu, where several selections can be made. Some of the available functions are shown in the collection below:

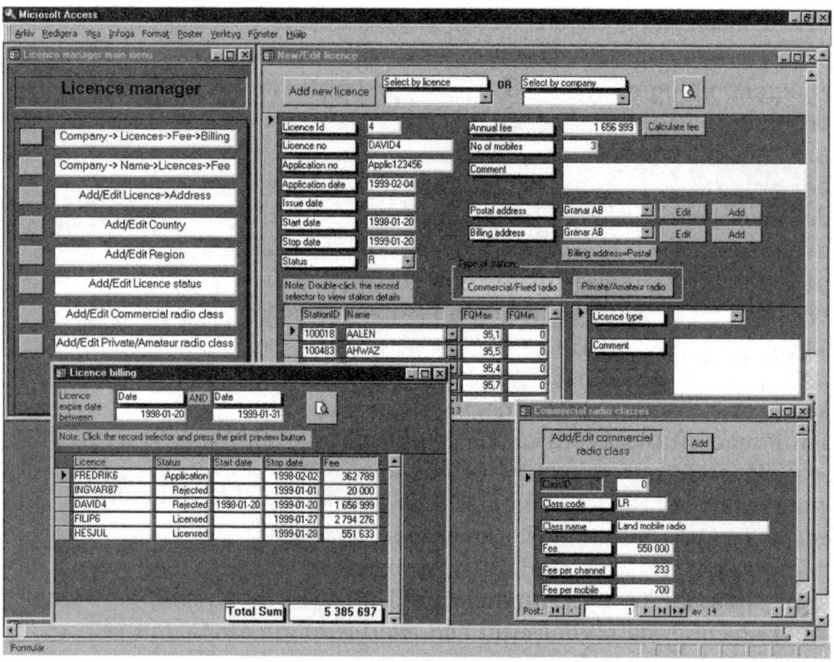

> Click on the box left of **Add/Edit License -> Address**. This opens a window, which is used to generate a new license and, if needed, a new licensee address.

Click on **Add new license**. This clears the entry fields and allows new data. Fill in reasonable data (details are not important for this exercise). Note that an entry first has to be made in the **License no** field. Otherwise it is not possible to enter data in the other fields.

All of the stations in the new network generated in the first exercise can be licensed in one form. Selection of stations is done by clicking on the **drop-down arrow** just right of the entry field "**Name**." This opens a list of all nonlicensed stations in the database. Selection of station is easily performed either by scrolling the list or by entering the first few characters of the station name. This causes automatic search for these first characters.

Select all of the stations in the new network.

Licensee addresses that already exist can be selected by clicking on the **drop-down arrow** just right of the **Postal Address/Billing Address fields**. New addresses can be entered by clicking on **Add**.

When all stations have been entered you can click on **Calculate fee**. This will calculate the total license fee for this license, based on the fee structure that has been defined. The fee structure can be edited in the **Edit commercial radio class** function that can be selected from the main menu.

Having entered complete data into the form gives (for example) the following view:

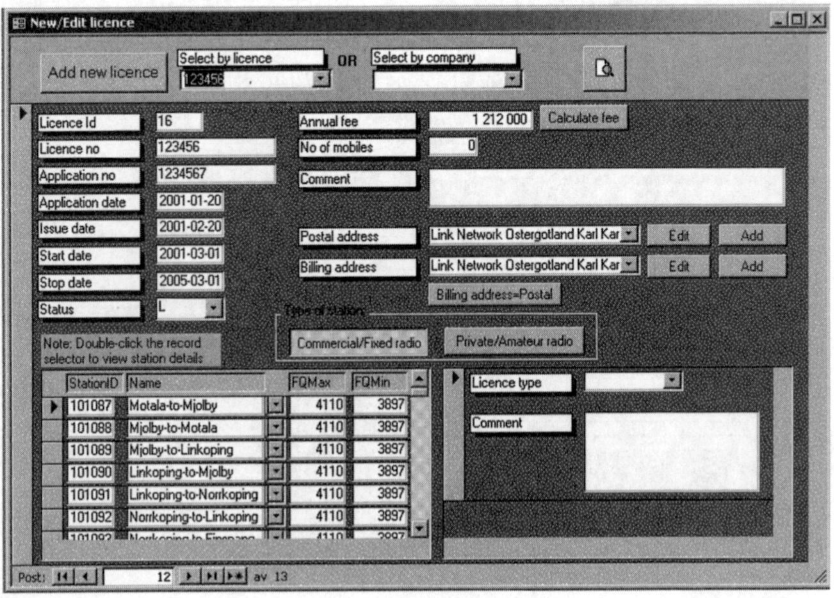

Now click on the **Print Preview** button in the upper right part of the form. This opens a preview window, where the license information as it may be printed appears.

There are several functions available for editing basic data that then can be selected from drop-down lists. These are available in the first window—the Main Menu.

Data generated in the licensing process are stored in the station database and can be retrieved later.

5.5.5 Exercise A4: coordination and notification

This exercise deals with the coordination and notification process that is necessary in many instances. The rules for when coordination and notification are required can be found in the ITU Radio Regulations.

A result of the first exercise was that the coordination contour (receiving) around the Norwegian earth station NIT 5 extended to include all the radio link stations that were planned in the exercise. Some of them transmit on an overlapping frequency assignment. This means that coordination should be performed with the Norwegian telecommunications administration. After successful coordination, a notification on the coordinated link stations should be sent to the ITU BR (Bureau of Radiocommunications) for entry in the master international frequency register.

The handling of the notification process is supported by the WRAP License Manager. Perform the following to illustrate how this is done.

> Start the License Manager. Select the item **Coordination** on the main menu.
>
> Study the form that opens. The available T-series forms can be seen by clicking on the **drop-down arrow** for the **Forms** field. Click on the **drop-down arrow** for the **Not coordinated stations**. This opens a list of stations in the database. Enter the first letter of the station to be coordinated (in this case one of the stations that operate on 3897 MHz) to assist in finding it in the list. Select that station.
>
> The proper form is **T11**. Select this one. If this is going to be a notification to be sent to the ITU you should select **ADD**. However, before performing the notification there may be a need to coordinate with Norway in this case. This procedure can be initiated using the **COORD** button instead. Let's assume that the coordination with Norway is ready, so select the **ADD** button. Data that were already available in the database are filled automatically in the form. Other data must be entered manually.

The Coordination function supports both the coordination process with affected countries and the notification process with the ITU. Each stage in these processes can be tracked with its status, date and comments that can be added. All generated forms are stored in historical layers in the database.

Some notes referring to the form that now is displayed:

- Shaded fields are only applicable in certain cases
- Fields that are automatically filled from the database are read-only. Data in these can only be changed from the technical WRAP system.
- Fields with a drop-down button are filled from selections that are made from the drop-down list.
- The T11 form can only handle the notification of one station with one frequency
- The Administration Unique Identifier is an important field. The format here is obviously defined by the administration submitting the notification.
- Field 7b takes values A, B and C but is only mandatory for the fixed service in the HF frequency range.
- Field 9ea is not filled automatically in the current version. The altitude above sea level can be found in the technical system by placing the cursor at the station and reading the altitude in the status bar.
- S6.7 should be entered in Field 11 for this particular case. **NOR** (Norway) should be selected from the drop-down list, since we assume that the station was successfully coordinated with the Norwegian administration.
- Field 9c is not filled automatically in the current version. The correct value can be found by opening the **Edit Station** window, select **Antenna** – **Edit** and reading the 3 dB-beamwidth from the table of values. It is in this case about 2°.
- Data for fields 12a and 12b may be a little complicated to find. Information is available in, for instance, ITU documents and software. Enter A in field 12b and for instance 660 in field 12a.
- Field 5g is not entered automatically in the current version. This is the distance between the transmitter and its associated receiver at the other end. It can be found by using the **Calculation/Search area** – **Line** in the technical system. Position the start of the line at the transmitter and the end at the receiver. Read the distance in the status bar.

Detailed information on the fields in the notification forms can be found in Ref 27 in the WRAP References document in the **UsersGuide\References** folder.

Click on **Save** when all desirable information has been entered. This will save the entry for later retrieval for printing or continued work in the notification process.

The Coordination form now should display a line for the just completed T11 notification. Marking this line and clicking on **Show** will display the stored form for viewing and printing.

A change to a notification that has been submitted as an **Addition** (i.e., for a new entry to the ITU BR) is notified by using the **MOD (Modification)** function.

Deletion of an existing entry in the master international frequency register is performed by the **SUP (Suppress)** function.

Withdrawal of a notification (ADD or SUP) is done with the **WITHDRAW** function.

The history of submissions for a particular station can be viewed by selecting a station of interest from the drop-down list that is seen when selecting that station in the list **Previously coordinated stations**.

The just completed T11 notification may look as follows when viewed with the **SHOW** function:

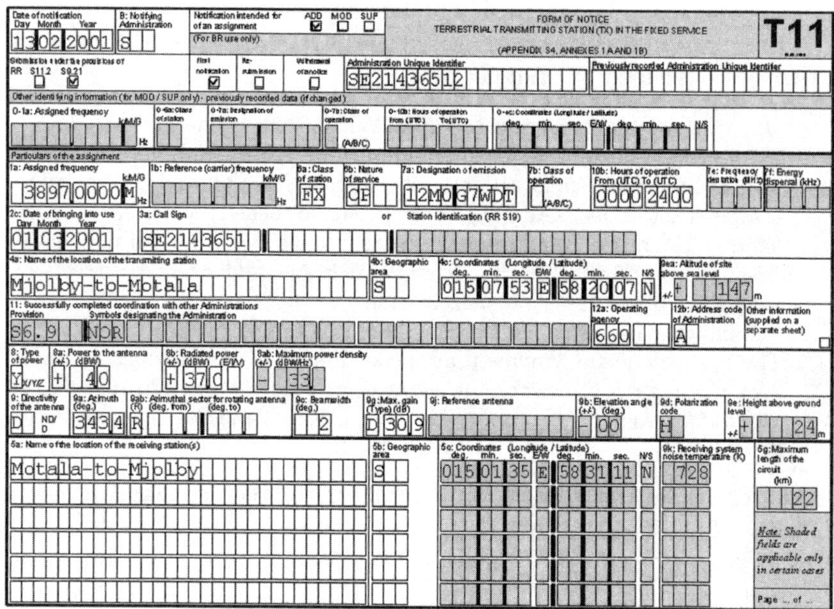

FORM OF NOTICE
TERRESTRIAL TRANSMITTING STATION (TX) IN THE FIXED SERVICE
(APPENDIX S4, ANNEXES 1 A AND 1B)
T11

Date of notification (Day Month Year): 13 02 2001
B: Notifying Administration: S
Notification intended for an assignment: ADD ☑ MOD ☐ SUP ☐
(For BR use only)
Submission under the provisions of RR S11.2 ☐ S9.21 ☑
First notification ☑ Re-submission ☐ Withdrawal of a notice ☐
Administration Unique Identifier: SE2143651 2
Previously recorded Administration Unique Identifier:

Other identifying information (for MOD / SUP only) - previously recorded data (if changed)
0-1a: Assigned frequency (k/M/G Hz) | 0-6a: Class of station | 0-7a: Designation of emission | 0-7b: Class of operation (A/B/C) | 0-10b: Hours of operation From (UTC) To (UTC) | 0-4c: Coordinates (Longitude / Latitude) deg. min. sec. E/W deg. min. sec. N/S

Particulars of the assignment
1a: Assigned frequency (k/M/G Hz): 3897.0000 M
1b: Reference (carrier) frequency (k/M/G Hz):
6a: Class of station: FX
6b: Nature of service: CP
7a: Designation of emission: 12M0 G7WDT
7b: Class of operation (A/B/C):
10b: Hours of operation From (UTC) To (UTC): 0000 2400
7e: Frequency deviation (MHz)
7f: Energy dispersal (kHz)
2c: Date of bringing into use (Day Month Year): 01 03 2001
3a: Call Sign: SE2143651 or Station identification (RR S19)
4a: Name of the location of the transmitting station: Mjolby-to-Motala
4b: Geographic area: S
4c: Coordinates (Longitude / Latitude) deg. min. sec. E/W deg. min. sec. N/S: 015 07 53 E 58 20 07 N
8ea: Altitude of site above sea level: +/- + 147 m
11: Successfully completed coordination with other Administrations — Provision: S6.9 Symbols designating the Administration: NOR
12a: Operating agency: 660
12b: Address code of Administration: A
Other information (supplied on a separate sheet) ☐
8: Type of power (X/Y/Z): Y
8a: Power to the antenna (+/-) (dBW): + 4.0
8b: Radiated power (+/-) (dBW) (E/V): + 37.0
8ab: Maximum power density (+/-) (dBW/Hz): − 33
9: Directivity of the antenna (ND/D): D
9a: Azimuth (deg.): 343.4
9ab: Azimuthal sector for rotating antenna (R) (deg. from) (deg. to): R
9c: Beamwidth (deg.): 2
9g: Max. gain (Type) (dB): D 30.9
9j: Reference antenna
9b: Elevation angle (+/-) (deg.): − 0.0
9d: Polarization code: H
9e: Height above ground level: +/- + 24 m
5a: Name of the location of the receiving station(s): Motala-to-Mjolby
5b: Geographic area: S
5c: Coordinates (Longitude / Latitude) deg. min. sec. E/W deg. min. sec. N/S: 015 01 35 E 58 31 11 N
9k: Receiving system noise temperature (K): 728
5g: Maximum length of the circuit (km): 22
Note: Shaded fields are applicable only in certain cases
Page ... of ...

5.5.6 Exercise B1: planning a microwave link network

Input data. The objective in this exercise is to plan and design a small 2 Mbit/s microwave link network in the 1.6 GHz band. Standard stations and equipment data available in the demo version of WRAP's station database should be used. Assume the following:

- Maximum height above ground for antenna towers is 24 m.
- Maximum single-hop distance is 25 km.
- The network shall link the following villages in the county of Östergötland in Sweden:
  - Sturefors
  - Kisa

- Towers can be placed at the highest points within or near the urban areas of the towns.
- The frequency band 1500 to 1690 MHz, channel separation 2 MHz, duplex distance 100 MHz shall be used.
- The nonavailability requirement is an SESR (Severely Errored Seconds Ratio) in accordance with ITU-R Recommendation F.697-2, which is less than 1.5×10^{-4}.

Entry of stations. The **Samples** folder contains a project file that sets the map view to suit this exercise. Perform the following after starting WRAP as described in Endnote 54.

> Open the project **OstergotlandMap. WPR**. This is done by first selecting **File – Open** in the upper text menu bar. Then find the ...**\WRAP \Samples** folder, which contains this project file and other files used for exercises in the User's Manual. When you have opened the file, save it again (change the name), using **Save as**... to allow saving under a new name. Saving now and then is wise if errors are made, and a saved, correct version of the project needs to be retrieved.
>
> Right-click in the map and select **Scale – 1:250 000**. Select **Properties**.... The first tab of the window that opens shows the available geographical vector themes. Left-click on the square for the theme **Text, built-up area 2000-9999** to check this theme for display. Accept and close the window by selecting **OK**.

In order to see the full text descriptions of the vector themes you may need to adjust the column widths. This is done by placing the cursor on the short vertical line that separates the columns and double-left-clicking to automatically adjust the column width to accommodate the full text of every vector theme in the column.

> Note: If you are using a display with only 800 × 600 resolution, you will not be able to see both villages to be connected at the same time in scale 1:250000. If you have such a display you should perform the following:
>
> - Right-click in the map and select Scale – 1:500000
> - Right-click in the map and select **Properties – Vector Data**. Mark the text of the theme **Text, built-up area 2000-9999** and left-click on the **Change** button.
> - Change **Visible scales – Max** to **500000**
> - Close the windows with **OK**.

The map now displays the names of a number of villages/towns. The center position of the map should be adjusted by right-clicking at the desired new center position and selecting **Center position** in the menu. The map display is derived from the height and terrain classification databases and gives an impression of the topography through the relief type of presentation. You may want to try other types of map displays, for instance as follows:

Right-click in the map and select **Properties…**, **Image Data**. Check the box for **Sweden** and close the window by pressing **OK**.

You now have a traditional "paper-type" map display. Try different scales, for instance 1:100 000 to allow a clear presentation of the elevation contours, names, and so on . You can use the **"Zoom in/Zoom out"** functions that are available on right-click in the map to adjust the scale between the selectable set of fixed scales, or set any scale of your own.

This type of map contains a lot of information and may actually make it difficult to see objects of real interest, such as the station names and in this case the names of the villages to be linked. It is usually more convenient to show a cleaner background map with less level of detail, so you may select to hide the "paper-type" of map display again, by right-clicking in the map, select the Image Data tab and deselect the 1:250000 scale map.

Right-click in the left geographical viewer and select **Profile**. This will later conveniently allow the display of the terrain profiles along the path. Take the opportunity now to set the earth radius factor to 0.6 instead of the nominal 1.33. This is done by right-clicking, selecting **Properties** and entry of 0.6.

Note. An earth radius factor of 0.6 is commonly used when planning radio links for high time availability.

Right-click in the map and select **Snap to highest**. This will cause the selected station positions to automatically be placed at the highest point within a certain area around the cursor (scale-dependent size of the area).

Position the cursor at Sturefors. Right-click without moving the cursor and select **Tx Position**. Notice that a cross and circle appears near the cursor position.

Then position the cursor at Kisa. Right-click and select **Rx Position**. Notice that the Profile Viewer now displays the terrain profile between the two selected points.

The end points of the link have now been set. The next step is to create stations at these points. This is performed as follows:

Select **Station – New…** in the top text menu bar. A window with a number of selectable folders appears. Choose **Fixed service, digital links**. Select **Wrap Fixed station 1 to 19 GHz/2 Mbit/s.**

The **Edit Station** window opens. Give a name to the first station in the link hop (Sturefors), for instance **Sturefors**.

Select the **Frequencies** tab. Enter **1500** in the **TX Fq (MHz)** box and **1600** in the **RX Fq (MHz)** box. Left-click on the left-arrow to move the frequencies into the table. The TX and RX equipment have been selected automatically as defined by the template for this type of station. Click on the **Tx Equipment** and **Rx Equipment** tabs. Notice that the antenna heights above ground are pre-defined to 24 m. Press **OK**, which closes the window and automatically opens the **Edit Station** window again, this time for entry of data for the opposite station in Kisa.

Note. Observe that the Edit Station window will open twice, once for each end of the link, and you have to enter station names for both stations (and frequencies for the first station) and close each window with OK. Otherwise the link will not be created properly.

> Enter the station name, for example, as **Kisa**. The frequencies have been set automatically to match those that were assigned to the Sturefors station, i.e., the TX and RX frequencies are reversed. Press **OK** to accept the entries.

Notice that there now are two lines of information in the upper right list view: One line for the Sturefors station and one line for the Kisa station. Display them on the map as follows:

> Mark both stations by left-clicking on the first station, press **Shift** and left-click on the second station. Then right-click and select **Show In Map**. Notice that a red/green-colored line appears, connecting the two stations. The red color indicates which station is transmitting on the low frequency in the duplex pair.

Check for clearance, find a good relay site. You need to check if the propagation path is sufficiently clear of obstructions. This is performed by displaying the terrain profile between the stations, including the display of the first Fresnel zone. Perform the following:

> Left-click on the Sturefors station in the list view to mark it. Press on the **Profile** icon. This results in the display of the first Fresnel zone and the ray path between the two stations in the left geographical viewer.

You can clearly see that this link path will not work—there are too many terrain obstructions. An intermediate relay station is needed. A convenient way of finding an optimum site for this station is to calculate the radio-optical coverage for both of the existing stations, thereby finding out which areas both stations will cover under the assumption of using a certain antenna height. Perform the following:

> Right-click in the map, select **Calculation/Search Area** and **Circle**
>
> Position the cursor on the radio link line about midway between the two stations. Press the left mouse button, hold down and move the cursor to draw a circle with a diameter to just enclose both stations within the circle.
>
> Mark both stations in list view by left-clicking on the first one, press **Shift** and left-click on the second station. Select the **Coverage** icon. This opens the Coverage window.
>
> Make the following selections
>
> - **Calculation: Minimum Req Antenna Height**
> - **Clearance: 0 percent** (to allow obstruction of the first Fresnel up to the line-of-sight)
> - Check **Above Ground Level**. *Note: If you are using version 3.3.1 or earlier, there is no such selection. Instead, make sure that the "Above sea level" is not selected.*

- **Calculation Area: Circle, Defined in Map Viewer**
- Stay with the **Average Resolution**
- Set the **Effective Earth Radius Factor** to **0.6**.
- Edit the **Description** field to give a suitable name to the calculated result.

Press **OK** to start the calculation and close the window.

Go to the upper left section of the main window. Left-click on the + sign in front of the **This Project** folder to open the underlying folder structure. Left-click on the + sign in front of the **Results** folder. The + sign in front of the **Coverage Results Area** folder indicates that there is a result available. Left-click on the + sign to open the folder.

Double-click on the result name. The **Edit Result** window opens. The first tab contains some numerical results that among other information give the minimum antenna height above ground that will give the required 0 percent first Fresnel zone clearance to both Sturefors and Kisa. Select the **Presentation** tab. Mark the ellipsoid shown and change the **Value** field to 24 m (this is the maximum available tower height). Close the window with **OK**.

Right-click on the result name and select **Visible**. Right-click again and select **Show Cursor Value**

You will now see a few small, colored areas in the map. These areas show where you can place the relay station, with an antenna height of 24 m above ground, and achieve a first Fresnel zone clearance of more than 0 percent. Now you can use a handy feature to zoom in exactly on these areas:

Find the magnifying glass icon in the tool bar. Left-click to select, position the cursor in the map near the colored result area, left-press and hold while moving the cursor to generate a rectangle that encloses the area of interest. When you release the left button, the map will zoom in to this area and you will be able to clearly see the result areas.

To get a nice detailed orientation on the map you may also put on the "paper map" by right-clicking in the map, select **Properties**..., **Image Data, Sweden 1:250000** and pressing **OK**.

Move the cursor within the result areas and read the result value to the right in the bottom status bar.

Perhaps you want to make a more detailed calculation, just around this smaller area? This is a good idea to really make sure that you can find the optimum spot where to place the antenna. Perform as follows:

Right-click in the map, select **Calculation/Search Area, Circle**. After selecting this area type you should position the cursor in the map where you want to place the center of the circle. Then left-press and hold, while moving the cursor until the circle has the desired size. Release the left button.

Now mark both stations in list view. Select the **Coverage** icon. Make the following selections in the Coverage window:

- **Calculation: Minimum Req Antenna Height**
- **Clearance: 0 percent** (to allow obstruction of the first Fresnel up to the line-of-sight)

- Check **Above Ground Level**. *Note: If you are using version 3.3.1 or earlier there is no such selection. Instead, make sure that the "Above sea level" is not selected.*
- **Calculation Area: Circle, Defined in Map Viewer**
- Stay with the **Average Resolution**

Note. You can select a higher calculation resolution instead of making the calculation area smaller. This is another way of getting a more detailed calculation.

Go to the upper left section of the main window. You will now see one more result name.

Right-click on the new result name and select **Visible**. Right-click again and select **Show Cursor Value**. You should also right-click on the previous result name and de-select the **Visible**.

Now you will now see a similar result as before, but with more details owing to the higher resolution of the calculation. These areas indicate where you should place the relay station.

Defining the relay stations. A convenient way to generate stations for the relay point is to duplicate the link between the two end points and then move the appropriate stations to the relay point. Perform as follows:

Mark the last station in the station list. Right-click and select **Duplicate**. Notice that a pair of new stations appears, with the name "Copy of xxx," where "xxx" is the name of the station which was duplicated. These new stations have exactly the same properties as the old stations (apart from the name).

What you should do now is to change the names of the appropriate stations and move them to the relay position. This means that one of the old stations and one of the new copy stations shall be moved to the relay position in a way to form two link hops, connecting *Sturefors—Relay Point—Kisa*. Continue as follows:

Double-left-click on station **Kisa** to open the **Edit Station** window. Edit the station name to **Relay-to-Sturefors** (this will be one of the stations to move to the relay point). Close the window with **OK**.

Double-left-click on station **Copy of Sturefors** to open the **Edit Station** window. Edit the station name to **Relay-to-Kisa** (this will be the other station to move to the relay point). Close the window with **OK**.

Double-left-click on station **Copy of Kisa** to open the **Edit Station** window. Edit the station name to **Kisa** (this will be the new Kisa station) Close the window with **OK**.

Right-click in the map at the position where you want to place the relay station. This should be within one of the areas calculated in the previous section to give more than 0 percent clearance with 24 m high antennas to both Sturefors and Kisa. Select **Tx Position**.

Right-click on the station name **Relay-to-Sturefors** and select with left-click **Move To Tx Position.**

Right-click on the station name **Relay-to-Kisa** and select with left-click **Move To Tx Position.**

Mark all stations in the list view and select with left-click **Show In Map.** You can see that the relay stations are placed at the selected position. Now return to the previous scale (1:250000) by right-click in the map and selecting **Previous Scale** or selecting **Scale, 1:250000.**

You may want to deselect the zoom function for the map in the tool bar. Otherwise you may accidentally place a zoom area in the map (to re-set the map you just select **Previous Scale** if this happens).

You will probably notice that the red/green-colored lines indicating the low/high transmission frequencies of the links correspond to the assignments for the original stations. This means that the relay site will transmit on both the high and the low frequency band, which is not common practice and may result in interference within the site. The frequency assignment should thus be changed as follows.

Double-left-click on the **Sturefors** station in the list view. Select the **Frequencies** tab. Mark with left-click the **Tx** frequency in the table. Left-click on the arrow pointing right to move the frequencies into the editing boxes. Edit the frequencies to reverse the Tx and Rx frequencies. Then left-click on the arrow pointing left to move the new frequencies to the frequency table. Close the window with **OK.**

Mark all stations in the list view, select with right-click **Show In Map.**

Now you can see that the red/green colors are properly set at the relay point.

The links can now be checked for Fresnel zone clearance:

Left-click on the Sturefors station in the list view to mark it. Press on the **Profile** icon to display the first Fresnel zone and the ray path between the two stations in the left geographical viewer.

Perform the same action with the Kisa station.

You should be able to notice that both links now operate with a clear line-of-sight. There are however obstructions of the lower half of the first Fresnel zone, and these will result in some diffraction losses.

Calculate the link performance. Continue now to check the performance of the links and compare with the requirements as they were defined initially. This is done by marking one link at the time and performing a *Radio Link Performance* calculation as follows:

Mark the **Sturefors** station in the list view. Select the **Radio Link Performance** tool with the icon .

The Link Performance Analysis window that opens contains a number of calculated values. The initial calculation is always performed with the free-space propagation model, which is not suitable for these particular paths since the Fresnel zone was about half obstructed. A more suitable, terrain-dependent propagation model should be selected:

Click on **Settings**. Make the following selections:

- Model: **Method II**. This is the one to be used in all cases when a terrain database is used for the propagation calculations.
- **Terrain type** to describe the path in general terms. "**Open**" is the most suitable selection for this path.
- **Rain Intensity** and **Layer Presence** should be set to values to represent the worst case planning conditions. You can stay with the default values, which are suitable for this area.
- Select **Propagation Model – Models – DETVAG90/FOA** with its default settings.
- **Atmospheric Attenuation** has parameters for temperature and water vapour density. Stay with the default values.
- The **Effective Earth Radius** factor should be set to 0.6. Close the windows with **OK** to accept the settings.

The result is calculated. Mark the line for **BER 0.001** in the **Performance** table. The window should now appear similar to the following (the exact values may be a little different):

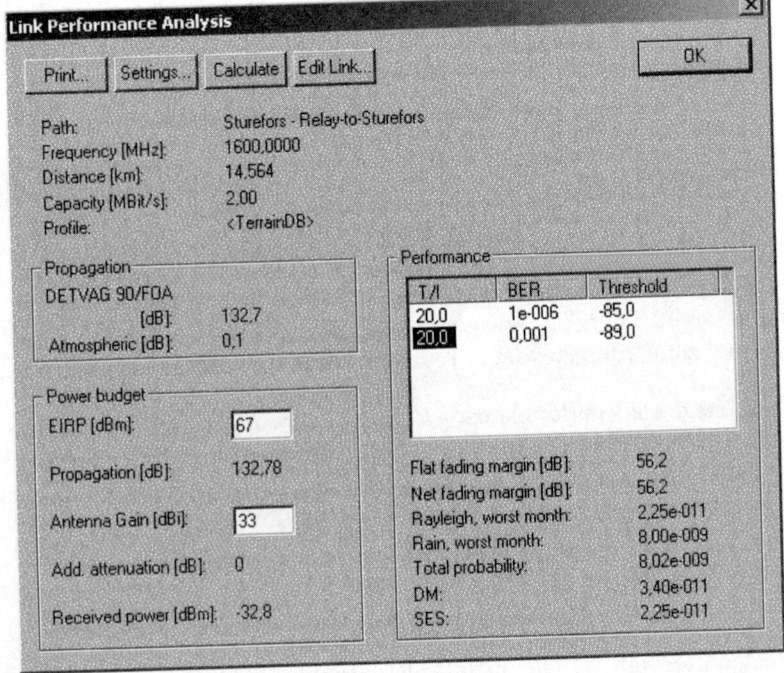

Comparing the values with the requirement for SES (Severely Errored Seconds) probability of less than 1.5×10^{-4} indicates that there is ample margin, and that for instance the radiated power can be reduced significantly.

Try different power levels by editing the field **EIRP [dBm]**. Left-click on **Calculate** between each change of the radiated power.

The calculated performance of the link fulfils the requirement even at quite a low radiated power.

The transmitter radiated power is changed at the **EIRP [dBm]** field. You may also change the receiver antenna gain in the **Antenna Gain [dBi]** field. It is set to 33 dBi in the template station that was used, which may be a little high for transportable links in this frequency range. A suggestion is to try with about 20 dBi of receiver antenna gain, and the corresponding reduction in radiated power (–13 dB), thus reducing the total power budget with 26 dB.

Check the other link in the same way.

5.5.7 Exercise B2: coverage for radio access points

This exercise complements the previous one with radio access points at two of the link locations. These access points are used to provide area coverage for mobile units that need to access the fixed network, similar to a mobile telephone system.

Entry of stations. Perform as follows to place radio stations at the Kisa and Relay link locations.

Mark the **Kisa** station in the list view with right-click and select **Set As Tx**. This means that you now have used the geographical coordinates of the Kisa station for the Tx Position.

Select the **New Station** icon in the tool bar. Open the **Land Mobile Service** folder in the **Select Type Of Station** window that opens. Select **Wrap Base station, analogue**, then **OK**.

Enter a suitable name in the **Edit Station** window, such as **RAP Kisa**. Notice that the coordinates are already set to the same as for the Kisa link station. Enter the **Tx** and **Rx** frequencies of 60 MHz in the **Frequencies** tab, then left-click on the left-pointing arrow to enter the frequencies into the table.

Select the **Tx Equipment** and **Rx Equipment** in turn and edit the antenna heights to 10 m, which is a more common antenna height for this type of station. Accept the data and close the window with **OK**.

The new station has been created as can be seen in the list view. Now perform the following for the next radio access point:

Mark the **Relay-to-Kisa** station in the list view with left-click. Right-click and select **Set As Tx**. This means that you now have used the geographical coordinates of the Relay-to-Kisa station for the Tx Position.

Right-click on the **RAP Kisa** station and select **Duplicate** to create a copy that then can be moved to the desired location. Double-left-click on the **Copy of RAP Kisa** station and edit its name to **RAP Relay**. Choose the **Frequencies** tab and edit the Tx and Rx frequencies to 59.8 MHz instead of 60 MHz. Close the window with **OK**.

Right-click on the **RAP Relay** station and select **Move To Tx Position**.

Mark all stations in the list view, right-click and select **Show In Map**.

You will notice that the names of stations at the same location overprint. Clicking on or just below the station symbol opens a window **Select station to be edited**. There you can select any station by double-left-click, which opens the **Edit Station** window for that station.

You have now created in total six stations in the WRAP project file. Please remember to save the file now and then!

Note that the demonstration version of WRAP has a limitation set to exactly six on how many stations that can be created in a project.

Coverage calculation. Now proceed to calculate the coverage areas for the radio access points. Selecting just one of stations will allow the calculation of the individual coverage area for this station. In this case it is convenient to select both stations and perform the calculation of the composite coverage, that is, the total area covered by any one of the two stations.

Draw a circle in the Map Viewer to define the calculation area by right-clicking in map, select **Calculation/Search Area – Circle**. Position the cursor about half-way between the two RAP stations, left-press and hold while moving the cursor to generate a circle that encloses all three sites. Release the left button when ready.

Mark both **RAP Kisa** and **RAP Relay** in the station list view.

Select the **Coverage** icon . This opens the Coverage window.

Make the following selections

- **Calculation: Received Power**
- **Mobile – Select – Analogue mobile**
- Check **Above Ground Level**. Note: If you are using version 3.3.1 or earlier there is no such selection. Instead, make sure that the "Above sea level" is not selected.
- Select **Downlink** (to calculate from the RAP base to the mobile station)
- **Calculation Area: Circle, Defined in Map Viewer**
- Stay with the **Average Resolution**
- Select the **DETVAG90/FOA** propagation model, default settings.
- Edit the **Description** field to give a suitable name to the calculated result.

Press **OK** to start the calculation and close the window.

Go to the upper left section of the main window. Left-click on the + sign in front of the **This Project** folder to open the underlying folder structure.

Left-click on the + sign in front of the **Results** folder. The + sign in front of the **Coverage Results Area** folder indicates that there are results available. Left-click on the + sign to open the folder.

Double-click on the result name of the RAP coverage. The **Edit Result** window opens. Select the **Presentation** tab. Left-click on the + sign to add one more result level for the presentation. Mark the red ellipsoid shown and change the **Value** field to –100 dBm. Mark the yellow ellipsoid and change to –90 dBm. Close the window with **OK**.

Right-click on the result name and select **Visible**.

You should now have a main window with the following appearance:

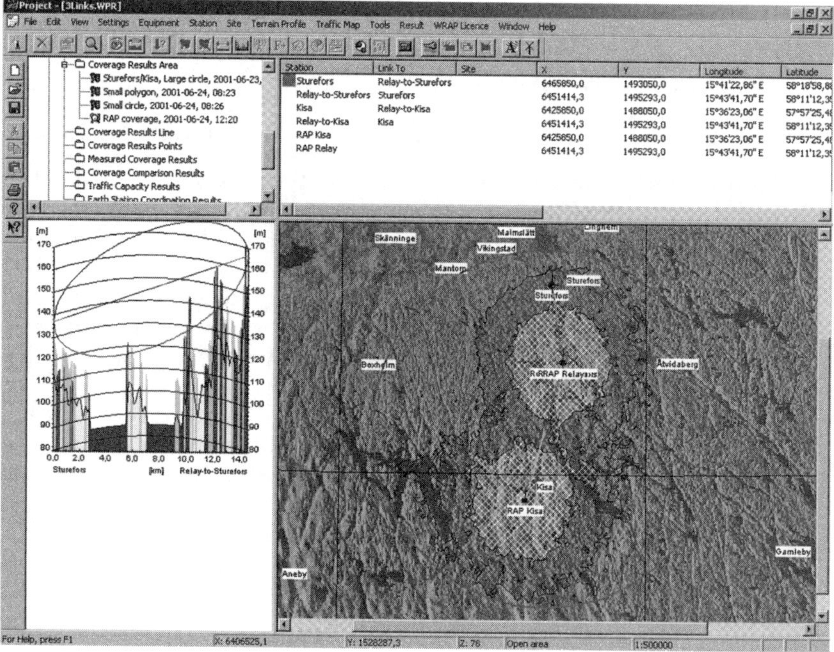

Note. If you are using the demonstration version of WRAP, your coverage diagrams may look different in areas where there is no terrain data available.

The map now displays the coverage of the RAP stations, with higher than –100 dBm received signal level in the mobile receivers indicated by the red area. The yellow areas indicate higher than –90 dBm.

5.5.7 Exercise B3: coverage under jamming

WRAP can be used for analysis of interference, both unintentional and intentional in the form of hostile jamming. This exercise demonstrates a way of using WRAP to see the impact of a jammer on the coverage of the previously entered radio access point stations around 60 MHz.

Entry of jamming station

Note. If you are using the demonstration version, you can only enter up to six stations in a WRAP project. Since this exercise is based on the project created in Exercises B1 and B2, now containing six stations, you must perform the following (only for demo users):

> Open the previously created project. Open the folder **This Project – Stations in Project**. Mark the radio link stations by first marking the first radio link station in the list view, then move the cursor to the last radio link station and, while holding down the Shift key, mark this one. Then right-click and select **Delete**.

Continue now to enter the jamming station at a suitable location:

> Put the cursor about 25 km west of the **RAP Relay** station. Right-click and select **Tx Position**. Note: A convenient way to measure 25 km is to right-click in the map, select **Calculation/Search Area – Line**. Then position the cursor at **RAP Relay**, left-click and move the cursor while looking at the status line at the bottom of the window. The length and direction of the line is shown. Left-click to put a breakpoint on the line, then right-click to end the line, then right-click again and select **Tx Position**.
>
> Select **Station – New – Jammers – Jammer** in the top text tool bar.
>
> Enter a suitable name, such as **Jammer**. Select the **Frequencies** tab and enter **Tx Frequency 59.9 MHz**. Enter the frequency into the table by left-click on the left-pointing arrow.
>
> You may study the characteristics of this jammer by selecting the **Tx Equipment** tab. The radiated power is 40 dBW (10 kW), and the antenna height above ground is 25 m. Left-click on the **Transmitter – Edit** button to open the Edit Transmitter window, where you can select the **Frequency Characteristics** tab. Now you can study transmitter power spectrum and see that it is about 300 kHz wide between the –3 dB points.
>
> Select the **Tx Equipment** tab. Left-click on **Antenna… Edit** to open the **Edit Antenna** window. Notice that the polarization is set to **Horizontal linear**. This is not a suitable polarization for the jammer, since the stations that it intends to jam use vertical polarization. Therefore, perform as follows to change to another antenna for the jammer:
>
> - Close the **Edit Antenna** window by left-click on **Cancel**.
> - Left-click on **Antenna – Select** in the **Edit Station – Tx Equipment** tab.
> - Select the antenna **WRAP ND, 0/Vert** from the list of antennas. Left-click on **OK**. Close **Edit Station** with **OK**.

Placing the jammer on 59.9 MHz, right in between the two RAP stations at 59.8 and 60.0 MHz will thus allow it to jam both stations. It will, however, suffer about 13 dB of loss in jamming efficiency since its spectrum is about 20 times wider than the receiver bandwidth of the RAP receivers. Its power and antenna height are still so much higher that it will be quite successful in jamming the communication links.

Calculate coverage. Draw a circular calculation area in the map by right-click in the map, select **Calculation/Search Area – Circle**, place the cursor about midway between the two RAP stations, left-press and move the cursor until the circle has a radius of about 40 km (see the status line at the bottom, where the radius is shown). Releasing the left button will set the circle.

Mark the two RAP stations in the list view.

Select the **Coverage** icon . This opens the Coverage window.

Make the following selections

- Calculation: No of Servers, Interference limited
- Mobile – Select – Analogue mobile
- Check **Above Ground Level**. *Note: If you are using version 3.3.1 or earlier there is no such selection. Instead, make sure that the "Above sea level" is not selected.*
- Select **Uplink** (to calculate from the mobile station to the RAP base)
- **Margin: 0 dB**. This means that the database value of the required S/I value (signal-to-interference ratio) will be used as the limit for acceptable S/I. In this case the required S/I value is 10 dB.
- Calculation Area: Circle, Defined in Map Viewer
- Stay with the **Average Resolution**
- Select the **DETVAG90/FOA** propagation model, default settings, for both the wanted and the interfering signals.
- Edit the **Description** field to give a suitable name to the calculated result.

Press **OK** to start the calculation and close the window.

Go to the upper left section of the main window. Left-click on the + sign in front of the **This Project** folder to open the underlying folder structure. Left-click on the + sign in front of the **Results** folder. The + sign in front of the **Coverage Results Area** folder indicates that there are results available. Left-click on the + sign to open the folder.

Double-click on the result name of the just-performed coverage calculation. The **Edit Result** window opens. Select the **Presentation** tab and notice that there are 3 levels: 0, 1 and 2 representing those numbers of servers (i.e., RAP stations with acceptable S/I values) to be presented. Close the window with **OK**.

Right-click on the result name and select **Visible**.

You can now see the areas of operation for the mobiles in the map within which the RAP receivers will have a higher S/I ratio than 10 dB from the mobile transmitters. There are some very small areas where both stations will give coverage and larger areas where there will be one-station coverage.

Note. Save the project now, if you have not done that before! Use a separate name from the previous project, which also contained the radio link stations.

You can compare this coverage to the noninterfered coverage by performing exactly the same calculation, but this time you must first delete the jamming station or reduce its power to a very small value. You will notice that the coverage is much larger without the jamming, with large areas covered by both of the RAP stations.

5B.8 Exercise B4: using the site concept

The concept of "Sites" is used in WRAP to allow the formation of groups of stations that are geographically related to each other and that would be convenient to copy, move, and so on as a complete entity. An example of when this is useful is in a military context, where for instance a mobile staff headquarters normally has a well-defined set of radio stations, but the staff headquarters may be moved to other locations in a geographically dynamic scenario.

The following exercise illustrates this by the formation of the stations at the Relay location into a Relay Node, which then can be conveniently moved to other locations.

Note. The described function and procedure are supported by WRAP version 3.3. WRAP version 4.0, released in September 2001, has a much more extensive and refined site/network/military unit handler.

Forming the site. Open the project containing all six stations. Perform the following:

Double-left-click on the **This Project** folder.

Double-left-click on the **Stations in Project** folder.

Right-click in the list view on one of the stations that are at the Relay location. Select **Set as Tx** to use its geographical coordinates to define the site position.

Go to **Site** in the text menu bar, left-click and select **New**.

The **Edit Site** window opens. Write a name for the site, for instance **Relay Node**.

Select the **Stations** tab. This shows an empty list of stations.

Left-click on **Add**. This opens a list view with all the stations in the project.

Mark the first station to be added to the site, for instance **RAP Relay**. Select **Next**, which opens a window where the coordinates (x, y, z) in metres for the station relative to the site origin shall be entered. Enter offset coordinates **x = 20, y = 20, z = 0**. Press **Finish** and notice that the station now appears in the list, with its just entered (x, y, z) coordinates.

Continue to add the next station by pressing **Add** and selecting **Relay-to-Kisa.** Let this station have coordinates (0,0,0). Repeat this for **Relay-to-Sturefors**. When the three Relay site stations appear in the list you can close the window with **OK**.

You will now see the list view containing one line with the site that was just created. It can be edited by left-double-click on the site name to open the **Edit Site** window. You can also right-click on the site name to open the list of available commands. Notice that you can select **Duplicate**, which is a convenient way of creating one or more identical sites, at the same position. These will all have the same station content. They can be moved individually to other locations, or be edited in other site and station properties.

Note. If you are using the demonstration version you will not be able to duplicate the site, as this would give more than six stations in the project.

You should save the project file. Do that under a new name, so you can go back to the previous file if you want to later.

Moving the site. The sites can be moved to other locations. The stations within the sites will then maintain their relative positions referenced to the site origin. Perform the following as a demonstration of this feature:

Left-click on the **Stations in Project** folder. Mark all the stations in the list view, right-click and select **Show in Map**.

Left-click on the **Sites in Project** folder to show the site in the list view.

Position the cursor in the map at a location to where you would like to move the **Relay Node**. Right-click and select **Tx Position**.

Right-click on the site name in the list view and select **Move to Tx Position**. Notice that the site moves to this position, and that the red/green lines showing the radio links are moved accordingly. This also means that the antennas have automatically been redirected to maintain the correct pointing direction.

When moving the radio links like this, you may obviously conflict with the need for a clear line-of-sight. Before moving you can make a coverage calculation as described on page 225, that is to find areas where there is line-of-sight to both of the opposite link stations. The coverage result for this calculation should still be available in the **Results** folder, so you can select it and show in the map to find appropriate areas.

Select **Snap to Highest** to automatically place the site at the highest point in the vicinity of the cursor position.

Check the Fresnel zone clearance by opening the station list view, mark the two links in turn and select the **Profile** icon.

More nodes can be of course be created in the project (not in the demonstration version). When a complex network is designed, there may be a need to disconnect link pairs and reconfigure to connect individual link stations with other link stations. Perform the following to demonstrate this.

Mark the **Sturefors** link in the list view. Right-click and select **Unlink**.

Mark the **Kisa** link in the list view. Right-click and select **Unlink**.

Mark both the **Sturefors** and the **Kisa** stations. Right-click and select **Link**. In the window that opens you can select which frequencies that should be used for the link.

You can continue to link the remaining unlinked stations in the project. In this case it is not very meaningful, since they are both located at the same position. You can however easily move one of them to Kisa or Sturefors and then perform the linking. Through this procedure you will have changed the connectivity of the network.

5.5.9 Exercise B5: checking for collocation interference

There may be a risk for interference if a number of stations are located near each other. This may give rise to interference such as intermodulation, harmonics, adjacent channel radiation, and so on. Analysis of this can be performed with the **Collocation Interference** tool. A normal procedure would be to search for other stations in the vicinity of each of the new installations and add these to the project. In this case we will just perform the analysis on the site that includes three stations, that is the Relay site.

Run the **Collocation Interference** tool to illustrate the procedure Perform the following:

Mark the stations at the Relay site in the project. This is done by first left-clicking on one of the Relay stations, then moving the cursor to the second Relay station and holding down the **Ctrl** key while left-clicking. Do this for the third station also.

Start the **Collocation Interference** tool.

The main window opens. Press **Calculate** (you may naturally study the settings).

Maximize the window after the calculation has been performed. Mark the first receiver in the top list box. Then expand the view in the bottom left list box by clicking on the + sign. Mark the line as shown below. The "Margin" value indicates the sensitivity degradation, in this case because of interference from the other transmitter at the same location.

An example of how the result may look like is shown below.

Collocation Interference Result

Receiver	f (MHz)	B (kHz)	Max block	Total	Step 1	Step 2	Ant Illum	Harmonics	IF	Rec Blocking	Co/Adj	Image	IM
Relay-to-Sturefors	1600,0000	1300	0,00	2	1	1	0	0	0	1	1	0	0
Relay-to-Kisa	1600,0000	1300	0,00	2	1	1	0	0	0	1	1	0	0

OK | Calculate | Settings... | Warnings... | Print selected... | Print all...

Receivers | Transmitters

- Relay-to-Sturefors (1600,0000 MHz)
  - Rec Blocking (1)
  - Co/Adjacent Channel Interference (1)
    - Relay-to-Kisa (1500,0000 MHz)

Transmitter	f (MHz)	B (kHz)	EIRP (dBW)	Margin (dB)	Dist (m)	Lb (dB)	Duty cycle
Relay-to-Kisa	1500,0000	1500	37,0	-46,6	0	21,9	1,00

The sensitivity degradation is not acceptable. Remedies to the situation can be tested within the Collocation tool, by for instance separating the antennas of the interfering stations to achieve a higher attenuation between the antennas. There is also the possibility to put selective filters on the transmitters and receivers that are not interference-compatible.

The way to do it in the Collocation tool is to left-double-click on the station name in the lower right list view to open the **Edit Station** window for that station, and correspondingly on the station name in the upper list box. This is however not described here, since the same procedure is demonstrated in detail in the next exercise on Interference calculations.

5.5.10 Exercise B6: interference check

Interference within the network. The new network should be checked for potential interference within the network and against other use of the assigned frequencies. This check is performed using the **Interference** tool. Perform the following:

Open the project with all stations. Mark the **Relay-to-Sturefors** station. Select the **Interference** tool.

Select **Settings… Propagation Model – Wanted – ITU-R P.526** and **…Interferer – ITU-R P.452.** Use the default settings for other options. Press **OK** to start the calculation and close the window.

The P.526 propagation model accounts for diffraction and is used here just as a slightly faster alternative to the more comprehensive and accurate DETVAG90/FOA model. There is no particular benefit in using the P.526 model, but it is selected here to demonstrate its availability in WRAP. The P.452 model is the appropriate one to use for the interference path, since it is designed to calculate the transmission loss that is not exceeded for a specified low percentage of time. This is just what you want from a propagation calculation for the interfering signals.

Mark the line for the frequency that is shown in the top list box. The following window appears (the calculated values may be different):

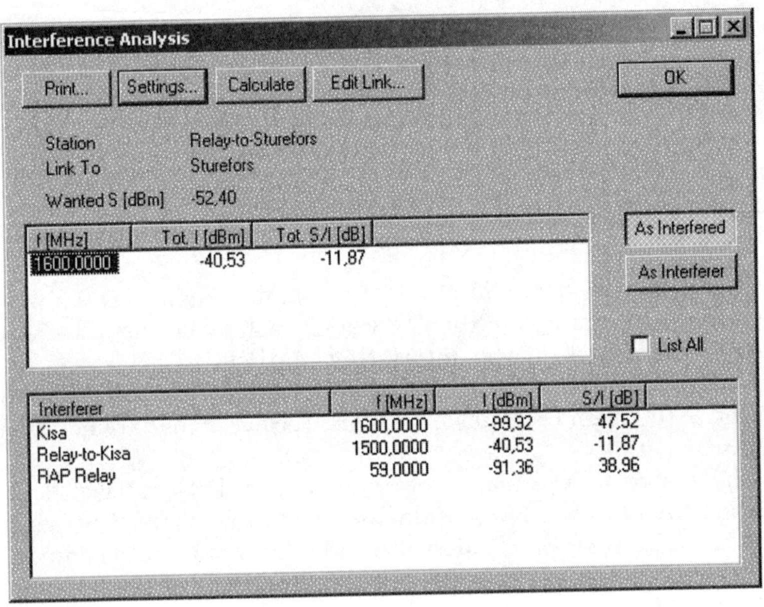

The frequency entry in the top list view has been marked to display the individual interference contributions as listed in the lower box. The total interference as received by the **Relay-to-Sturefors** station is shown in the upper list, together with the signal-to-interference ratio. The S/I value is not acceptable. Selecting the **As Interferer** function will show the interference that the transmitter of **Relay-to-Sturefors** causes in the other receivers in the network.

You may continue and check the remaining stations for interference in the same way.

The bottom list can be sorted by left-clicking on the column header. In this case there are just a few individual interference results, but in large networks there may be quite a number of cases that will need investigation.

The worst interfering station is the other link transmitter at the Relay location. The reason for the interference is that the transmitter and receiver antennas are located in exactly the same position, thus causing a fairly low attenuation between the antennas. The transmitter noise sidebands and the limited receiver selectivity will then cause interference.

The interference situation can be solved by for instance adding selective filters to the radio link receivers and the transmitters at the relay site. Continue as follows to do this:

Double-left-click on the most interfering station in the bottom list (the one showing the lowest S/I value). This opens the **Edit Station** window.

Select the **Tx Equipment** tab and the **Filter – Select** button. This opens the **Select Filter** window. Select **Add**... and notice in the **Select Equipment** window that opens that there are a few filters in the list.

Mark the **Band Pass, 900 MHz** filter (don't mind that the frequency of operation of the link actually is around 1500 MHz—the filter will automatically be "tuned" to the assigned frequency). Have a look at the filter characteristics by left-clicking on **Edit** to open the **Edit Filter** window. Select the **Frequency Characteristics** tab to display the filter curve.

Notice that the **Reference Frequency** can either be defined by the assigned **Tx/Rx Frequency** or be **Fixed**. The "Fixed" frequency can be entered if this alternative is selected.

Close the window with **OK.** You may notice a message: **This equipment is not yours to edit**. This means that you cannot change the parameters, but you can use it for your purposes (if you want to use and change the equipment, you must first make a copy of it).

Close the windows with **OK** until you are back in **Edit Station**. Select the **Rx Equipment** tab and add the same filter to the receiver. Close **Edit Station** with **OK**.

Now you are back in the **Interference Analysis** window. Left-click on **Calculate** to perform a new calculation with the changed station. Notice that the S/I value has improved significantly, but it may still not be acceptable (this link needs better than 20 dB S/I).

The main reason for the interference situation now is that the receiver of the **Relay-to-Sturefors** link has not been protected by additional filtering. Perform as follows:

Left-click on **Edit Link**.... This opens the **Edit Station** window for the **Relay-to-Sturefors** link. Perform the same actions as described previously to add the filters to both the receiver and transmitter.

When you close the **Edit Station** window, a new **Edit Station** window is automatically opened, this time for the opposite station (Sturefors). This station needs no filters to operate properly, so just close this window.

The result will now indicate a high S/I ratio for the receiver of Relay-to Sturefors. Check how its transmitter interferes by left-click on **As Interferer** and marking the frequency in the upper list box.

The interference conditions within this network are now acceptable. To make sure you can check by closing the Interference Analysis window and mark, in turn, the other stations in the project and repeat the analysis for them.

Interference from other stations. Interference to and from other users of the assigned frequencies should be checked. This requires knowledge of the overall frequency use in the area, such as a national master station database and a military station database. WRAP supports this, but the demonstration version is limited in functionality to a maximum of six stations in a project, and stations may not be placed outside the county of Östergötland. To demonstrate some of this, it is possible to search the IFL (ITU's International Frequency List, to be replaced with the BR IFIC—International Frequency Information Circular). This requires that the IFL in WRAP format is installed, which will show up as an additional station database named IFL in the upper left database folder view.

In this particular case it is of little interest, since Sweden does not have any entries in the IFL for these frequencies.

Instead, we can demonstrate the procedure by searching in the demonstration WRAP database that is provided in the standard installation. The search should be made in the area and frequency range of interest. Perform as follows:

> First make sure that the project with the radio link network that you have created earlier is open.
>
> Draw a circle in the map around the radio link network by right-clicking in the map, select **Calculation/Search Area – Circle**, position the cursor at the desired center position, left-press and hold while moving the cursor until you have a radius of about 100 km. Release the left button.
>
> Double-left-click on the **WRAPdB** database name in the upper left folder view. This opens the underlying folder structure. Open the **Stations in Database** folder and double-left-click on the **Search Engine**... line. This opens a window where entries to define search criteria can be made.
>
> Make sure that the **Area** is checked and notice that the circular area just drawn in map is shown. Check the **Tx** and **Rx Frequency** and enter **1400** to **1700 MHz**.
>
> Perform the search by left-clicking on **OK**. Now all stations in the WRAPdB database within the circle AND within the range 1400 to 1700 MHz are shown in the upper right list view.
>
> The marked stations can be shown in the map by right-clicking on the marked stations and selecting **Show In Map**. Do that and zoom in to a suitable scale so you can see the links properly.

The continued procedure is now to add these stations to the project, to allow calculations including these stations.

Note that stations must exist in the WRAP "Project" in order to perform calculations. No calculations can be performed on stations that are just shown in the list view directly from the databases.

Note. The following procedure cannot be performed in the demonstration version, since it will result in more than six stations in the project.

> Mark the stations that were searched from the database as shown in the list view. Right-click on the marked stations and select **Add To Project**.
>
> Click on the **Stations In Project** folder in the upper left folder view. This results in a list of all stations in the upper right list view. Note that all radio link stations now appear on two lines, to be able to present information about both transmission directions.
>
> Mark all stations and select **Show In Map**.

Calculations can now be performed in the same way as before for the stations in the new radio link network, to see if their receivers are interfered with by existing transmitters, or if their transmitters interfere with existing receivers. In this particular case there will be no harmful interference. Instead, the just-created project is interesting to use for the following exercise on Frequency Assignment.

Save the project with all these stations under a new name for the following exercise!

5.5.11 Exercise B7: Frequency Assignment

Note. This exercise cannot be run in the demonstration version, as it requires more than six stations in the project.

This exercise will demonstrate the use of the very efficient automatic frequency assignment algorithms in WRAP.

The project that was created in the previous project will be used, so you should open it first of all. The 3-node radio link network in that project works properly, with low interference and creating low interference levels to the other links operating in the area in the same frequency range. But the existing links are in fact not compatible, and as the new link network uses the same frequency range there is a need to optimize the frequency use. This can be done with the **Frequency Assignment** tool. This tool also gives a convenient graphic display of the interference situation for each station to support manual frequency assignment. Perform as follows:

> Mark all radio link stations. Start the **Frequency Assignment** tool.
>
> First an appropriate channeling plan must be selected. Click on **Allotment – Select Allotment.** Choose the **Link Raster, 2 MHz** allotment and click on **OK.**

Note. The terminology was changed between versions 3.3 and 4.0. In 4.0 the term "Allotment" is used instead of "Raster" as is used in 3.3. This text uses the 4.0 terminology.

Now select **Blocking – Settings**. Select the **ITU-R P.526** propagation model. Click on **OK**.

If you still have the Radio Access Point stations at around 60 MHz in the project you must select a **Default Mobile**, because these stations take part in the assignment process and need to be accounted for as potential interferers and victims of interference. Select the **Analogue mobile**. Left-click on **OK** in the Blocking Settings window.

Select **Blocking – Calculate**. The list views become white when the calculation is ready. Mark the **Station Name** for each station in the upper left list view. This will display the margin diagram that gives information on the assignment margins. The keyboard Up/Down arrows can be used to quickly scroll through the list of stations.

The colors in the Margin diagram indicate the following:

- Blue: Already in use by the marked station
- Yellow: Assignable
- Red: Not assignable (owing to nonfulfilment of signal-to-interference criteria).

The assignment can be changed manually by clicking on the vertical lines.

You will notice that many of the existing frequencies indicate a red color, and are quite heavily interfered. There is obviously a need to make a complete new assignment to these links, to try to improve the situation.

Frequencies can be removed by right-clicking in the upper right list box and selecting whether to delete just this frequency or all frequencies. Before performing a complete automatic assignment it is common practice to delete all frequencies. So left-click on one of the frequencies in the upper right list box, select **Delete All** and respond **No** to the question if you want to remove the frequencies for this station only.

Perform the automatic assignment by selecting **Automatic – Start**. There are several settings to control the assignment algorithm in the **Automatic – Settings** menu. Just use the default settings.

When the assignment has been performed, you mark each station in turn in the upper left list box and check that the assignment margin diagram indicates a good assignment with acceptably low interference.

The detailed interference levels for a particular assigned frequency are displayed in the **Worst Interfered Stations** and **Worst Interfering Stations** fields. To do that you need to mark the **Used** frequency in the lower left list box, for the particular station that is marked in the upper left list box. See the following window for an example of this.

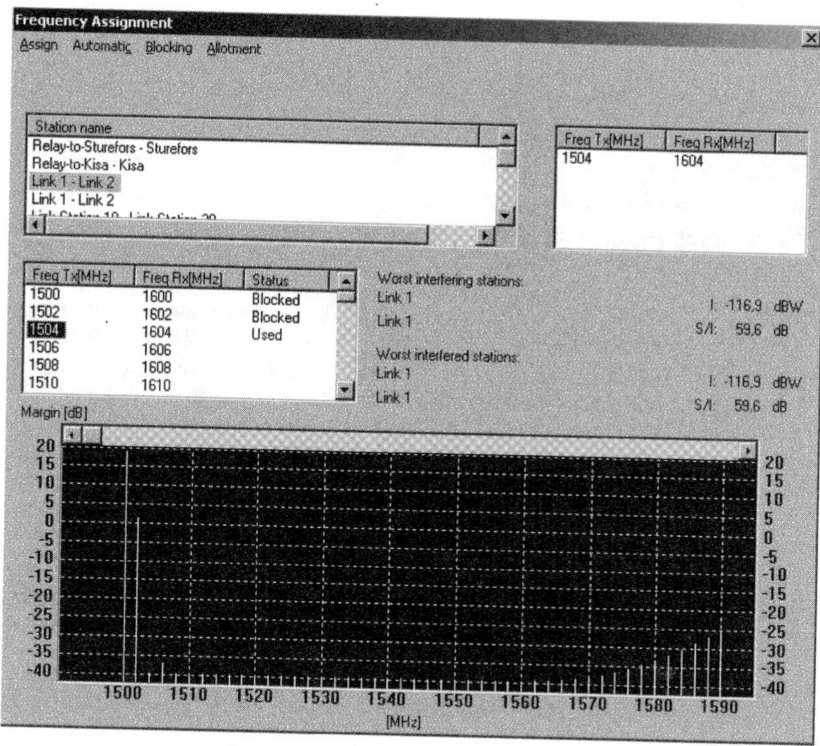

The new assignment can be saved to the project file by left-clicking on **Assign** and selecting **Save**. This closes the Frequency Assignment window.

Don't do the following! This is just for information purposes! The frequency assignment to the stations in the project has now been changed. These changes can be saved to the station database by marking all stations to be saved and selecting **Save**.

5.6 Endnotes

[1] The term "frequency spectrum management" is used to describe various administrative and technical procedures that are intended to ensure the operation of radio stations of different radiocommunication services at any given time without causing or receiving harmful interference. It takes place at two levels: national and international.

[2] Professor Dr. R. G. Struzak of the ITU Regulation Board, "Access to Spectrum/Orbit Resources and Principles of Spectrum Management," http://www.ictp.trieste.it/~radionet/2000_school/lectures/struzak/AcceSpctrOrbICTP.htm

[3] As referenced by Struzak, *Ibid.*

[4] "Managing the Radio Spectrum—Rapid Response Unit" The World Bank Group, http://rru.worldbank.org/Resources.asp?results=true&stopicids=53

[5] The laws of nature allow various applications of radio waves to interfere with each other and nullify their benefits, unless they are correctly designed or operated. Such

interference is avoided by allowing each application some amount of radio frequency spectrum for exclusive use. Sometimes, however, special arrangements are made to share a particular radio frequency spectrum. The terms "radio waves," "radio frequency spectrum" and "spectrum," have the same meaning.

[6] As referenced by Struzak, *Ibid.*, 2.

[7] Struzak, *Ibid.*, 2.

[8] Managing the Radio Spectrum, *Ibid.*, 4.

[9] The 1947 Atlantic City Radio Conference made foundations for the present international spectrum management by copying, to some degree, the US national spectrum management system of that time. Struzak, *Ibid.*, 7.

[10] National spectrum management began in the early 1920s with the record keeping—logging out frequencies to applicants essentially on a first come, first served basis. *Ibid.*

[11] Brown, C., "Spectrum Analysis Basics," Hewlett-Packard Company, http://we.home.agilent.com/upload/cmc_upload/tmo/downloads/E206WIRELESS_SABASICS.pdf.

[12] The international Table of Frequency Allocations is contained in Article 5 of the Regulations. It specifies the way frequency bands are to be shared among different radiocommunication services in the three regions 1, 2 and 3 identified and specified in Fig. 6-1, Chap. 6.

[13] The Radio Regulations are part of the Administrative Regulations complementing the provisions of the ITU. These regulations specify the basic conditions that are relevant for the international radio-regulatory arrangement. Some of these conditions are specified in rather general terms (e.g., the international Table of Frequency Allocations), while some other conditions are specified in a more detailed manner (e.g., procedures for mandatory coordination, notification and recording of frequency assignments). Other elements, such as: procedures for issuing licenses; availability of frequency bands for specific applications; type of approval procedures; use of radio equipment by foreign persons; and so on, are governed by national legislation instruments. The national legislation instruments (e.g., national tables of frequency allocations) are not available from ITU, but may be obtained from the regulatory authorities of the concerned Member State. The addresses of the regulatory authorities are contained in the ITU Global Directory: http://www.itu.int/GlobalDirectory/ . Conventions that govern the use of telecommunications are binding on all Members.

The international Radio Regulations are based on the use of two main concepts:

- The concept of frequency block allocations is intended for use by defined radio services (Table of Frequency Allocations as contained in Article 5 of the Radio Regulations). This concept generally provides common frequency allocations to mutually compatible services operating with similar technical characteristics in specific parts of the spectrum. It also provides stable planning environment for administrations, for equipment manufacturers and for users.
- The concept of voluntary or obligatory regulatory procedures (for coordination, notification and recording) is adapted to the allocation structure.

[14] The term "frequency planning" is a carry over from the early days of radio, when only operating frequency of radio station and its geographic location could vary.

[15] The present system has received criticism almost from the very beginning, but a better system is yet to be agreed upon. As discussed earlier, the developing countries fear that there will be no spectrum left to satisfy their future needs. They would also like to exploit their old equipment until it ceases to function. The developed countries argue that the regulatory barriers are inhibiting them from implementing new technologies, including new applications.

[16] The governments represent the consumers/users, service providers, and equipment manufacturers in the ITU decision process.

[17] Professor Dr. R. G. Struzak presented these proposals, Struzak, *Ibid.* 10.

[18] They further point out that reliance upon administrative decision-making could force us to live with decision that are arbitrary and often wrong in determining the best interest of users. Market forces, on the other hand, can raise the prices of wireless applications and tilt the balance between the further developments of wired- and wireless communication services.

[19] Kwerel, E., "Auctioning Spectrum Rights," Walt Strack, Office of Pland and Poilicy, Wireless Telecommunications Bureau, U.S. Federal Communications Commission, February 20, 2001.

[20] Zimmermann., as referenced in Struzak, *Ibid*, 17.

[21] "Spectrum Management Regulations and Related References," http://www.army.mil/spectrum/library/regulations.htm. Materials for this section also heavily draws upon "Science and Spectrum Management," European Science Foundation (ESF) Committee on Radio Astronomy Frequencies (CRAF) 2002. The author wishes to thank the European Science Foundation for granting permission to reproduce in full or in part various text, charts, tables, and figures. A special word of thanks is directed to Dr. Wim van Driel, Observatoire de Paris, CRAF Chairman, and Dr. Titus Spoelstra, CRAF Frequency Manager/Secretary for helping the author to improve the contextual coverage of this chapter.

[22] The ITU Radio Regulation defines an Administration as "any governmental department or service responsible for discharging the obligations undertaken in the Constitution of the International Telecommunication Union, in the Convention of the International Telecommunication Union and in the Administrative Regulations," ITU, 1993, *Final Acts of the Additional Plenipotentiary Conference (Geneva, 1992)—Constitution and Convention of the International Telecommunication Union*, International Telecommunication Union, Geneva, as referenced in "Science and Spectrum Management," *ibid.*

[23] Public entities place fundamental importance to radio frequency use for the country's defense system, uninterrupted operation of critical infrastructures (e.g., public telecommunication facilities), police and fire brigade, publicly funded scientific research, and public broadcasting. Private entities include companies active as telecommunication operators, private broadcasting facilities, private safety organizations, as well as general public using nonlicensed devices (remote control devices, microwave ovens, and so on).

[24] National Telecommunications and Information Administration (NTIA) Manual of Regulations and Procedures for Federal Radio Frequency Management, January 2000 Edition with January/May/September 2001 Revisions.

[25] A White Paper on Future Federal Communications Commission Spectrum Policy, Motorola, ET Docket 02-135, August 30, 2002.

[26] Federal Spectrum Management: How the Federal Government Uses and Manages the Spectrum, NTIA, March 2001.

[27] Kwerel, E., and Williams, J., "A Proposal for a Rapid Transition to Market Allocation of Spectrum," Federal Communications Commission, November 2002.

[28]Statement of Peter F. Guerrero, "Telecommunications: History and Current Issues Related to Radio Spectrum Management," Testimony Before the Committee on Commerce, Science, and Transportation, U.S. Senate, U.S. General Accounting Office report, GAO-02-906, Washington, DC: June 11, 2002, provides an extensive discussion of the organization of spectrum management in the United States.

[29] General Accounting Office Report to Congressional Requesters, "Telecommunications: Comprehensive Review of US Spectrum Management with Broad Stakeholder Involvement is Needed," GAO-03-277, January, 2003.

[30] Science and Spectrum Management, *Ibid*, 21.

[31] EN, 24.4.2002, "Decision No 676/2002/EC of the European Parliament and of the Council of 7 March 2002 on a regulatory framework for radio spectrum policy in the European Community (Radio Spectrum Decision)," pp. L 108/1-L 108/6.

[32] EC, 1999, Directive 1999/5/EC of the European Parliament and of the Council of 9 March 1999 on radio equipment and telecommunication terminal equipment and the mutual recognition of their conformity, as referenced in *ibid.*, 3.

[33] "European Commission's Green Paper on Radio Spectrum Policy," http://europa.eu.int/ISPO/spectrumgp/sgpcom/aer.htm.

[34] As defined in Chap. 4, Section on Transport of Information and Energy.

[35] *Ibid.*

[36] The EC policy is again based on the worldwide plans established, under the auspices of the ITU, for the terrestrial services, and the worldwide plans form part of the Radio Regulations:

- The frequency allotment plan for coast radiotelephone stations operating in the exclusive maritime mobile bands between 4000 and 27500 kHz (Appendix 25 to RR);
- The frequency allotment plan for the aeronautical mobile (OR) service operating in the exclusive bands between 3025 and 18030 kHz (Appendix 26 to RR);
- The frequency allotment plan for the aeronautical mobile (R) service operating in the exclusive bands between 2850 and 22000 kHz (Appendix 27 to RR).

The following "Regional" plans, established under the auspices of the ITU, are still relevant to the terrestrial services:

- Frequency assignment plans for VHF and UHF Television Broadcasting annexed to the Regional Agreement for the European Broadcasting Area, Stockholm, 1961 (ST61), including a frequency assignment plan for FM sound broadcasting in the band 41 to 68 MHz;
- Frequency assignment plans for LF and MF broadcasting annexed to the Regional Agreement on LF/MF Broadcasting (Regions 1 and 3), Geneva, 1975 (GE75);
- Frequency assignment plan for MF broadcasting annexed to the Regional Agreement on MF Broadcasting, (Region 2), Rio de Janeiro, 1981 (RJ81);
- Frequency assignment plan for VHF/FM sound broadcasting annexed to the Regional Agreement concerning FM Sound Broadcasting Stations (Region 1 and part of Region 3), Geneva, 1984 (GE84);
- Frequency assignment plan for stations of the maritime mobile and aeronautical radionavigation service in the MF bands in Region 1 annexed to the Regional Agreement concerning the MF maritime mobile and aeronautical radionavigation services in Region 1, Geneva, 1985 (GE85- MM-R1);
- Frequency assignment plan for stations of the maritime radionavigation service (radiobeacons) for the European Maritime Area in the band 283.5 to 315 kHz annexed to the Regional Agreement concerning the planning of the maritime radionavigation service (radiobeacons) in the European Maritime Area, Geneva, 1985 (GE85-EMA);
- Allotment plan for the broadcasting service in the band 1605 to 1705 kHz in Region 2 annexed to the Regional Agreement for the use of the band 1605 to 1705 kHz in Region 2, Rio de Janeiro, 1988 (RJ88);
- Frequency assignment plan for VHF and UHF Television Broadcasting annexed to the Regional Agreement concerning planning of the VHF/UHF Television Broadcasting stations in the African Broadcasting Area and neighbouring Countries, Geneva 1989 (GE89).

[37] Space service includes:

- Fixed Satellite Service, which refers to radiocommunication between two earth stations at given positions, where one or more satellites are used.
- Mobile Satellite Service, which refers to radiocommunication involving mobile earth stations, space stations and/or fixed stations.
- Radio-determination Satellite Service, which refers to radiocommunication service for purposes of radio-determination involving one or more space stations.
- Earth Exploration Satellite Service, which refers to radiocommunication involving one or more space stations for collection of information relating to the characteristics of the earth and its natural phenomena from active or passive sensors on earth satellites, and/or airborne or earth-based platforms.

- Broadcasting Satellite Service, which refers to radiocommunication in which signals are transmitted or retransmitted from space stations, intended for direct reception by the general public.
- Amateur Satellite Service, which refers to amateur radiocommunication involving the use of one or more space stations.
- Space Operations Service, which refers to radiocommunication in which space craft or other objects in space are used for scientific or technological research purposes.
- Inter-satellite Service, which refers to radiocommunication for providing communication links between artificial earth satellites.

[38] National Sovereignty Implies a Specific National Articulation of the Telecommunication Legislation, refer to Scherer, J., (ed.), 1995, *Telecommunications Laws in Europe*, Kluwer Law International, The Hague/London/Boston, 3rd revised (ed.), as referenced in "Science and Spectrum Management," *Ibid.*, 3.

[39] Schwarz, T., and Satola, D., *Telecommunication Legislation in Transitional and Developing Economies*, World Bank Technical Paper 489, World Bank, Washington, DC, 2000, http://rru.worldbank.org/documents/Telecommunications_legislation.pdf.

[40] Legislation addressing spectrum policy introduced in the 107th Congress included S. 2869, H.R. 5638, H.R. 4738, and H.R. 4641.

[41] Schwarz, T., and Satola, D., *Ibid.*, 39.

[42] Materials for this section are heavily drawn from "Science and Spectrum Management," European Science Foundation (ESF) Committee on Radio Astronomy Frequencies (CRAF) 2002. The author wishes to thank the European Science Foundation for granting permission to reproduce in full or in part various text, charts, tables, and figures. A special word of thanks is directed to Dr. Wim van Driel, Observatoire de Paris, CRAF Chairman, and Dr. Titus Spoelstra, CRAF Frequency Manager/Secretary for helping the author to improve the contextual coverage of this chapter.

[43] The primary focus here is on the concept of "technical efficiency"—that is, getting the most use, or "output," out of a portion of spectrum, given the mission or market context of its use.

[44] Economic efficiency relates to whether spectrum is allocated across various uses in a way that maximizes society's welfare.

[45] Refer to Appendix 1 for an extract from: "Economic aspects of Spectrum Management," Report ITU-R SM.2012.

[46] In free markets, National Regulatory Authorities must evaluate economic incentives that give signals to firms and consumers that help to ensure that resources flow to their most valued use. The GAO report on Telecommunications points out that with spectrum, this free flow of resources is not fully functional. Refer to GAO-03-277 Telecommunications, *Ibid.*, 29. National Regulatory Authorities must, therefore, undertake periodic analysis to determine whether spectrum is allocated in an economically efficient manner.

[47] Refer to Appendix 1, Paragraph 1.6.4 Spectrum control (Enforcement inspections and monitoring), *Ibid.*, 30.

[48] As defined in Chap. 4, Transport of Information and Energy.

[49] ITU, 1995b, *Handbook on National Spectrum Management*, ITU-Publications, Geneva, as referenced in "Science and Spectrum Management," *Ibid.*, 30.

[50] ITU, 1995c, *Handbook on Spectrum Monitoring*, ITU-Publications, Geneva, as referenced in "Science and Spectrum Management," *Ibid.*

[51] *Ibid.*

[52] International Telecommunication Union's World Radiocommunication Conference:

Main Results of World Radio Conference 2000, http://www.itu.int/ITU-R/conferences/wrc/wrc00/results/index.html.

African Telecommunications Union: Guidelines for World Radiocommunication Conference 2003.

Asia-Pacific Telecommunity: Preparatory Group for World Radio Conference 2003, http://www.aptsec.org/radio/apt-wrc.htm.

Inter-American Telecommunication Commission: Preparations for World Radiocommunication Conference 2003, http://www.citel.oas.org/WRC/wrc.asp.

These sites provide a summary of results of the most recent World Radiocommunication Conference (2000), and preparations for the next conference (2003) in Africa, Asia-Pacific, and Latin America. The principles, concerns, and industry-wide consultative process of a mature economy leading up to 2003 can be gleaned from the U.S. Federal Communications Commission website (http://www.fcc.gov/wrc-03).

[53] Struzak, R., "Access to Spectrum/Orbit Resources and Principles of Spectrum Management," http://www.ictp.trieste.it/~radionet/2000_school/lectures/struzak/AcceSpctrOrbICTP.htm#_Toc474458659.

[54] Carlsson, O., "An Integrated System for Computerized Spectrum Management, AeroteckTelub.

[55] Reproduced in full with permission from AerotechTelub, Sweden. The author wishes to thank Dr. Olov Carlsson, Head, Spectrum Management, AerotechTelub for his helpful suggestions.

Chapter

6

Management Process

The radio frequency spectrum is being used by a growing variety of radiocommunication applications. Although the traditional uses of broadcasting, maritime and aeronautical communications, point-to-point fixed links, land mobile radio, and numerous navigation services remain, in almost all of their categories new applications are being introduced at a growing pace, and novel uses of radio spectrum are being exploited. Recent years have seen a dramatic change to digital modulation techniques, in both fixed and mobile services, and the next generation of digital broadcasting services is close to implementation. There is increasing interest in the field of satellite communications to provide mobile services, and more generally, radio is offering the means to rapidly introduce new telecommunication services, to provide competition in developed countries and to establish basic telephone services in developing countries. All of these applications have to coexist alongside scientific and other, for example, amateur, uses of radio as well as against a background of radio frequency emissions from non-radiocommunication sources.

The task of accommodating all of these radio services and systems in the finite usable range of radio frequency spectrum comes under the generic title of *spectrum management*. This process is mainly the responsibility of government administrations (although some degree of delegation is possible) and it is imperative that those administrations coordinate their efforts internationally. There are several reasons for this, as discussed in Chap. 7.

This chapter describes the framework in which spectrum is managed internationally and nationally.[1]

6.1 International Spectrum Management

The international administrative cooperation body having the responsibility for coordinating spectrum management at the global level is the International Telecommunication Union (ITU). The ITU is the oldest specialized agency of the United Nations, and its origins extend back to the International Telegraph Union, which was founded in 1865; that is, before the invention of the telephone and the demonstration of the practical application of radio transmission. The ITU currently has a membership of about 190 sovereign countries, the member states, and about 500 nongovernmental entities, including equipment manufacturers and operators/service providers, the sector members, and a permanent headquarters in Geneva, Switzerland. A list of other U.N. organizations deploying activities related to spectrum regulations and scientific use of radio frequencies is given in Table 6-1.

The importance of the ITU in the field of spectrum management can be judged by the prominence given to radiocommunication in Article 1 of the ITU Constitution. In that text, the Union is required to effect allocation of bands of the radio-frequency spectrum, the allotment of radio frequencies and registration of radio frequency assignments, and any associated orbital position in the geostationary-satellite orbit in order to avoid harmful interference between radio stations of different countries and to coordinate efforts to eliminate harmful interference

TABLE 6-1 The United Nations and Some of Its Organizations

UN organization	Administrative headquarters	Scientific interest
International Civil Aviation Organization (ICAO)	Montreal, Canada	
International Telecommunication Union (ITU)	Geneva, Switzerland	
International Trade Centre UNCTAD/WTO	Geneva, Switzerland	
Office for Outer Space Affairs	Vienna, Austria	Outer space studies
United Nations Educational, Scientific and Cultural Organization (UNESCO)	Paris, France	Scientific committee on the allocation of frequencies for radio astronomy and space science, IUCAF
World Meteorological Organization (WMO)	Geneva, Switzerland	Remote sensing, meteorology

SOURCE: Ibid.

between radio stations of different countries and to improve the use made of the radio-frequency spectrum and of the geostationary-satellite orbit for radiocommunication services.

6.1.1 The international radio regulations

The ITU sets the overall international spectrum management framework through the International Radio Regulations. This body of text has international treaty status, and thus the Regulations are binding for all members of the ITU. The radio regulations contain, in Article S5, the international frequency allocation table. For the purpose of the international frequency allocation table, the world has been divided into three regions. Region 1 covers the whole of Europe, the Middle East and Africa; Region 2 comprises the Americas, and Region 3 Asia and Australia (see Fig. 6-1).

An extract of one page of Article S5 is shown in Table 6-2. This article shows the allocation of the radio spectrum, broken down into a large number of discrete bands and to a number of defined radiocommunication services. The frequency range covered is from 9 kHz to 1000 GHz. The radiocommunication services defined in the Radio Regulations and for which there are allocations include broadcasting, mobile radio, the fixed services, amateur radio, radionavigation and radiolocation, a number of science services, and, in nearly all cases, the corresponding satellite based transmissions of a space service. The Radio Regulations do not in general subdivide the basic radio services into detailed applications. For example, although the mobile service is subdivided into its land, maritime and aeronautical variations, there is no provision in the regulations relating to paging, private mobile radio or cellular radiotelephones.

For many radio services it is necessary to have common worldwide allocations, for example, in the high frequency (short-wave) bands where signals propagate over vast distances, in bands used for international maritime and aeronautical communications and navigation, and where satellite-delivered services are involved. In other cases, global allocations may be desirable to minimize incompatibilities in border regions or to create large markets for equipment. However, historic differences in usage and subsequent difficulties in negotiating changes have resulted in some significant variances in the allocations from region to region. Another important element is the use of different allocations to create and maintain "exclusive" regions of political and economic influence and closed monopolistic markets.

In addition to the broad applications of the international table, which may provide for more than one radio service in any given band, the table contains a large number of footnotes. Some of these specify constraints on the use of the radio service or frequencies in question, while others provide additional or alternative frequency allocations to individual countries or groups

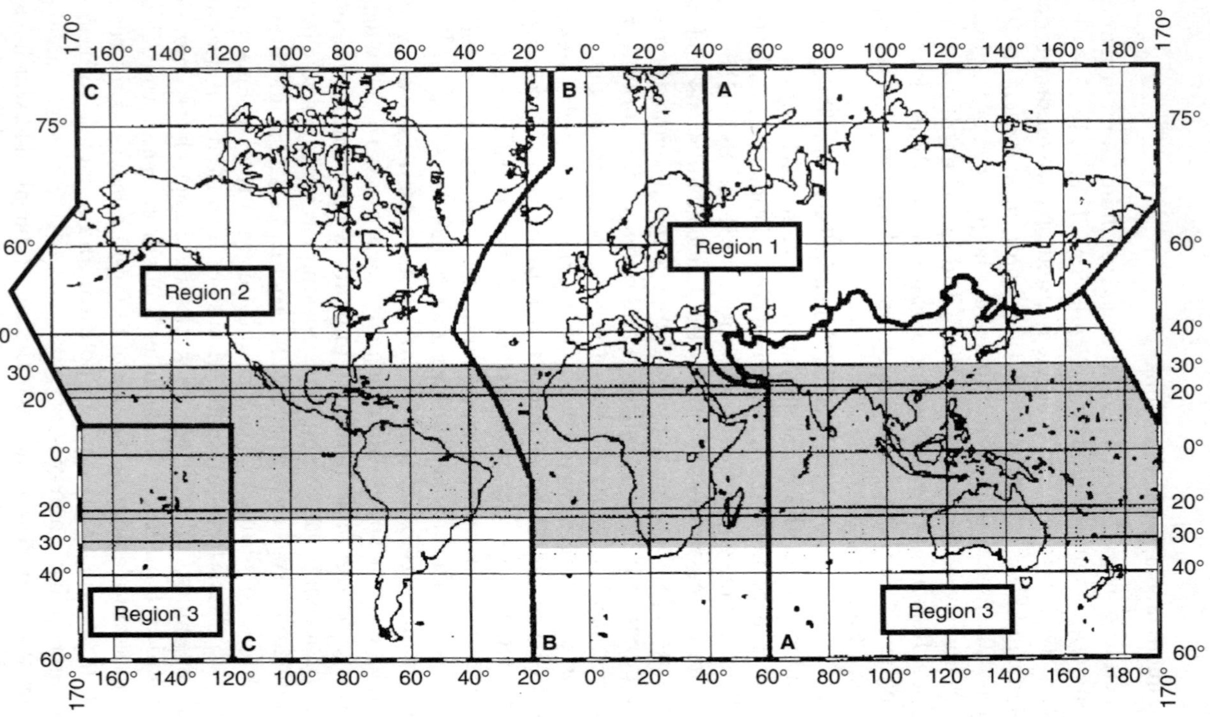

Figure 6-1 For the allocation of frequencies the world has been divided into three Regions as shown in this map. (*Source: "Science and Spectrum Management," European Science Foundation (ESF) Committee on Radio Astronomy Frequencies (CRAF) 2002, Ibid., 1.*)

TABLE 6-2 Extract of a Page of Article S5 of the ITU Radio Regulations. (1610.6–1631.5 MHz)

Allocation to Services		
Region 1	Region 2	Region 3
1610.6–1613.8 Mobile satellite (Earth-to-Space) Radio astronomy Aeronautical radionavigation	1610.6–1613.8 Mobile satellite (Earth-to-Space) Radio astronomy Aeronautical radionavigation Radiodetermination satellite (Earth-to-Space)	1610.6–1613.8 Mobile satellite (Earth-to-Space) Radio astronomy Aeronautical radionavigation Radiodetermination satellite (Earth-to-Space)
S5.149 S5.341 S5.355 S5.359 S5.363 S5.364 S5.366 S5.367 S5.368 S5.369 S5.371 S5.372	S5.149 S5.341 S5.364 S5.366 S5.367 S5.368 S5.369 S5.371 S5.372	S5.149 S5.341 S5.355 S5.359 S5.364 S5.366 S5.367 S5.368 S5.369 S5.372
1613.8–1626.5 Mobile satellite (Earth-to-Space) Aeronautical radionavigation Mobile satellite (Space-to-Earth)	1613.8–1626.5 Mobile satellite (Earth-to-Space) Aeronautical radionavigation Radiodetermination satellite (Earth-to-Space) Mobile satellite (Space-to-Earth)	1613.8 to 1626.5 Mobile satellite (Earth-to-Space) Aeronautical radionavigation Mobile satellite (Space-to-Earth) Radiodetermination Satellite (Earth-to-Space)
S5.341 S5.355 S5.359 S5.363 S5.364 S5.365 S5.366 S5.367 S5.368 S5.369 S5.371 S5.372	S5.341 S5.364 S5.365 S5.366 S5.367 S5.368 S5.370 S5.372	S5.341 S5.355 S5.359 S5.364 S5.365 S5.366 S5.367 S5.368 S5.369 S5.372
1626.5–1631.5 Maritime mobile satellite (Earth-to-Space) Land mobile satellite (Earth-to-Space) S5.532 S5.341 S5.351 S5.354 S5.355 S5.359	1626.5–1631.5 Maritime mobile satellite (Earth-to-Space) S5.341 S5.351 S5.354 S5.355 S5.359 S5.373A	

SOURCE: Ibid., 1.

NOTE: The cells in this table apply for a specific frequency band and region as indicated. Radiocommunication services having a primary allocation are printed in capital characters. General footnotes applying to a specific frequency band are added at the bottom of each cell. Footnotes applying to a specific service only are added behind this service. The footnotes are explained elsewhere in the radio regulations.

of countries. Some footnotes represent real operational needs or usage; others are the results of compromises in international negotiations.

The Radio Regulations contain much more than the international frequency table alone. They contain rules for the use and operation of frequencies; they specify operating procedures for stations, especially in the maritime and aeronautical services; and they lay down the procedures for the coordination of frequencies. The latter is the mechanism used to check if the use of frequencies in one country will cause interference to, or suffer interference from, other existing frequency assignments of other countries. If not, the frequency can then be registered and afforded protection from other, future, users. This procedure can be very complex and time-consuming, but it forms the core of the regulations and achieves order in what would otherwise be utter chaos. Not all frequency assignments need to be cleared internationally in this way: existence of low power applications, those well inside a country's border or cases where special bilateral or other arrangements are in force, are exempt. But coordination is necessary in order to be able to claim international recognition and hence protection. It is therefore essential for many applications, including virtually all satellite-based ones.

For some radio applications, the frequency requirements of each country are met in a predetermined *frequency assignment plan*. This approach is most common in the broadcasting and broadcasting-satellite services. In many cases the plan is on a regional or sub-regional basis. Once the plan has been agreed upon, a country may use its assignments without further formal coordination. Normally, a plan-modification procedure provides the mechanism for bringing into use assignments with different characteristics from those specified in the plan.

A variation of the planning process is used in the *allotment plan*. Here, the plan may specify a particular frequency or frequencies to be used by a country in a particular area (as compared to a specific location for an assignment plan). Allotment plans are used in the aeronautical and maritime mobile services. A special case of allotment plan has also been used for certain fixed satellite bands in which each country has been allotted a range of frequencies for use over a specified portion (arc) of the geostationary-satellite orbit. In this case, the allotment is converted into a specific assignment before bringing it into use.

In spite of the official recognition of science services, such as the Radio Astronomy Service, the Earth Exploration-Satellite Service and the Space Research Service, the ITU does not fully recognize the extent to which their characteristics are different, and sometimes even very different, from those of the other radiocommunication services. This holds especially true for the passive services. The Radio Regulations place all services on an equal footing and do not provide preferential treatment to the more vulnerable *passive services* as is clear from the Radio Regulations Articles S4.5, S4.6, and S4.7.

Article S4.5.

> The frequency assigned to a station of a given service shall be separated from the limits of the band allocated to this service in such a way that, taking account of the frequency band assigned to the station, no harmful interference is caused to services to which frequency bands immediately adjoining are allocated.

Article S4.6.

> For the purpose of resolving cases of harmful interference, the radio astronomy service shall be treated as a radiocommunication service. However, protection from services in other bands shall be afforded the radio astronomy service only to the extent that such services are afforded protection from each other.

Article S4.7.

> For the purpose of resolving cases of harmful interference, the space research (passive) service and earth exploration-satellite (passive) service shall be afforded protection from different services in other bands only to the extent that such services are afforded protection from each other.

Although it is the mission of the ITU, that is, of its radiocommunication sector, to ensure the rational, equitable, efficient and economical use of the radio-frequency spectrum by all radiocommunication services (see Section 6.1.3), the specific case of passive services is at present not handled on an equitable basis in the physical and technical sense. It is common practice in the design of active services to raise the power of emitted signals to a point where the level of natural, additive noise onto the received signal becomes negligible. In such a context, where active spectrum users may raise their transmitting powers beyond such a level, spectrum management is reduced to ensuring each user has its required signal-to-interference ratio, that is, to handling relative signal power levels. Passive services, on the other hand, are based on measurements of natural radiation, sometimes of very low levels; hence, they need protection in absolute terms.

The Radio Astronomy Service has suffered a number of harmful interference cases during the last years, a situation which continues at present. This interference is mainly caused by satellites with inadequate protection for radiocommunication service(s) in an adjacent frequency band. Some radio astronomy operations in protected frequency bands have thereby been made very difficult or even impossible for a number of years to come, although the Radio Regulations specifically permit no harmful interference in these bands.

6.1.2 The International Radio Regulations in the context of global regulation

On a global scale, the international regulatory framework for frequency regulation and spectrum management is the ITU Radio Regulations. These regulations have the status of an international treaty.

In terms of international law, the national administrations play a key role in spectrum management. In some local situations, where coordination between private users of the radio spectrum is required or desired, agreements between these private users can be obtained.

These agreements should be reached in proper coordination with the national administration; otherwise the private users undermine their own case (see Chap. 7, Section 7.2). Furthermore, such agreements or "Memoranda of Understanding" should obey the legal principles as given above and conform to the current national and international legislation, that is, the ITU Radio Regulations. The legal status of such agreements is very limited and even nonexistent in terms of international law.

The ITU Radio Regulations apply to terrestrial, aeronautical and space radiocommunication services. As concerns the space-borne systems, the following comments should be made.

A treaty with a status higher than the ITU Radio Regulations is the *Treaty on Principles Governing the Activities of States in the Exploration and Use of Outer Space, including the Moon and other Celestial Bodies*, the Outer Space Treaty (OST) (United Nations Treaties and Principles on Outer Space, 1994). An even higher status for the OST is being considered, based on the fact that it was formulated within the most comprehensive world organization, the United Nations, as a sort of Magna Carta for Space. But this interpretation is subject to dispute.

OST articles relevant to the protection of terrestrial radio stations in general and scientific radio stations in particular are[2]:

Article I.

> The exploration and use of outer space, including the moon and other celestial bodies, shall be carried out for the benefit and in the interests of all countries, irrespective[3] of their degree of economic or scientific development, and shall be the province of all mankind.
>
> Outer space, including the moon and other celestial bodies, shall be free for exploration and use by all States without discrimination of any kind, on a basis of equality and in accordance with international law, and there shall be free access to all areas of celestial bodies.

There shall be freedom of scientific investigation in outer space, including the moon and other celestial bodies and states shall facilitate and encourage international cooperation in such investigation.

Article VI.

> States Parties to the Treaty shall bear international responsibility for national activities in outer space, including the moon and other celestial bodies, whether such activities are carried on by governmental agencies or by nongovernmental entities, and for assuring that national activities are carried out in conformity with the provisions set forth in the present Treaty. When activities are carried on in outer space, including the moon and other celestial bodies, by an international organization, responsibility for compliance with this Treaty shall be borne both by the international organization and by the States Parties to the Treaty participating in such organization.

Article VII.

Each State Party to the Treaty that launches or procures the launching of an object into outer space, including the moon and other celestial bodies, and each State Party from whose territory or facility an object is launched, is internationally liable for damage to another State Party to the Treaty or to its natural or juridical persons by such objects or its component parts on the Earth, in air or in outer space, including the moon and other celestial bodies.

Article VIII.

A State Party to the Treaty on whose registry an object launched into outer space is carried shall retain jurisdiction and control over such object, and over any personnel thereof, while in outer space or on a celestial body. Ownership of objects launched into outer space, including objects landed or constructed on a celestial body, and of their component parts, is not affected by their presence in outer space or on a celestial body or by their return to the Earth. Such objects or component parts found beyond the limits of the State Party to the Treaty on whose registry they are carried shall be returned to that State Party, which shall, upon request, furnish identifying data prior to their return.

Article IX.

In the exploration and use of outer space, including the moon and other celestial bodies, States Parties to the Treaty shall be guided by the principle of cooperation and mutual assistance and shall conduct all their activities in outer space, including the moon and other celestial bodies, with due regard to the corresponding interests of all other States Parties to the Treaty. States Parties to the Treaty shall pursue studies of outer space, including the moon and other celestial bodies, and conduct exploration of them so as to avoid their harmful contamination and also adverse changes in the environment of the Earth resulting from the introduction of extraterrestrial matter and, where necessary, shall adopt appropriate measures for this purpose. If a State Party to the Treaty has reason to believe that an activity or experiment is planned by it or its nationals in outer space, including the moon and other celestial bodies, would cause potentially harmful interference with activities of other States Parties in the peaceful exploration and use of outer space, including the moon and other celestial bodies, it shall undertake appropriate international consultation before proceeding with any such activity or experiment. A State Party to the Treaty which has reason to believe that an activity or experiment planned by another State Party in outer space, including the moon and other celestial bodies, would cause potentially harmful interference with activities in the peaceful exploration and use of outer space, including the moon and other celestial bodies, may request consultation concerning the activity or experiment.

A second specific convention based on the Outer Space Treaty 1967, in particular on its Articles VI and VII, is the Liability Convention 1971 concerning international responsibility and liability of states for their national activities in space.

Articles 1, 2, 5.1, and 5.3 of this *Convention on International Liability for Damage Caused by Space Objects* are especially relevant for the protection of scientific radio stations.[4] These articles address the definition of damage and the liability of the responsible state if a space station causes damage.

The third specific convention based on the Outer Space Treaty 1967 is the *Convention on Registration of Objects Launched into Outer Space* of 1974, in particular its Articles VIII, X, and XI. These articles deal, respectively, with the obligation of states where a vehicle is launched into outer space and is registered, to retain jurisdiction and control over such an object and over any personnel thereof (Art. VIII), the opportunity to observe the flights of space objects (Art. X), to inform the Secretary-General of the United Nations, the public and the international scientific community, of the nature, conduct, location and results of such activities (Art. XI).

Given the increasing threat of harmful interference to radio astronomy and other passive scientific applications of radio frequencies by transmissions from satellites, and the fact that satellites used for international direct broadcasting contribute significantly to this, it should be noted that in the *Principles Governing the Use by States of Artificial Earth Satellites for International Direct Television Broadcasting* (1972) it is stated clearly that:

> In order to promote international cooperation in the peaceful exploration and use of outer space, States conducting or authorizing activities in the field of international direct television broadcasting by satellite should inform the Secretary-General of the United Nations, to the greatest extent possible, of the nature of such activities. On receiving this information, the Secretary-General should disseminate it immediately and effectively to the relevant specialized agencies, as well as to the public and the international scientific community (item 12).

And:

> With respect to the unavoidable over-spill of the radiation of the satellite signals, the relevant instruments of the International Telecommunication Union shall be exclusively applicable (item 15).

The ITU itself, its constitution and convention, and the ITU Radio Regulations are considered "Related International Agreements." This implies that international law at its highest level should in the context of the current problem be the Outer Space Treaty (OST), while the ITU documents, treaties, and agreements act as appendices to this law. Therefore, the ITU Radio Regulations and related documents should be read in the context of the OST as far as space applications are concerned.

For the protection of radio frequencies relevant for scientific research, the key articles are *Articles VI and VII* of the OST. It should be noted that in the OST "damage" is a generic term and understood in the sense that the victim defines the damage, just as in the case of physical damage. However, in the strict juridical sense, the definition of damage is often subject to the general interpretation intended by the drafters and the participating states, subject to reasonableness and ultimately also subject to a decision of the judicial body called upon to judge a particular case.

Working on the spectrum management and in relation to the interest of the scientific community, scientists should base their arguments not only on the ITU Radio Regulations and related ITU documentation, but they must also be aware of the protection argumentation based on the OST. The OST contains no restriction concerning the kind of exploration of outer space, including the moon and other celestial bodies: this can be done by launching space vehicles, but also done by various different techniques used in scientific research. This treaty uses the term "exploration" only in a generic way. The same holds for "damage."

6.1.3 The ITU–R sector

The ITU organization has a somewhat federative structure with three *sectors* including *radiocommunication* (ITU-R).

The functions of the radiocommunication sector are to ensure the rational, equitable, efficient, and economical use of the radio-frequency spectrum by all radiocommunication services, including those using geostationary-satellite orbit and to carry out studies without limit of frequency range and to adopt recommendations on radiocommunication matters. Close coordination is carried out (mainly at national level) between the ITU-R, ITU-T, and ITU-D sectors.

The radiocommunication sector works through world and regional radiocommunication conferences, the Radio Regulations Board, radiocommunication assemblies, and radiocommunication study groups. The ITU-R activities are supported by the Radiocommunication Bureau, headed by a director.

Among its various activities, the Radiocommunication Bureau has the responsibility for operating a database containing the declared (not the actual or real) frequency use and related parameters for all stations of a radiocommunication service and for radio astronomy stations. The national administrations notify this bureau about these stations according to a well-defined procedure. Such notification enables the ITU to serve the international community and the administrations in their spectrum management, for example by providing information about the characteristics of a station for which measures to protect it from interference must be taken. A station, for which the notification has not been done properly, may find difficulties in getting due attention for protection requests. This notification process is done for all kinds of stations, both terrestrial and space-borne.

World Radiocommunication Conferences (WRCs) are the only conferences that have the authority to change the International Radio Regulations. WRCs are held every two or three years. Each WRC will develop and propose an agenda for the next WRC, as well as a provisional agenda for the WRC after that. The final decision on each WRC agenda rests with the ITU Council.

The main issues of the *WRC-95* conference were: the simplification of the ITU Radio Regulations, and new allocations to the Mobile Satellite Service (MSS), including feeder links.

Important issues at the *WRC-97* conference were: allocations for multimedia applications, satellite broadcasting, maritime issues, the problem of paper satellites, and continued pressure to consolidate more spectrum allocations for specific satellite applications.

Serious coordination efforts still need to be made to allow the implementation of many planned (mostly non-GSO) satellite systems and to allow the peaceful coexistence of new and existing systems. It appears that sharing in certain bands between different satellite systems and various terrestrial applications will be very difficult if not impossible. The demand for spectrum has risen dramatically with the possibility of terrestrial and satellite-based high-density data systems. Specific assignments were made for such applications up to frequencies of 66 GHz. Although many of these systems could still be far away in the future, astronomers are warned that the (currently interference-free) mm-wave spectral regions may soon have active applications. Other services, like those of terrestrial fixed and aeronautical radio navigation, are getting seriously worried about "harmful interference," which was strongly expressed at the conference.

Among the important results of the *WRC-2000*, we noted: the global assignment of 160 MHz to IMT-2000 (in Europe known as UMTS) in the frequency range 2.5 to 2.7 GHz to facilitate future developments of mobile communication; allocation of frequencies to the Radionavigation-Satellite Service (RNSS) in the frequency ranges 1.2, 1.3, and just above 5 GHz. These frequencies are intended to be used by the European civil satellite navigation system GALILEO. The Conference also adopted regulatory measures for sharing between the Fixed-Satellite Service (FSS) and terrestrial services in some frequency domains. The conference adopted resolutions asking for further study to conclude on this issue in the next WRC (i.e., WRC 2003). High-density applications in the Fixed Service (FS) and FSS were also a key issue and led to several decisions. For scientific research, the conference's main result was a complete reallocation of the frequencies between 71 and 275 GHz. The new allocations imply that frequencies allocated to science services have been identified in the most optimum way to comply with the propagation conditions of the terrestrial atmosphere and to minimize potential coordination and sharing issues with nonscience services. Also, to the greatest possible extent, space-to-Earth transmissions have been moved as far as possible from frequencies allocated to the science services.

A major result of the conference was the complete replanning of the Satellite Broadcasting Plan for the ITU-R Regions 1 and 3, that is, Europe, Africa, Asia, and the Pacific region.

The WRC-2000 also adopted provisional limits for transmissions from space to protect terrestrial science services, for example, radio astronomy. The resolution of this important issue for scientific research will also be completed by the next WRC.

The agenda of the *WRC-03* contains a number of items that are important for scientific research.[5] Among these, the most important are to:

- Finalize the work on spurious emission criteria in Appendix S3 of the ITU Radio Regulations for space services with regard to passive science services;
- Consider possible extension of the allocation to the mobile-satellite service (Earth-to-space) on a secondary basis in the band 14 to 14.5 GHz to permit operation of the aeronautical mobile satellite;
- Consider regulatory provisions and possible identification of existing frequency allocations for services which may be used by high altitude platform stations;
- Determine regulatory measures to protect radio astronomy against interference from the radionavigation-satellite service operating just above 5 GHz, consider allocations on a worldwide basis for feeder links in bands around 1.4 GHz to the non-GSO MSS with service links operating below 1 GHz, consider additional allocations on a worldwide basis for the non-GSO MSS with service links operating below 1 GHz;
- Consider technical and regulatory provisions concerning the band 37.5 to 43.5 GHz, that is, to protect radio astronomy from interference resulting from space-to-Earth transmissions from FSS applications in adjacent frequency bands.

The Radiocommunications Assembly (RA) is responsible for the organization and work program of the ITU-R study groups (SGs) to approve and issue ITU-R Recommendations and Questions developed by the study groups and suggest suitable topics for the agenda of future WRCs. The two main tasks of the study groups are:

- To prepare the technical basis for the Radiocommunication conferences;
- To develop ITU-R Recommendations on technical characteristics and operational procedures relating to the various radiocommunication

systems and services, and on associated issues of spectrum management. Thus, the role of the RA is to provide the technical basis for effective use of the spectrum and geostationary-satellite orbits, to recommend performance standards for radio systems and to ensure the effective and compatible interworking of systems, and to disseminate technical information.

6.1.4 ITU-R study groups

The study groups of the radiocommunication sector are responsible for carrying out the work in each area of activity identified in Table 6-3. This work includes the study of questions and the preparation of draft recommendations on the matters referred to them. Those draft recommendations are submitted for approval to a Radiocommunication Assembly or, between such conferences, by correspondence to administrations. In practice, most of the work is carried out in working parties (WPs) and task groups (TGs).[6] There is one study group for each main radio service or group of services, plus specialized study groups on spectrum management, interservice sharing, and radio wave propagation. Table 6-3 shows the different study groups of this sector with their working parties.

Much of the output of the Radiocommunication Assembly and the study groups provides technical input to the WRCs. To assist in this process, the radiocommunication sector requires an assembly to be held "associated in time and place" with each WRC. *A Special Committee on Regulatory/Procedural Matters*, SC, supports the preparation of a WRC.

In addition, a series of *Conference Preparatory Meetings* (CPMs) have been established to focus the work of the study groups in preparation for each WRC. The pattern is that a first CPM will be held soon after each WRC to ensure that the work program of the study groups takes into account the needs of the next WRC as set by its agenda. Some time before the next WRC takes place, a second CPM will be held to synthesize the study group's work into a report which administrations and others can take into account in making proposals to the WRC. The CPM itself is not expected to carry out technical studies. The final report of the CPM provides the technical basis for the WRC.

The duties of the Radio Regulations Board (RRB) consist of:

1. The approval of Rules of Procedure, which include technical criteria, in accordance with the ITU Radio Regulations and with any decision which may be taken by competent radiocommunication conferences;
2. The consideration of any other matter that cannot be resolved through the application of the Rules of Procedure;
3. The performance of any additional duties, concerned with the assignment and use of frequencies, in conformity with the function of the ITU-R Sector.

TABLE 6-3 Study Groups of ITU-R Sector

ITU-R Study Group	Subject
1	**Spectrum management** Working parties: *WP1A*: Principles and techniques for the effective use and management of the radio frequency spectrum *WP1B*: Spectrum sharing criteria and methods to enable the efficient use of the spectrum *WP1C*: Techniques for spectrum monitoring and related issues
3	**Radiowave propagation** Working parties: *WP3J*: Propagation fundamentals *WP3K*: Point-to-area propagation *WP3L*: HF propagation *WP3M*: Point-to-point Earth-space propagation
4	**Fixed-satellite service** Working parties: *WP4A:* Efficient orbit/spectrum utilization *WP4B*: Systems, performance, availability and maintenance *JWP4-9S*: Frequency sharing between the Fixed-Satellite Service and the Fixed Service *WP4SNG*: Satellite news gathering (SNG), outside broadcast via satellite *RG WP4B*: Performance requirements and asynchronous transfer mode technology (ATM)
6	**Broadcasting services** Working parties: *WP6B*: Digital coding *WP6E*: Terrestrial emission *WP6M*: Interactivity and multimedia *WP6P*: Broadcasting systems, production, baseband signals, and so on *WP6Q*: Quality assessment *WP6R*: Recording for broadcasting *WP6S*: Satellite broadcasting
7	**Science services** Working parties: *WP7A*: Time signals and frequency standard emissions *WP7B*: Space radio systems *WP7C*: Earth exploration satellites systems and meteorological elements *WP7D*: Radio astronomy *WP7E*: Interservice sharing and compatibility
8	**Mobile, radiodetermination, amateur and related satellite services** Working parties: *WP8A*: Land mobile service excluding IMT-2000; amateur and amateur-satellite service *WP8B*: Maritime mobile service including Global Maritime Distress and Safety Systems (GMDSS), aeronautical mobile service, and radiodetermination service

TABLE 6-3 Study Groups of ITU-R Sector (*Continued*)

ITU-R Study Group	Subject
	WP8D: All mobile satellite services except the amateur satellite service, and radiodetermination satellite service *WP8F*: IMT-2000 and systems beyond IMT-2000
9	**Fixed service** Working parties: *WP9A*: Performance and availability, interference objectives and analysis, effects of propagation, and terminology *WP9B*: Radio-frequency channel arrangements, radio systems characteristics, interconnection, maintenance and various applications *WP9C*: HF systems *WP9D*: Sharing with other services (except for the fixed-satellite service).
SC	Special Committee on regulatory/procedural matters
CCV	Coordination committee for vocabulary
CPM	Conference preparatory meeting

SOURCE: Ibid., 1.

6.2 U.S. Spectrum Management

6.2.1 Present arrangements and limitations

This section is not intended to be exhaustive, although it is intended to be sufficiently comprehensive to support the topics the readers would like to know about during the course of their research and investigations on U.S. Spectrum Management. In several cases, this section touches on topics but does not completely detail those topics. This occurs where the author did not delve too deeply into issues not considered vital to the discussion of his main inquiry. Readers desiring more information on specific units or agencies not covered here should contact those agencies directly. The information is only as accurate and as current as could be obtained from published sources describing some of the organizations, since the author did not specifically contact every organization addressed herein to verify the accuracy and currency of the information presented. The materials collected and the analyses conducted for the federal and independent agencies are reflected in Tables 6-4 and 6A-1.

6.2.2 Federal—Executive Branch

The National Telecommunications and Information Administration (NTIA) is the Executive Branch agency, with principal responsibility for developing and articulating domestic and international telecommunications policy.[7] NTIA conducts studies, makes telecommunications policy recommendations, and presents Executive Branch views on telecommunications issues to the Congress, the Federal Communications Commission (FCC), and the public.

The National Telecommunications and Information Administration manages the federal government's use of the radio spectrum, while FCC manages the spectrum for the state and local governments and the private sector.[8] At times FCC and NTIA's attempt to resolve key policy issues relevant to allocation of adequate spectrum to support both commercial interest and the critical missions of the federal agencies, such as national defense, protection of the president and foreign officials, assuring public safety of air and water transportation, federal law enforcement, disaster relief, protection of national resources, ensuring the security of power generation and nuclear material, the health and well-being of the military veterans, and the efficient operation of our postal service becomes protracted and contentious.

In the present regulatory environment, neither FCC nor NTIA has the ultimate decisionmaking authority over all spectrum in the United States. Instead, as depicted in Fig. 6-2, some 20 federal agencies rely on the Interdepartmental Radio Advisory Committee (IRAC) to receive frequency assignment that will allow them to develop solutions to key spectrum-management issues and fulfill their established missions. The mission staff in various agencies must identify and justify the need for a frequency assignment and complete all the required engineering and technical specifications for the applications. NTIA, with the approval of IRAC, authorizes spectrum use by agencies, while ensuring, among other things, that the assignment will not interfere with other users.

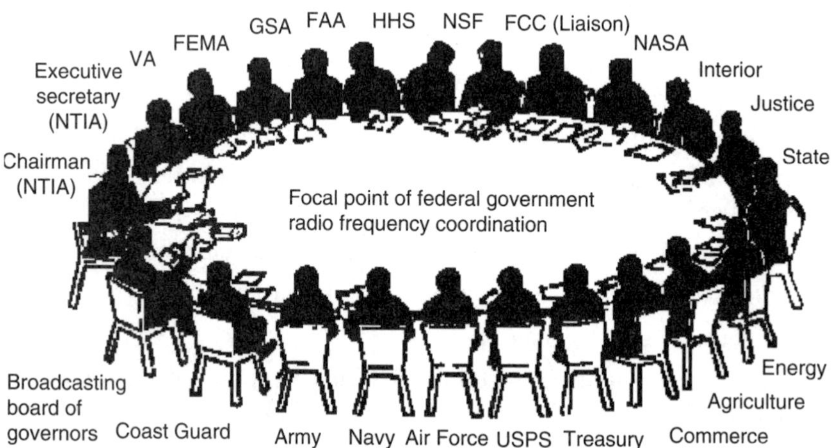

Figure 6-2 Interdepartment Radio Advisory Committee (IRAC).

Appendix 6A includes two summary tables, Table 6A-1 and Table 6A-2, that lists the different agencies' varied missions, their spectrum needs, and the way they must justify and review their assignment needs.[9]

6.2.3 State and local

There are several issues that concern the municipal, local, county, state, and tribal governments, including public rights-of-way, facilities siting, universal service, removal of barriers to competitive entry, public safety communications, and other issues regarding the management and/or implementation of the Telecommunications Act of 1996 that explicitly or inherently share intergovernmental responsibilities or administration with local, county, state, or tribal governments.

For example, the state and local public safety entities use land mobile radio (LMR) systems as a primary means of communication. LMR systems, like other wireless technologies, require radio spectrum to operate. As discussed in the preceding section, the FCC manages the nonfederal government use of the radio spectrum, including the spectrum used by state and local public safety entities. Therefore, the state and local agencies with public safety responsibilities must deal with the FCC to secure the necessary approvals for operating in portions of the radio spectrum. As Fig. 6-3 illustrates, these interactions could pose a serious challenge to persons who are not experts in spectrum management.

Therefore, underlying the FCC processes associated with frequency assignment, frequency administration, and spectrum allocation that help to ensure that state and local radio spectrum use is consistent with established spectrum policy is a formalized FCC structure of the Local and State Government Advisory Committee (LSGAC).[10] The LSGAC has a diverse representation of municipal, local, county, state and tribal governments.

1. Six elected municipal officials (city mayors and city council members).
2. Three elected county officials (county commissioners or council members).
3. One elected or appointed local government attorney.
4. One elected state executive (governor or lieutenant governor).
5. Two elected state legislators.
6. One elected or appointed public utilities or public service commissioner.
7. One elected or appointed Native American tribal representative.

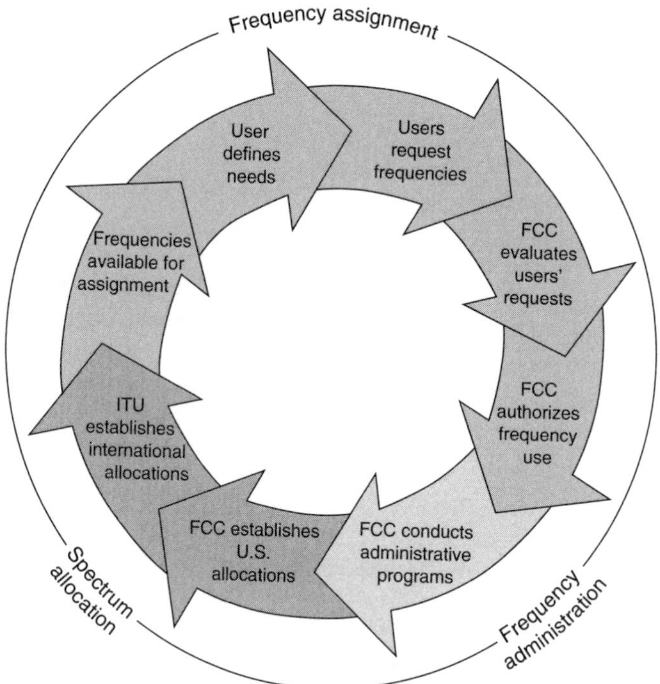

Figure 6-3 Cyclical Process of Spectrum Management. (*Source: State and Local Spectrum Management Process Report, Final, November 1998, FCC.*)

The underlying purpose of LSGAC is not to address individual consumer issues or complaints about telecommunications services. Instead, the members of the LSGAC are there to serve as key information providers to state and local government officials who need answers to questions on the FCC's rules and proceedings or who have comments about certain aspects.

As regards allocating spectrum to state and local public safety entities and to commercial industry, the FCC must consider several factors, including U.S. spectrum policy, international spectrum policy, and congressional influence before allocating spectrum. After completion of the processes depicted in Fig. 6-4, the FCC determines spectrum allocations for state, local, and commercial entities. The National Table of Frequency Allocations lists these allocations. Appendix F of the National Table of Frequency Allocations lists the spectrum bands allocated for use by state and local public safety entities.

Another federal government entity that also addresses the public rights-of-way, universal service, public safety communications, and other municipal, local, county, state, and tribal government concerns is

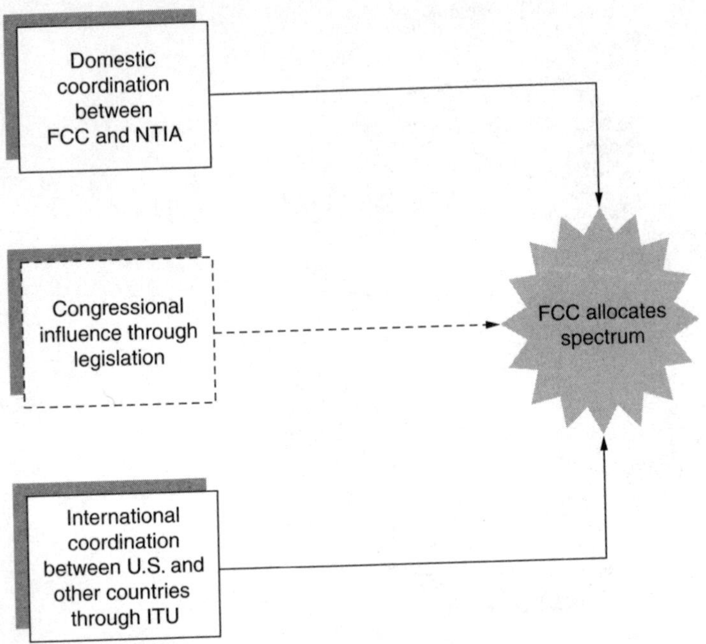

Figure 6-4 Spectrum Allocation Process. (*Source: Ibid.*)

the NTIA. For example, the Public Safety Program (PSP) organized under Vice President Gore's National Partnership for Reinventing Government (NPRG) is a program designed to help save lives by finding solutions to give people better, faster and more effective hazard warnings. To bring to fruition the PSP goals, federal government agencies involved in weather forecasting and disaster management are engaged in a cooperative venture with private sector telecommunications firms to develop alternative means of delivering weather and other disaster warnings leveraging modern telecommunications technologies, for example, advanced television receivers, wireline and wireless telecommunications devices, and the Internet. In this venture, NTIA has played a crucial role.

On July 17, 2000, in cooperation with the interagency working group, NTIA hosted a roundtable event. The invitees included consumer advocacy groups, state and local public safety officials and a cross section of the telecommunications industries. As a result of this roundtable event, a Hazard Ready Internet Service Provider (ISP) pilot project was undertaken, followed by a subsequent demonstration of the prototype of this system at an NTIA press conference in November 2000. Since that time NTIA has been enhancing its public safety program staff to:

1. Identify the long-range spectrum requirements for the next 10 years.
2. Ensure compatible sharing and interoperability among federal, state, and local entities where required.
3. Ensure that technical standards are available to ensure interoperability.
4. Identify the most appropriate options to fund state and local upgrade of their telecommunications systems.
5. Provide leadership and federal liaison with various public safety groups.
6. Increase the responsiveness to critical public safety issues.
7. Ensure adequate access in coordination with the FCC to allocate spectrum for public safety communications for the safe, effective, and efficient protection of life and property, consistent with the National Performance Review objectives.

The PSP has been primarily directing its efforts toward leadership, policy, and technical support to the Public Safety Wireless Network (PSWN) program and the Federal Law Enforcement Wireless Users Group (FLEWUG).[11] PSP helped coauthor and edit a number of PSWN program reports on:

1. State and local partnerships to address current public safety issues.
2. Spectrum.
3. Interoperability.
4. Pilot projects.
5. Advanced technology.

As a member of the federal public safety agencies, FLEWUG addresses issues relevant to federal public safety. The PSWN, as a program, supports much of the activity of the FLEWUG. Efficient and effective use of spectrum resources in a shared and interoperable environment of the future requires continued direct support to the PSWN/FLEWUG and other similar programs of concern to the municipal, local, county, state, and tribal governments.

6.2.4 Pressures from Congress and from industry

The wireless sector has witnessed tremendous growth and evolution in recent years. This growth has caused the U.S. Congress and industry to raise issues about the current spectrum management structure that relies on policies wherein government dictates the use of spectrum.

Their primary concern is whether the current structure is able to adequately address the needs of both the commercial and government users. In the past, while facing rapid growth and competition in wireless markets, commercial wireless providers had access to spectrum. Commercial wireless markets are still growing and developing new services and technologies that require additional spectrum. At the same time, the events of September 11, 2001, underscored the:

1. Importance of wireless communications.
2. Growing needs for spectrum for homeland security, national defense, emergency management, law enforcement, and public safety.

Many parties believe there are significant challenges to meeting the growing demands for spectrum within the current regulatory framework. Consequently, there are pressures from industry and Congress, as evidenced by the growing number of Congressional hearings, to address the spectrum issues. The author can hardly be expected to review these concerns in such details as would allow the readers to think that the subjects have been thoroughly discussed. Those interested in the comprehensive review of U.S. spectrum management with broad stakeholder involvement will find the statement of Peter Guerrero, Director of Physical Infrastructure Issues, General Accounting Office, particularly timely, relevant, and interesting.[12]

6.3 Telecom and Broadcasting by Regions[13]

Between the broad framework established at the global level by the ITU and the detailed frequency planning necessary for national administrations, there has always been, in several other parts of the world, a need for regional coordination. The widespread introduction of pan-regional radio services depends on harmonization of the radio frequency allocations, the adoption of common approaches to the conformity assessment of radio equipment and agreement on mutually acceptable procedures for cross-border licensing. Even when radio services are not intended for operation on a pan-regional basis, there are advantages for users, manufacturers and regulators in harmonizing frequency usage and radio regulatory regimes. The forum for achieving such harmonization in Europe is the *Electronic Communications Committee* (ECC), of the *Conference of European Posts and Telecommunications Administrations* (CEPT). In the Americas it is CITEL and in the Asia Pacific region, the Asia Pacific Telecommunity (APT). In other regions of the world similar organizations are emerging. This section describes the regional organizations that are most developed in Asia, Africa, Europe, and other continents.

6.3.1 Africa

Pan African Telecommunications Union (PATU). Structured along the lines of the ITU, Pan African Telecommunications Union[14] represents the telecommunications division of the Organization of African Unity (OAU). PATU contributes to the standardization of telecommunications networks and the coordination of telecommunications services among member states. It conducts studies in telecommunications and in other relevant fields with a view to making recommendations and reports to member states. PATU publishes the information and research materials and also to benefit the member states, it encourages the exchange of information and staff between the administrations of member states.

Pan African Telecommunications Union harmonizes the position of member states during international meetings, while coordinating the program planning, feasibility studies, and the adoption of the international standards recommended by the ITU that are of interest to member states or assigned to it by the Organization of African Unity (OAU). South Africa has observer status in PATU, and its full membership is under consideration.

6.3.2 Americas

Inter-American Telecommunication Commission (CITEL). CITEL is an entity of the *Organization of American States* (OAS), and its purpose is to use all the means at its disposal to facilitate and further the development of telecommunications in the Americas to contribute to the development of the region. CITEL aims to provide a forum for discussion and coordination on developing technical, operational, and service standards for global systems and for preparing for international conferences which allocate spectrum to these services. At present, there is no forum in CITEL for harmonizing other elements of policy and regulation that would normally be expected to govern the operations of these systems (e.g., licensing, interconnection, competition policy, tariffs, and so on).

CITEL now brings together 34 nations of North, Central, and South America and the Caribbean. The headquarters of the OAS and the executive secretariat of CITEL are in Washington, DC.

Until now, CITEL's fundamental mission has been technical. CITEL is the only regional organization open to all OAS member states with a thorough technical knowledge of telecommunications, not only because member countries are represented by telecommunication administrations, but because recently the door has been opened to active participation by the private sector, which can participate in CITEL as associate members. CITEL's goal is to continue to develop and strengthen its role in standards coordination, in radiocommunication, and in telecommunication development activities, bearing in mind the differences between the ITU and CITEL. The ITU is a global telecommunication body and as such it addresses problems from a global perspective.

CITEL on the other hand was established to address regional issues as they impact and are impacted by global issues.

To obtain these objectives in CITEL, the administrations work in close collaboration with the private sector and coordinate with regional and international organizations. CITEL provides the private sector with the opportunity to participate in the definition and execution of its program of activities. CITEL is actually providing the telecommunications industry and member countries with a forum where they can take an active role in establishing standards in the Americas, harmonizing equipment certification procedures, liberalizing regulatory structures in member countries, promoting competition and privatization and coordinating spectrum allocation and use. Participation in CITEL also provides the private sector with additional access to the markets within the Americas.

CITEL fulfills its objectives through the following organs: the CITEL Assembly, the Permanent Executive Committee (COM/CITEL), the Permanent Consultative Committees (PCCs), and the Secretariat. The organs shall include such committees, subcommittees, working groups and ad hoc groups as may be established when required.

The CITEL Assembly establishes the Permanent Consultative Committees (PCCs) that it considers necessary to attain the CITEL objectives together with specific terms of reference for each PCC. A PCC shall continue in force until such time as the CITEL Assembly itself, or COM/CITEL deems its functions and purpose to be concluded.

CITEL Committees. At present, CITEL has four Committees: COM/CITEL, PCC.I, PCC.II, and PCC.III. The objectives and mandates of these PCCs are summarized as follows:

PCC.I: Public Telecommunications services. PCC.I acts as a technical advisory body within the Inter-American Telecommunication Commission with respect to standards coordination, planning, financing, construction, operations, maintenance, technical assistance, equipment certification processes, rate principles, and other matters related to the use, implementation, and operation of public telecommunications services in the CITEL member states.

In accordance with the ITU Regulations PCC.I undertakes the coordination of regional preparations for major ITU Conferences and meetings, including the preparation of common regional proposals (IAP) and positions when deemed appropriate.

PCC.II: Broadcasting. PCC.II acts as a technical advisory body within the Inter-American Telecommunication Commission with respect to standards coordination, planning, operation, and technical assistance for the broadcasting service in its different forms.

PCC.III: Radiocommunications. PCC.III acts as a technical advisory body within the Inter-American Telecommunication Commission with respect to standards coordination, planning, and full and efficient use of the radio spectrum and satellite orbits, as well as matters pertaining to the operation of radiocommunication services in the member states.

In accordance with the ITU Radio Regulations and taking into account ITU Recommendations, PCC.III has the mandate to:

- Promote harmonization in the use of the radio-frequency spectrum and the operation of radiocommunication services in the member states, bearing especially in mind the need to prevent and avoid, as far as possible, harmful interference in radiocommunication services;
- Foster the development and implementation of modern technologies and new services in the field of radiocommunication to meet the needs of member states, in conjunction with a more efficient use of the spectrum;
- Undertake a coordinated effort with the different CITEL Groups in those areas that by their very nature lend themselves to joint action;
- Undertake the coordination of regional preparations for major ITU conferences and meetings, including the preparation of common regional proposals (IAP) and positions when deemed appropriate.

CITEL and standardization. CITEL has a *Working Group on Standards Coordination*, WGSC. This working group is not a standards-making body. Instead, the terms of reference for the WGSG are to facilitate network interconnectivity and interoperability on a regional as well as on a global basis.

Recently, ETSI has offered to establish a Memorandum of Understanding, MoU, with CITEL on cooperation between the two organizations. A decision of the assembly of CITEL on this issue was deferred because of strong opposition by the United States. The United States considered a formal structural relationship between CITEL and ETSI unnecessary and stated that the mandate of the WGSG allows them to consider standards from all recognized standards bodies with a preference for ITU standards.

North American National Broadcasting Associations (NANBA). The North American Broadcasters Association (NABA), founded in 1972, was formerly known as the North American National Broadcasters Association.[15] The association is actively involved in issues related to spectrum allocation, broadcasters' use of digital technology, intellectual property, as well as distribution technologies including satellites and fiber. NABA represents North American broadcasters as a nongovernmental organization at the World Intellectual Property Organization (WIPO), and the ITU. It is one of the eight members of the World Broadcasting Unions (WBU).[16] One of NABA's five key areas of activity is spectrum related issues: preservation of broadcasters' spectrum both terrestrial and satellite.[17]

6.3.3 Arab

The Arab States Broadcasting Union (ASBU) is a professional, non-profit organization that develops, coordinates, and investigates all matters likely to impact the broadcasting sector.[18] Its primary goals[19] are to:

- Prepare broadcasting radio frequency tables based on international technical regulations, systems, and standards.
- Provide necessary training, in cooperation with international organizations, to the ASBU members to the use of broadcasting radio frequency tables.
- Promote and defend Arab countries' radio frequency needs in international forums.

It organizes the following types of engineering seminars, symposiums, and workshops to train the ASBU members: High-Frequency coordination, the World Radio Conference, Planning of Digital Télévision Terrestrial Broadcasting—just to name a few.[20]

6.3.4 Asia-Pacific

Asia Pacific Telecommunity. The APT, established in 1979, is a Regional Telecommunication Organization established by an Intergovernmental agreement. Membership in the APT is open to any state within the Asia-Pacific region which is a member of the United Nations or the Economic and Social Commission for Asia and the Pacific (ESCAP). Associate membership in the APT is open to any associate member of the ESCAP. In addition, affiliate membership in the APT is open to any provider of telecommunication services to the public within the region that is nominated for affiliate membership by a member or associate member. The APT now has a strength of 32 members, 4 associate members, and 47 affiliate members.

In accordance with a resolution of its General Assembly, APT invites the private sector and government-owned companies, academia, research organizations, consulting companies, to participate in the activities of APT such as seminars, fora, exhibitions, study groups activities. Under this resolution, 45 companies/organizations have joined APT since 1993.

In the past few years, the APT has very successfully facilitated the representation of the Asia-Pacific region at international fora. It has effectively expressed and represented the collective views of the region. As a consequence, the APT's profile and stature has been raised considerably in recent years at events such as the ITU-R World Radiocommunication Conference, the World Telecommunication Policy Forum and the ITU Plenipotentiary Conference. The APT will strengthen its role as coordinator of regional positions for major ITU meetings and processes and serve as a valuable information resource for member countries as well as coordinating regional issues on developments in the international arena.

Within the region, the APT will coordinate policy issues relating to the development of the Asia-Pacific Information Infrastructure. In May 2000, the APT and CEPT reached a cooperation agreement in the spirit of the purposes of the ITU in promoting and providing a framework for:

- Exchange of information and documentation.
- Coordination of positions. Strengthening the relations between APT and CEPT and with the ITU.
- The exchange of views during preparations for ITU conferences and meetings.
- Development of telecommunications in the two regions.

Objectives of APT. Among the objectives of the Telecommunity, the following items are noted:

14317. To correlate planning, programming and development of telecommunication networks;

14318. To promote the implementation of all agreed networks;

14319. To assist in development of national components of efficient networks;

14320. To foster coordination within the region of technical standards and routing plans;

14321. To seek adoption of efficient operating methods in regional telecommunication service.

In furtherance thereof, the telecommunity may:

14332. Undertake, in coordination with the International Telecommunication Union, when pertinent, technical and other studies relating to developments in telecommunication technology of common interest to its members and associate members;

14333. Encourage the exchange of information, technical experts and other specialized personnel amongst the telecommunication organizations of its members and associate members;

14334. Study the feasibility of transfer of technology in the field of telecommunications amongst its members and associate members;

14335. Arrange the provision of short-term technical assistance to its members and associate members, when so required;

14336. Advise its members and associate members in the assessment of their needs with respect to telecommunication personnel and programs for training;

14337. Promote, in cooperation with appropriate international organizations concerned with telecommunications in the region, the establishment within the region of telecommunication training institutes of a regional and multinational character;

14338. Promote and assist in the formulation and implementation of bilateral or multilateral telecommunication programs within the region in cooperation with appropriate international or regional organizations.

The general principles and policies for the fulfillment of the objectives of the Asia Pacific Telecommunity. The Telecommunity aims to be a vehicle for the sharing of information and for giving support to governments in the process of deregulation and liberalization, expanding telecommunication networks, development of information technologies and changing policies to accommodate global trading and commerce initiatives in which telecommunications is increasingly a part of or at the center. The telecommunity can be the vehicle for clarifying international telecommunications issues, forming regional policies and advancing these positions in global fora.

An active APT role in the interchange of information and experiences, the development and distribution of techniques via seminars, dialogues, courses, databases, compendia and expert systems will assist these developments throughout the region. The need to tap the expertise and resources of the private sector will be a key mechanism in this process of transition. The development of monitoring techniques is part of this process.

The Asia Pacific Telecommunity will continue to promote APT standardization, research and development efforts and technology development in the APT region accompanied by technology transfer and joint projects for telecom equipment production, consultative services and execution of telecom projects. There are possibilities of considerable cooperation amongst APT countries in meeting technical challenges and many other issues associated with software development and software modifications. This cooperation is seen in the active participation of the private sector in the recently created *Asia-Pacific Telecommunity Standardization Program* (ASTAP) forum. Since standardization is essential for realizing the interoperability of any information networks, the APT would:

- Strengthen its standardization activities through promoting research and development collaborative activity, joint projects and joint research among members including industry members with a view to promoting the integration of the research results into global standards;
- Make every effort to encourage industry members to participate more actively and at the same time take full advantage of the know-how they have to offer.

Technology transfer and human resources development are important activities in order to ensure a balanced development of telecommunication services in the Asia-Pacific region. The APT should pursue such activities including holding seminars and meetings taking account of the real needs of developing member countries.

APT work program. The APT General Assembly instructs the APT Management Committee to plan and implement the APT work program focusing on the distinguished priorities to deliver the maximum benefits to its members. In addition, the General Assembly instructs the Management Committee to develop guidelines in relation to the level of contributions of the industry members in the ongoing analysis of the financial viability of the telecommunity.

The APT pays special attention in promoting and facilitating the development of the telecommunication and information infrastructure in the Asia-Pacific region. Several AII pilot projects have been initiated which are expected to encourage the development and promotion of Asia-Pacific Information Infrastructure (AII). Thus, AII envisioned the promotion of construction of high quality network infrastructure through the National Information Infrastructure (NII) initiatives by the members, the advancement of applications for supporting various information and multimedia systems, the development of technologies for developing human resources, and greater information dissemination to APT members by publishing the AII Compendium. The recent selection of new AII Pilot Projects for implementation was conducted by the AII Expert Group of the APT. As of now, there are 29 AII pilot projects that involve telecommunication application, multimedia and online services of commerce, education, health and socio-political issues.

Another special interest of APT is the Industry and Users' Forum. There has been a dramatic change of telecommunication industry and user scenario since the birth of APT in 1979. The service provider originally owned by the member state has largely been liberalized and privatized, giving way to affiliate members and leaving the state as either a regulator or a political representative. The actual players of telecommunications are the private entities or the affiliate members. The ITU has a private sector forum in the ITU-D to recognize the role played by sector members. APT, with similar foresight, has formed an Industry and Users' Forum in line with Rule 23 of the Management Committee Rules of Procedures.

The APT has currently the four study groups that together with the key subjects are:

Study Group 1: Policy and regulation

Study Group 2: Network

Study Group 3: New services towards business creation

Study Group 4: Internet and E-Commerce

Questions dealing with a large variety of issues ranging from market liberalization via network technology and multimedia services to e-commerce issues are addressed in these study groups.

Asia-Pacific Economic Cooperation (APEC). APEC is a unique forum for facilitating cooperation, trade, investment, and economic growth in the Asia-Pacific region. Its membership includes 21 economic jurisdictions, with a combined GDP of 19 trillion U.S. dollars, accounting for 47 percent of world trade.

Since its inception in 1989, APEC has been facilitating practical policy formulation, including technical cooperation in the Asia-Pacific region. In 1990, APEC created the Telecommunications and Information Working Group (TELWG), with a set of *TEL Sponsorship Guidelines*. The business/private sector participates in all TELWG activities and initiates and drives many projects either independently or in cooperation with the public sector. The ongoing activities of TELWG, among other things, include:

- Maintaining the TELWG website on radio-frequency spectrum management policies and practices in member economies.[21]
- Updating compiled information on legal and regulatory environment for telecommunications in all economies.[22]

The Asia-Pacific Broadcasting Union (ABU). The ABU is an association of the television and radio networks in the Asia-Pacific region. It represents its members internationally on frequency management, standards-setting and other policy areas.[23]

Founded in 1964, the ABU now has 100 members in 49 countries and areas, including a number of associate members in Europe and North America. It undertakes wide-ranging activities pertaining to the current needs of the members. The Secretariat's Technical Department implements these activities[24] while the Technical Committee, which includes all ABU members, decides all policy matters.

A director assisted by two senior engineers heads the Technical Department.

The Technical Committee holds a yearly meeting, usually in the last quarter. Meeting participants include the ABU members, affiliates, and a number of invited observers from several International Organizations. At these meetings, decisions and recommendations are made based on proposals submitted by the members. Also, the meetings provide a forum for exchange of information on technological developments.

The Technical Committee addresses four issue areas: spectrum, resources and services, production, and transmission under the supervision of four topic chairmen. Table 6-4 identifies and lists the current technical activities.

6.3.5 EU spectrum management

European Conference of Postal and Telecommunications Administrations (CEPT) / European Telecommunications Standards Institute (ETSI). Established in 1959, the European Conference of Postal and Telecommunications Administrations' (CEPT) activities include, among other things, cooperation on technical standardization issues. In 1988, CEPT created the European Telecommunications Standards Institute (ETSI) and transferred all its telecommunication standardization activities.

The underlying European policy is to separate postal and telecommunications operations from policy-making and regulatory functions. Thus, CEPT has become a body of policy-makers and regulators while the postal and telecommunications operators have created their own organizations: Post Europe and ETNO. With its 45 members, CEPT now covers almost the entire geographical area of Europe and offers its members[26] the opportunities to influence developments within ITU and UPU in consonance with European goals.[27]

The CEPT has three committees:

1. Comité européen des régulateurs postaux (CERP) deals with postal matters.
2. European Radiocommunications Committee (ERC).
3. European Committee for Regulatory Telecommunications Affairs (ECTRA).

ERC and ECTRA address the telecommunications issues including harmonization and adopt recommendations and decisions.

In May 1991, the ERC established a permanent office in Copenhagen, the European Radiocommunications Office (ERO), which supports the activities of the committee and conducts studies for it and for the European Commission. In September 1994, ECTRA also established a

TABLE 6-4 Current ABU Technical Activities

Technical activities	Activity details
Coordination of shortwave frequencies	A High Frequency Coordination Group, ABU-HFC, provides a forum for international radio broadcasters in the region to mutually coordinate the use of the shortwave frequency bands on a seasonal basis. This activity minimizes interference in shortwave services.
Activities related to the ITU	Allocation of frequencies for broadcasting services and setting technical standards on behalf of the members.
Interunion activities	The ABU, as an active member of the Technical Committee established by the WBU[25], participates with eight other unions from around the world to cover broadcasting interests: HF Coordination, Digital Radio, Digital Terrestrial Television Transmission
Technical studies	Harmonization of technical standards and operational practices in the region. This activity is becoming increasingly important because of the rapid changes in technology.
Technical Advisory Service (TAS)	Assistance to the smaller ABU members in improving their standards of service and selecting technologies and formats most suitable for their particular applications. This assistance includes short-term consultancies and expert advisory services from the more advanced members for, free of charge. A set of rules established by the Technical Committee governs the TAS' mission undertaken each year.
Technical information	Dissemination of technical information to the members, including the publication of the ABU Technical Review and regular distribution of clippings containing a selection of the most useful papers published in technical journals worldwide.
Publications	Publication of monographs and manuals on subjects of particular importance to ABU members.

permanent office in Copenhagen: the European Telecommunications Office (ETO) for the same purpose.

Since September, 2001[28], the CEPT has been more active as a forum for strategic planning, decision-making, and preparing for the ITU conferences.[29]

The European Broadcasting Union (EBU). The EBU, founded in February, 1950, by Western European radio and television broadcasters, has 71 active members in 52 countries of Europe, North Africa, and the Middle East, and 45 associate members in 28 other countries. In 1993, the EBU merged with the OIRT—its counterpart in Eastern Europe.[30]

Working on behalf of its members in the European area, the EBU provides a full range of operational, commercial, technical, legal and strategic services. The EBU also works in close collaboration with sister unions on other continents.

In matters pertaining to technical and/or legal challenges, such as harmonization of the use of the radio frequency spectrum to avoid interference and to guarantee the free circulation of receivers, the Technical Department promotes and defends the members' interests vis-à-vis the relevant international organizations and in professional forums. The EBU considers that the appropriate forum for obtaining agreements for radio spectrum policy for broadcasting allocations and assignments is the ITU, with initial European level coordination via the CEPT. The EBU Technical Department reviews position papers on radio spectrum policy and makes recommendations on how to achieve coordination of both allocations and assignments throughout an area which is larger than that represented by the European Union.[31]

6.4 Appendix A: US Spectrum Management

TABLE 6A-1 US Spectrum Management—Executive Branch

Federal agency	Missions	Present arrangements and limitations
United States Department of Commerce (DOC)		
National Telecommunications and Information Administration (NTIA)	NTIA's mission includes distribution of spectrum to federal agencies; engineering and analysis relative to interference resolution; and management of the Interdepartment Radio Advisory Committee (IRAC).[32]	The NTIA Manual of Regulations & Procedures for Federal Radio Frequency Management is the guidebook for frequency authorization in the United States and Possessions. Within the manual the required information is defined and the standards and guidelines are provided. The process for filing with the NTIA is provided in detail in Chapter 9 of the NTIA manual.[33]
	The NTIA is also responsible for maintaining the National Table of Frequency Allocations.	Coordination between nongovernment and Government users of the RF spectrum is accomplished by joint meetings of the FCC and the NTIA.
Interdepartment Radio Advisory Committee (IRAC)	IRAC's mission includes maintenance of the National Table of Frequency Allocations.	NTIA performs its functions through the assistance of the Interdepartment Radio Advisory Committee (IRAC).[34] Coordination between nongovernment and government users of the RF spectrum is accomplished by joint meetings of the FCC[35] and the NTIA.
United States Department of Defense		
Army	The Army Spectrum Manager (ASM) directs Army-wide spectrum management activities, develops and implements spectrum management policy, and allocates frequency resources (frequency assignment) to support the Army. The ASM serves as the principal advisor	Spectrum assignments to be used within the Army include: a. Air traffic control. b. Amateur frequencies. c. Land mobile radio (LMR). d. Citizen band (CB) radios.

	to the Director of Network Enterprise Technology Command/9th Army Signal Command in regard to radio frequency spectrum management and radio regulatory matters.[36]	e. Special considerations for CONUS HF requests. f. Maritime mobile (MM) frequencies. g. Specialized mobile radio (SMR) service. h. Trunked land mobile radio (LMR) system. i. Satellite communications.[37] The Army is authorized by Chap. 7 of the NTIA Manual[38] to use frequencies in certain nongovernment bands to meet peacetime tactical and training requirements, as well as certain other bands for test range requirements. Frequencies are assigned by supporting AFCs only when spectrum requirements cannot be satisfied in government bands and when operation will not cause interference to nongovernment service.[39] The Army accepts interference caused by authorized nongovernment users. Military use of a particular frequency in these bands does not preclude new nongovernment assignments on that frequency.
Navy	The Naval Electromagnetic Spectrum Center (NAVEMSCEN) is the Department of the Navy's primary organization responsible for implementation of electromagnetic spectrum policy. NAVEMSCEN utilizes current telecommunications and information technology to effectively and efficiently provide Navy and Marine Corps customers worldwide assured access to the electromagnetic spectrum, and Spectrum Support both nationally and	A frequency assignment is required for all government owned Communications-Electronic (C-E) systems or C-E equipments that require the use of the radio frequency spectrum. The assignment authorizes the system or equipment to operate on a discrete frequency or group of frequencies, within a specified set of constraints, such as power, emission bandwidth, location of antennas, and operating time of day. Authority for approval of the assignment authorizing the use of radio

(*Continued*)

TABLE 6A-1 US Spectrum Management—Executive Branch (*Continued*)

Federal agency	Missions	Present arrangements and limitations
	internationally, through spectrum certification and frequency assignment processes.[40]	frequencies by Navy and Marine Corps activities within the United States and Possessions (US&P) rests with the Administrator, NTIA. NAVEMSCEN, as delegated by the Chief of Naval Operations (CNO) and Commander Naval Network Operations Command (CNNOC), is the only Naval agency authorized to coordinate frequency assignment issues with the NTIA. International law recognizes that the use of the radio frequency spectrum within its territory is a sovereign right of each nation. Navy and Marine Corps activities operating within the territory of a Host Nation must obtain frequency assignments from the Host Nation through the respective Theatre Commander-in-Chief (CINC)/Joint Frequency Management Office (JFMO). U.S.-based activities that have operational deployment or training requirements within another nation's sovereign territory must request frequency assignments as indicated above unless otherwise directed by Theatre CINC or National Command Authority (NCA). These assignments must comply with the Host Nation's rules, regulations and procedures. All Navy and Marine Corps commands, organizations, and activities are required to follow the spectrum management policies and procedures promulgated by the CNO and CNNOC.[41]

Air Force	Air Force Frequency Management Agency (AFFMA) "plans, provides, and preserves access to the radio frequency spectrum for the Air Force and selected DoD activities in support of national policy objectives, systems development, and global operations through analysis and negotiation with international and national, civil and military organizations."[42]	DoD does not own any spectrum, but is allowed to use certain bands. Most US spectrum is neither government nor nongovernment, but is shared spectrum. FCC represents nongovernment users; Department of Commerce (NTIA) represents Government users. Equipment approved in U.S. cannot automatically be used overseas; frequency allocations are not the same around the world. All equipment and frequencies must be approved by host nation All AF equipment that uses the spectrum (including FCC "Part 15" devices) must be coordinated through AFFMA before it is purchased or used. "Landing Rights"—slang for official approval from a foreign country to use specific electromagnetic equipment and frequency in their country.
United States Department of State	The Communications and Information Policy (CIP) Division at the US Department of State has the authority and the ultimate responsibility for establishing foreign telecommunications policy. CIP's mission is to address spectrum issues that impact military readiness, national defense, and homeland security. To that end, CIP identifies the needs early so that the office may engage in an aggressive outreach effort to garner support for these important proposals from other countries.[43]	CIP develops positions and proposals[44] to the International Telecommunication Union's World Radiocommunication Conference. The CIP division takes the following steps when preparing for WRCs: Immediately following a WRC, the first session of the international Conference Preparatory Meeting (CPM) is held. Delegations review the agenda items adopted by the preceding WRC and identify areas needing further technical study before spectrum decisions can be supported at the upcoming WRC. These areas

(*Continued*)

TABLE 6A-1 US Spectrum Management—Executive Branch (*Continued*)

Federal agency	Missions	Present arrangements and limitations
		are then assigned to the ITU's radio sector study groups. To the extent possible, the outline of the technical report that the second session of the CPM will forward to the next WRC is developed. Coordinators for the various chapters are named. CIP establishes a "Principals" group to decide the U.S. position in cases where technical assessments alone may not suffice, or where national security issues mandate an early decision. CIP forms an intra-governmental Core Delegation to the next WRC. This group reports to the Principals group. CIP draws upon technical preparation by the International Telecommunication Advisory Committee-Radiocommunication Activity (ITAC-R). The General Services Administration (GSA) charters it to the Department of State as an Advisory Committee under the Federal Advisory Committee Act (FACA). CIP participates in the inter-agency preparatory process. Here, preliminary views followed by draft proposals are generated. Foremost among them are proposals going to national defense and homeland security issues. A Head of Delegation is named who becomes a Department of State employee until the WRC ends.

United States Department of Agriculture (USDA)	USDA's mission is to adhere to federal regulations and procedures for spectrum management when establishing and operating wireless communications systems and services. The USDA Forest Service represents USDA to the NTIA Office of Spectrum Management and on the IRAC.[45]	Requirements for frequency assignment are submitted to the USDA IRAC Representative. Frequency requirements for new systems, or major upgrades, are planned in advance and the request for frequency assignment is forwarded to the USDA IRAC Representative at least one year prior to the planned procurement. Sharing of USDA systems among agencies is considered in the development of new or in the expansion of an existing system.[46] Such sharing is required when spectrum availability is limited, unavailable, or the traffic does not justify a separate system. NTIA mandates that agencies consider the use of commercial services in any system planning.[47] Arrangements permitting cooperative communications, of mutual benefit, between agencies on each other's authorized radio frequencies may be made by a MoU (between federal units) or a Cooperative Agreement (between federal and non-federal units) signed by the responsible official having jurisdiction. A formal MoU between units within USDA agencies or staff offices is not required. Copies of the MoU are kept on file with the agency or staff office and the USDA IRAC Representative. Verbal or written authorization of use, within the limits of the holder's authorization or license, may be given by the holder for a short duration and specific purpose without prior coordination with the USDA IRAC Representative. Frequencies installed under

(*Continued*)

TABLE 6A-1 US Spectrum Management—Executive Branch (*Continued*)

Federal agency	Missions	Present arrangements and limitations
		this arrangement are removed from the radio equipment at the conclusion of the time authorized. Arrangements between agencies or staff offices within USDA, or with external agencies (federal or non-federal), are forwarded to the USDA IRAC Representative when authorizations or licenses are required. Requests for radio frequency assignments that are identified as Freedom of Information Act (FOIA)-exempt, are accompanied by a justification letter stating the specific exemption(s) under the FOIA.
United States Department of Energy (DOE)	The mission of DOE Spectrum Management is to lead the Department in managing and protecting radiocommunication and spectrum dependent resources by: (1) representing DOE in national and international forums; (2) developing and executing policy, plans, procedures, and technical criteria to provide advice, assistance, and guidance to the DOE Lead Program Secretarial Offices, Headquarters Program Offices, and field sites on the use of federal radio spectrum systems and services; (3) administering the effective planning and certification of spectrum dependent systems, and the assignment and efficient use of radio frequency authorizations that support DOE radiocommunication facilities and operations; and (4) reengineering	DOE has a large investment in radiocommunication systems that depend on protected use of the spectrum. This investment is represented by more than 9500 frequency assignments authorized by NTIA. All DOE radio frequency authorization requests must be approved by the office of the Chief Information Officer (CIO). The CIO is also responsible for appointing representatives to serve as points-of-contact to NTIA. These representatives serve on the various committees, subcommittees, ad hoc groups, and working groups which assist NTIA in assigning frequencies not only to DOE, but all other federal agencies as well. In addition, the CIO appointed representatives also help NTIA in developing and executing national policies, programs, procedures, and

	business processes for improved efficiency and effectiveness through the application of information technology.[48]	technical criteria affecting the allocation, management, and use of the radio spectrum in the best interests of DOE, the federal government, and the nation. Requests for DOE radio frequency authorization (RFA) assignments are submitted to the Office of the CIO. The DOE Spectrum Management Program office, under the Office of the CIO, will validate the RFA proposal for regulatory and technical compliance, and forward the request to NTIA. The request will be scrutinized by NTIA and the IRAC's Frequency Assignment Subcommittee (FAS), to insure that the RFA proposal meets national spectrum use guidelines, and does not cause, or is subject to receive, interference in the existing electromagnetic environment. If the RFA proposal meets NTIA regulatory standards, and is unanimously voted for approval by the FAS, then NTIA will grant a frequency assignment to the applicant.
United States Department of Interior (DOI)	The National Park Service (NPS) obtains and maintains effective wireless telecommunications systems, which comply with all relevant standards and authorities. These standards and authorities permit government stations to "use such frequencies as shall be assigned to each or to each class by the president … and shall conform to such rules and regulations designed to prevent interference with other radio stations and the rights of others as the (Federal Communications Commission) may prescribe."[49]	The regulations governing radio frequency management are in the Department of Commerce publication, "Manual of Regulations & Procedures for Federal Radio Frequency Management." Requirements for NPS facilities are found in that manual and in the Radio Handbook accompanying Departmental Manual Part 377 (377 DM). The DM states: The formulation and enunciation of national telecommunication policies designed to ensure achievement of the national objectives is an

(Continued)

TABLE 6A-1 US Spectrum Management—Executive Branch (*Continued*)

Federal agency	Missions	Present arrangements and limitations
		essential element of the role of the federal government. Telecommunication policies are made by the Congress, by the Court, by the President and the Assistant Secretary of Commerce for Communications and Information with respect to the agencies and establishments of the federal government, and by the FCC for the public. Policy is made through treaties to which the United States adheres with the advice and consent of the Senate, through executive agreements, by executive departments and agencies in the discharge of their telecommunication responsibilities, and by custom and precedent. These policies may be separated into three categories: (1) National Telecommunication Policy; (2) Telecommunication Policy applying to the agencies and establishments of the federal government; and (3) Federal Communications Commission Telecommunication Policy. Operations of the NPS on wireless devices are governed by any or all of the foregoing categories. When the authority is granted by the NTIA to operate on a frequency, a radio frequency assignment is issued to the NPS through DOI. Full technical compliance with the parameters of the authorization is required. The Service-wide Radio Program Coordinator forwards the

		assignment to the superintendent, spelling out the most salient features/ imitations of the authorization and stipulating additional terms and conditions relating to the assignment, if any.[50]
United States Department of Justice	The mission of the Department of Justice (DOJ) is ". . . to enforce the law and defend the interests of the United States according to the law; to provide federal leadership in preventing and controlling crime; to seek just punishment for those guilty of unlawful behavior; to administer and enforce the nation's immigration laws fairly and effectively; and to ensure fair and impartial administration of justice for all Americans."	Wireless communications play an integral role in carrying out the mission both in direct tactical communication between law enforcement officers and in critical investigative tools used in an environment of sophisticated adversaries. The Justice Management Division, Wireless Management Office (WMO) coordinates departmental spectrum usage and provides the national level representation to the Interdepartment Radio Advisory Committee and its various subcommittees and ad hoc Groups. DOJ components that develop, implement, or utilize radiocommunication systems are responsible to provide the WMO with the information necessary to ensure regulatory compliance and obtain authorization. Each component is responsible to designate a spectrum coordinator who, among other things, interfaces with the WMO spectrum management office. Each DOJ component spectrum coordinator is assigned to a member of the WMO staff who serves as their point of contact for spectrum requirements. The POC provides direct support to the component's unique requirements and provides guidance for their future spectrum

(Continued)

TABLE 6A-1 US Spectrum Management—Executive Branch (*Continued*)

Federal agency	Missions	Present arrangements and limitations
		dependent radiocommunication program development.
United States Department of Transportation		
Coast Guard	In matters pertaining to spectrum management[52], the Coast Guard's mission is to identify the cost and impact to Coast Guard operations of proposed spectrum reallocations of any portion of RF spectrum used by the Coast Guard.[53]	As directed by the Congress, the Coast Guard is required to fund a portion of the NTIA operating budget based on the number of radio frequency assignments it has. This fee is based on a percentage of the NTIA approved budget, currently about (dollars) 60 per assignment. The Coast Guard has approximately 17,000 assignments. NTIA spectrum fees cost the USCG about (dollars) 1 million annually. Among the potentially costly problems the Coast Guard faces is the use of communications and navigation equipment prior to obtaining spectrum authorizations required by REF B and C.[54] Under both domestic and international law, radio frequency-dependent equipment can be confiscated and/or monetary fines levied personally on the individuals involved if such equipment is used "illegally", that is, without authorization. Additionally, uncoordinated and unauthorized spectrum usage leads to increased harmful electromagnetic interference, which can severely degrade safety and operational mission capability. Inadequate management of radio spectrum resources or ineffective identification of the

		impacts associated with spectrum reallocation[55], can potentially result in the loss of Coast Guard authorization to operate systems using those resources, without recourse.
Federal Aviation Administration (FAA)	One of FAA's missions is to ensure that the FAA communications, navigation, and surveillance (CNS) and supporting systems receive the required FAA radio spectrum support prior to the appropriation of funds for systems that require the use of the radio spectrum. Its mission also includes the commitment to the use of new spectrum-efficient technologies and procedures to preserve the radio spectrum, especially aeronautical radio spectrum, that is reserved for exclusive worldwide use by international civil aviation.[56]	To ensure that adequate spectrum support is available to support operational requirements and projected growth patterns, radio spectrum support must be obtained prior to the submission of annual budget estimates to the OMB.[57] The specific procedures for obtaining spectrum support and the definition of systems requiring such support are contained in the NTIA manual.[58] The Program Director for Spectrum Policy and Management (ASR-1) has the responsibility within the FAA for obtaining the assurance of such support through the Spectrum Planning Subcommittee (SPS) of IRAC. Because of the importance of adequate radio spectrum to support civil aviation and other aeronautical systems, the Office of Spectrum Policy and Management (ASR) supports nearly all FAA acquisition programs. Such support includes representation as the radio spectrum subject matter expert on acquisition product teams and other activities that require decisions that could impact spectrum usage.[59] The FAA organizations planning or programming for a system, service, or equipment that will use the radio spectrum must contact ASR-1 so that spectrum support may be obtained as far in advance of any formal planning process as possible. In all cases,

(*Continued*)

TABLE 6A-1 US Spectrum Management—Executive Branch (*Continued*)

Federal agency	Missions	Present arrangements and limitations
		ASR-1 should be contacted prior to any budgetary actions. ASR-1 provides cognizant offices with copies of the SPS's decision in response to the FAA's submission. The time required for a response may vary depending on the stage and complexity of the system but averages 6 to 9 months from time of submission. Insufficient detail in the submission or the subsequent identification of EMC[60] problems could delay program approval for an extended period of time.[61]
United States Department of Treasury	In matters pertaining to spectrum management, the mission of the Department of the Treasury is to: a. Authorize radio frequencies only for justifiable operational requirements when acceptable alternatives are not available; b. Maintain strict supervision and oversight so that funds are not obligated for the development or procurement of equipment that uses the radio frequency spectrum until the requirements of paragraph 34.3, Office of Management and Budget (OMB) Circular A-11[62] are satisfied; c. Ensure that all radio frequency emitters under control of the Department of the Treasury is properly authorized/licensed; d. Ensure that any use of nongovernment public safety radio services frequencies for cooperative law enforcement, or other liaison activities, is	Departmental Offices (DO), bureaus, the Office of the Inspector General (OIG) and the Treasury Inspector General for Tax Administration (TIGTA) may use assigned/allocated federal radio frequencies as necessary to accomplish their respective missions pursuant to pertinent frequency authorizations, as outlined in the Treasury Directive 86-02, November 15, 2001.[64] Frequency authorizations are obtained by submitting proposals, in Spectrum XXI format[65], to the Wireless Programs Office (WPO)[66] staff, Office of the Chief Information Officer.[67] Frequency proposals are processed and forwarded to NTIA by the WPO staff in accordance with procedures outlined in Chapter 9 of the NTIA Manual. Upon approval of the applications, the frequency authorizations are recorded in the NTIA's Government Master File (GMF) and the requestors are legally authorized to use the:

	in accordance with applicable provisions of the NTIA Manual[63] as follows: (1) Part 7.12, "Use of Frequencies Authorized to Non-Government Stations under Part 90 of the Federal Communications Commission (FCC) Rules;" and (2) Section 8.3.3, "Coordination of Frequencies Used for Communication with Non-Government Stations Licensed Under Part 90 of the FCC Rules." e. Ensure that spectrum dependent devices are not be used outside of the United States and Possessions without prior authorization from the national spectrum regulatory body and; f. Ensure that personnel are protected from radiation levels that exceed generally accepted exposure criteria.	Radio frequency 166.4625 MHz (Treasury common) Nongovernment PSRS frequencies Mobile Satellite Service (MSS) radios pursuant to the technical and operational parameters listed on the frequency authorization.[68, 69]
United States Department of Health and Human Services	HHS is not a direct service agency.	The FDA has concerns about wireless devices and interference.[70]
United States Department of Veteran Affairs	The Radio Frequency Spectrum Management Office, Department of Veteran Affairs, procures and manages the wireless radio spectrum resources of the Department, and represents the Department at the National level in all radio frequency matters.[71] The other mission of the Office is to provide consulting services to VA organizations requiring new radio systems or intending to expand or modernize existing radio systems.[72]	VA users of radio frequencies are required to coordinate use of radio transmitters with the Office prior to purchasing or activating any radio service. The Office of Telecommunications oversees the nationwide implementation of VA and federal telecommunications policies and directives. The Office performs these functions through the Telecommunications Support Service which serves as VA's Radio Frequency Spectrum Management Program Office and Communications Security (COMSEC) Custodian.[73]

TABLE 6A-2 US Spectrum Management—Independent Agency

Federal agency	Missions	Present arrangements and limitations
Federal Communications Commission	The FCC, which is an independent regulatory agency, was established by the Communications Act of 1934 and is charged with regulating interstate and international communications by radio, television, wire, satellite and cable. The FCC's jurisdiction covers the 50 states, the District of Columbia, and U.S. possessions.	The FCC administers spectrum for nonfederal government use, and the NTIA, which is an operating unit of the Department of Commerce, administers spectrum for federal government use.[74] Within the FCC, the Office of Engineering and Technology (OET) provides advice on technical and policy issues pertaining to spectrum allocation and use. OET also maintains the FCC's Table of Frequency Allocations.[75]
Federal Emergency Management Agency (FEMA)	The Federal Emergency Management Agency is the lead agency for emergency planning and response activities in the federal government. FEMA's mission is to coordinate National Security Emergency Preparedness programs and plans among federal departments and agencies; coordinate the development of plans, in cooperation with the Secretary of Defense, for mutual civil-military support during National Security Emergencies; and guide and assist State and local governments in achieving preparedness for National Security Emergencies.	FEMA leverages the Radio Amateur Civil Emergency Service (RACES) program and the American Radio Relay League's Amateur Radio Emergency Service (ARES) program as well as thousands of local ham clubs to provide support to state and local government emergency operations. FEMA Regional Communications Officer liaisons with the Defense Commissioner, FCC, and State RACES Officer to provide planning guidance, technical assistance, and funding for establishing a RACES organization at the State and local government level.[76] There are two basic limitations on the use of RACES stations: A general limitation is that all messages transmitted by a RACES station must be authorized by the emergency organization for the affected area. Another limitation on the use of RACES stations is invoked in Wartime Emergency Situations.[77] The following frequency bands are available to stations transmitting communications in RACES on a shared

basis with the amateur service. In the event of an emergency that necessitates the invoking of the president's War Emergency Powers under the provision of Section 706 of the Communications Act of 1934, as amended, only RACES stations and amateur stations participating in RACES may transmit on the following frequencies:

Frequency or frequency bands
kHz:
1800–1825
1975–2000
3500–3550
3930–3980
3984–4000
7079–7125
7245–7255
10100–10150
14047–14053
14220–14230
14331–14350
21047–21053
21228–21267
MHz:
28.55–28.75
29.237–29.273
29.45–29.6
50.35–50.75
52–54
144.50–145.71
146–148
2390–2450

(*Continued*)

TABLE 6A-2 US Spectrum Management—Independent Agency (*Continued*)

Federal agency	Missions	Present arrangements and limitations
		In addition, 1.25 cm (220.0–225.0 MHz), 70 cm (420.0–450.0 MHz), and 23 cm (1240–1300 MHz) are available. Frequencies at 3.997.0 and 53.30 MHz are used in emergency areas to make initial contact with a military unit and for communications with military stations on matters requiring coordination.
General Services Administration	No information is available at the time of writing this book.	
National Aeronautics and Space Administration	The NASA Radio Frequency Spectrum Management Program's mission is to perform the overall planning, policy, coordination and implementation necessary to ensure adequate frequency spectrum in support of the agency's present and future programmatic goals.[78]	The Associate Administrator for Space Flight (AA/OSF) is designated the NASA Spectrum Manager and is responsible for ensuring compliance with pertinent international and national rules and regulations of all NASA RF spectrum users. Additionally, the AA/OSF nominates the Chairperson of the US Study Group 7 of the Radiocommunications Sector (RCS) of the ITU. The AA/OSF has delegated authority for the overall planning, policy and administration of the NASA Spectrum Management Program to the Agency Spectrum Policy and Planning Officer within the Office of Space Flight. Glenn Research Center (GRC) is the Lead Center for the Spectrum Management Program, and the GRC Center Director is assigned all programmatic implementation responsibilities for the program. The GRC Center Director has delegated responsibility for execution of the Spectrum Management Program

implementation responsibilities to the Agency Spectrum Program Manager/GRC.

Specifically, the Agency Spectrum Policy and Planning Officer (NASA HQ/OSF) establishes the policies, and the Agency Spectrum Program Manager/GRC implements the necessary procedures[79] to:

Obtain adequate frequency spectrum to support Agency programs.

Ensure Agency compliance with national and international rules and regulations pertaining to the use of radio frequencies.

Ensure the timely processing of spectrum allocation and frequency assignment requests for Agency programs.

Ensure timely dissemination of technical and regulatory changes, which have a bearing on Field Installation activities, to the Field Installation Spectrum Manager for evaluation and implementation.

Provide the means for Program Offices to provide guidance to project managers so that programs requiring the use of electromagnetic radiating devices are coordinated at the conceptual stage with the appropriate Spectrum Manager.

Ensure identification and mitigation of any RFI which might be caused by or suffered by agency operational programs.

Provide planning and implementation of actions required to obtain new allocations or enhanced radio regulations through national or international organizations.

(*Continued*)

TABLE 6A-2 US Spectrum Management—Independent Agency (*Continued*)

Federal agency	Missions	Present arrangements and limitations
National Science Foundation (NSF)	One of the missions of the Directorate for Mathematical and Physical Sciences, Division of Astronomical Sciences, is to manage electromagnetic spectrum, thereby ensuring the access of the scientific community to portions of the radio spectrum that are needed for research purposes.[80]	The Electromagnetic Spectrum Manager serves as the NSF Representative to the IRAC and its subcommittees and ad-hoc committees, such as the Spectrum Planning Subcommittee, responsible for the apportionment of spectrum space for established and anticipated radio services, and the Radio Conference Subcommittee, which prepares US Government proposals for the WRCs of the ITU.[81] The Spectrum Manager also serves as technical advisor to US delegations to WRCs when appropriate, and as representative to the US Radiocommunications Sector (RS) National Committee of the ITU. He or she works with Study Group 2 of the RS, to ensure that technical requirements for radio astronomy are adequately portrayed, and serves as a member of the US delegations to RS meetings. The Electromagnetic Spectrum Manager also: Serves as point of contact on specific radio regulation topics, as dictated by provisions of the NTIA Manual of Regulations and Procedures for Federal Radio Frequency Management,[82] and the FCC Rules and Regulations.[83] Provides telecommunications management guidance and serves as expert advisor and technical consultant to any element of the NSF on the use of the radio frequency spectrum for research purposes. Performs and arranges for pertinent studies on the scientific uses of the electromagnetic spectrum.

		Advises and assists US research scientists and engineers in obtaining licenses for radio transmitters or resolving cases of interference. Procures authorizations for, protects from harmful interference, and ensures the authorized operations of radio-frequency transmitters assigned to the NSF and used at the NSF National Centers.[84]
Broadcasting Board of Governors	The International Broadcasting Bureau (IBB), formed in 1994 by the International Broadcasting Act which also created a nine-member, bipartisan Broadcasting Board of Governors (BBG), was initially part of the U.S. Information Agency (USIA). When USIA was disbanded in October 1999, the IBB and BBG were established as independent federal government entities.[85] The IBB's primary mission is to broadcast the programs of the Voice of America, WorldNet, Radio and TV Martí, Radio Free Europe/Radio Liberty, and Radio Free Asia to transmitting stations and to AM, FM, short wave, and cable broadcasters worldwide.[86]	The Spectrum Management Division within the Office of Engineering and Technical Services of the IBB has frequency management responsibility for the Voice of America, Radio Free Europe/Radio Liberty, Radio Marti and Radio Free Asia. The Division provides *spectrum occupancy* data to the IBB's customers.[87]
United States Postal Service	No information is available at the time of writing this book.	

6.5 Appendix B: Spectrum Management Organization Links

Secure Sites

- AFFMA SCS Tracking Actions
- Air Combat Command Spectrum Management Office
- HNSWD

United States

- DISA—Office of Spectrum Analysis and Management, http://www.disa.mil/ops/os.html
- Air Force Frequency Management Agency, http://www.affma.hq.af.mil/public/index.html
- Office of the Army Chief Information Officer, http://www.army.mil/ciog6/offices/NETCCG/NETCESTV/spectrum_home.html
- U.S. Army Yuma Proving Ground (USAYPG) Radio Frequency Spectrum Management, http://www.yuma.army.mil/frequency%20management%20office/index.html
- Department of Commerce—Office of Radio Frequency Management, http://www.orfm.noaa.gov/
- Department of Energy Frequency Management, http://cio.doe.gov/spectrum/index.htm
- Federal Communications Commission, http://www.fcc.gov/
- NARTE—National Association of Radio and Telecommunications Engineers
- NASA Office of Space Communications
  - Deep Space Network Spectrum Management http://dsnra.jpl.nasa.gov/freq_man/index.html
  - Kennedy Space Flight Center, Radio Frequency Management http://vulture.ksc.nasa.gov/main.html
- Navy—Electromagnetic Spectrum Center, http://www.navemscen.navy.mil/
- Mid-Atlantic Area Frequency Coordination Office, http://spectrum.nawcad.navy.mil/
- NRQZ—National Radio Quiet Zone, http://astrosun.tn.cornell.edu/faculty/haynes/asat/nrqz.html
- NSF—National Science Foundation Frequency Management, http://www.aas.org/~light/pollution_nsf.html

- NSMA—National Spectrum Managers Association http://www.nsma.org/
- NTIA—Institute of Telecommunication Sciences, http://www.its.bldrdoc.gov/
- Telecommunications Industry Association, http://www.tiaonline.org/

International

- APCO International, http://www.apcointl.org/
- APT—Asia Pacific Telecommunity, http://www.aptsec.org/
- Australia—Australian Telecommunications Authority (AUSTEL), http://www.aca.gov.au/
- Brunei—Jabatan Telekom Brunei, http://www.brunet.bn/telecom/jtb/regu.htm
- Canada—Spectrum Management and Telecommunications, http://spectrum.ic.gc.ca/
- CITEL—Inter-American Telecommunication Commission, http://www.citel.oas.org/
- Denmark—National Telecom Agency, http://www.tst.dk/forside.asp
- European Radiocommunications Office (ERO), http://www.ero.dk/
- Hungary—Communications Authority, http://www.hif.hu/
- India—Group on Telecommunications, http://www.nic.in/pmcouncils/got/
- INMARSAT, http://www.inmarsat.com/about_inm.cfm
- ITU, http://www.itu.int/home/index.html
  - ITU Regions Allocation Tables, http://www.itu.int/brfreqalloc/
- Israel—Ministry of Communications, http://www.moc.gov.il/new/english/index.html
- Latvia—Telecommunication State Inspection, http://www.vei.lv/
- The Netherlands—Post and Telecommunications Department, http://www.minvenw.nl/cend/dvo/zoek/paginanietgevonden.html
- Singapore—Info-communications Authority of Singapore, http://www.ida.gov.sg/Website/IDAhome.nsf/Home?OpenForm
- South Africa—South African Telecommunications Regulatory Authority (SATRA), http://www.doc.gov.za/regulators/satra/
- Switzerland—Federal Office for Communications (OFCOM), http://www.bakom.ch/de/index.html
- United Kingdom—Radiocommunications Agency, http://www.radio.gov.uk/

6.6 Endnotes

[1]Material for this chapter is heavily drawn from "Science and Spectrum Management," European Science Foundation (ESF) Committee on Radio Astronomy Frequencies (CRAF) 2002. The author wishes to thank the European Science Foundation for granting permission to reproduce in full or in part various text, charts, tables, and figures. A special word of thanks is directed to Professor Dr. R. G. Struzak of the ITU Regulation Board, Dr. Wim van Driel, Observatoire de Paris, CRAF Chairman, and Dr. Titus Spoelstra, CRAF Frequency Manager/Secretary, as well as Dr. Joseph N. Pelton, Director, Space and Advanced Communications Research Institute, The George Washington University, for helping the author to improve the contextual coverage of this chapter.

[2]For additional comments, refer to *"CRAF Handbook for Radio Astronomy,"* European Science Foundation, Strasbourg, pp.108–116, 1997.

[3]*Ibid.*

[4]*Ibid.*, 2: idem, "*CRAF Handbook for Radio Astronomy*," pp.108–116, 1997.

[5]At the time of writing this document, the conference had not yet taken place.

[6]When work on a specific question has been completed the Task Group is discharged by an RA.

[7]Robbins, P., "National Telecommunications and Information Administration Structure," http://216.239.37.104/search?q=cache:WVxRwU31aYsJ:deepspace.jpl.nasa.gov/advmiss/docs/ntiastructure.pdf+%22National+Telecommunications+and+Information+Administration+Structure%22&hl=en&ie=UTF-8

[8]Under the current framework for managing spectrum, difficulties remain in resolving conflicts among existing spectrum users. For example, in its January 2003 Report to Congress on Telecommunications, U.S. General Accounting Office (GAO) noted, "considerable conflicts exist between incumbent spectrum users and potential new commercial providers, and no consensus exists on how best to balance the needs of private sector with those of the public sector." "*Telecommunications: Comprehensive Review of U.S. Spectrum Management with Broad Stakeholder Involvement is Needed*," GAO Report to Congressional Requesters, GAO-03-277, January 2003.

[9]Peter, F. G., Director, Physical Infrastructure Issues, U.S. General Accounting Office, "*Telecommunications: History and Current Issues Related to Radio Spectrum Management*," Testimony Before the Committee on Commerce, Science, and Transportation, U.S. Senate, GAO-020814T, June 11, 2002.

[10]Local and State Government Advisory Committee, http://www.fcc.gov/statelocal/.

[11]For more details, refer to the NTIA Annual Report 2000, http://www.ntia.doc.gov/ntiahome/annualrpt/2001/2000annrpt.htm.

[12]Peter F. Guerrero, "*Telecommunications: History and Current Issues Related to Radio Spectrum Management*," Testimony Before the Committee on Commerce, Science, and Transportation, U.S. Senate, Tuesday, June 11, 2002, GAO-02-814T; another good reference material is a second report, "*Telecommunications: Comprehensive Review of U.S. Spectrum Management with Broad Stakeholder Involvement is Needed*," January 2003, GAO-03-277.

[13]*Ibid.*, 1.

[14]http://www.doc.gov.za/department/int/patu.html.

[15]http://www.nabanet.com/about_naba/.

[16]The World Broadcasting Unions (WBU) acts as a coordinating body at the international broadcasting level to provide global solutions on key issues for its member unions. As the technical arm of the WBU, the Technical Committee (WBU-TC) is responsible for technical broadcasting issues of importance to the members including the preparation of guidelines for the application of new technologies and frequency planning.

[17]http://www.nabanet.com/wbuArea/members/WBU.tc.html.

[18]http://www.asbu.org.tn/.

[19]*Ibid.*

[20]*Ibid.*

[21]http://www.apectelwg.org/apecdata/telwg/27tel/dcsg/dcsg11.htm.

[22]http://www.apectelwg.org/apec/atwg/previous.html.

[23]http://www.abu.org.my/main.htm.

[24]http://www.abu.org.my/technical/technical.htm.

[25]*Ibid.*, 16.

[26]Administrations from the following 45 countries are members of CEPT: Albania, Andorra, Austria, Azerbaijan, Belgium, Bosnia and Herzegovina, Bulgaria, Croatia, Cyprus, Czech Republic, Denmark, Estonia, Finland, France, Germany, Great Britain, Greece, Hungary, Iceland, Ireland, Italy, Latvia, Liechtenstein, Lithuania, Luxembourg, Malta, Moldova, Monaco, Netherlands, Norway, Poland, Portugal, Romania, Russian Federation, San Marino, Slovakia, Slovenia, Spain, Sweden, Switzerland, The former Yugoslav Republic of Macedonia, Turkey, Ukraine, Vatican, and Yugoslavia.

[27]The role and purpose of CEPT was redefined at its plenary assembly on 5-6 September 1995 in Weimar. For more details on Weimar Agreement, refer to http://www.cept.org/.

[28]At its Plenary Assembly meeting in Bergen 20–21 September 2001 the CEPT amended the CEPT Arrangement and Rules of Procedure, created a Presidency, and adopted policy agenda, thereby strengthening the organization.

[29]In response to the convergence in the telecommunications sector and the requirements of the information society, the CEPT has created the new Electronic Communications Committee, replacing the two committees that deal separately with radiocommunicaitons and telecommunications. CERP, the committee that deals with postal services, remains unaffected by this change. Furthermore the Assembly endorsed the creation of a single permanent office to support the work of CEPT.

[30]http://www.ebu.ch/union/union.php.

[31]For an example of its function, refer to the EBU review of the "Green Paper on Radio Spectrum Policy," http://www.ebu.ch/departments/legal/pdf/leg_radio_spectrum_policy.pdf?display=EN

[32]http://www.ntia.doc.gov/.

[33]http://www.ntia.doc.gov/osmhome/redbook/redbook.html.

[34]http://www.ntia.doc.gov/osmhome/iracmemb.pdf.

[35]Refer to Table 6A-1.

[36]http://www.army.mil/ciog6/offices/NETCCG/NETCESTV/spectrum_home.html.

[37]http://www.usapa.army.mil/pdffiles/r5_12.pdf.

[38]NTIA has published REF B, available on the web at http://www.ntia.doc.gov/osmhome/redbook.html. This manual mandates the procedures required to fulfill its responsibilities in ensuring the effective and efficient management of all frequency spectrum available to the federal government, compliance with international treaty obligations and domestic law. NTIA procedures further ensure appropriate coordination with the activities of the Federal Communications Commission (FCC) and others. FCC has responsibilities similar to NTIA for all non-federal government spectrum use. REF C implements REF B, and is available by intranet from http://cgweb.comdt.uscg.mil/g-sct/programs/m2400.1f/m2400.1f.htm.

[39]For particular limitations concerning assignments in the above bands, refer to Paragraphs 5–7 Spectrum Requests in Non-Government Bands and 5–8. International Registration, http://www.usapa.army.mil/pdffiles/r5_12.pdf.

[40]http://www.navemscen.navy.mil/pages/mission.htm.

[41]http://www.navemscen.navy.mil/pages/assignment.htm.

[42]http://www.affma.hq.af.mil/public/index.html.

[43]http://www.state.gov/e//eb/rls/rm/2002/11239pf.htm.

[44]In its efforts, CIP is supported by other Executive Branch agencies through the Department of Commerce's NTIA and also by the FCC.

[45]http://ws.usda.gov/3300_1app_c.html.

[46]Prior arrangements for sharing USDA wireless systems with other non-USDA entities must be coordinated with the USDA IRAC Representative.

[47]All wireless operations require Radio Frequency Authorizations from NTIA or a license from the FCC.

[48]http://spectrum-basic.doe.gov/index.cfm?FUSEACTION=Mission.home.

[49]http://www.nps.gov/refdesk/DOrders/DOrder15.html.

[50]http://www.nps.gov/refdesk/DOrders/DOrder15.html.

[51]Source: U.S. Department of Justice.

[52]Coast Guard spectrum requirements are increasing as new systems such as the Integrated Deepwater System, The National Distress and Response System Modernization Project, Ports and Waterways Safety System and The Great Lakes Icebreaker are developed. The demand for spectrum necessary for enhanced Homeland Security and public safety is increasing on a daily basis.

[53]http://www.uscg.mil/reserve/msg02/coast312%2D02.htm.

[54]*Ibid.*, 38.

[55]Legislation and other factors have led to the reallocation of government spectrum to the non-government sector in the United States and abroad and have added to the complexity of the problem the Coast Guard faces.

[56]http://www1.faa.gov/ats/aaf/asr/library/rfi/6050.19/6050-19.pdf.

[57]OMB Circular A-11 requires that certification of spectrum support be obtained prior to the submission of annual budget estimates to the OMB. This applies to the appropriation of funds for either the development, procurement, or modification of CNS or supporting systems or equipment that require use of the radio spectrum.

[58]OMB Circular Number A-11 states in part, Estimates for the development or procurement of major communications electronics systems (including all systems employing space satellite techniques) will be submitted only after certification by the NTIA, Department of Commerce, that the radio frequency required for such systems is available. The NTIA manual provides that systems can be reviewed at four stages as it matures into operational status:

(1) Stage 1: Conceptual (no testing at this stage and radiation is not permitted).

(2) Stage 2: Experimental (new techniques or equipment/proof-of-concept).

(3) Stage 3: Development (preproduction testing).

(4) Stage 4: Procurement (for operational use).

These policies apply within the FAA:

Stage 1 provides guidance on the feasibility of obtaining certification of spectrum support at subsequent stages. Stages 2, 3, and 4 each receive their own certification of spectrum support. This is necessary for cases to begin experimentation and development with several frequency band options that are refined as the system matures and spectrum requirements are solidified. Not all systems are required to go through all four stages.

[59]In many cases, ASR participates as a core team member of the integrated product team; however, in some cases, ASR participates as an extended core team member.

[60]Electromagnetic compatibility (EMC) problems may be created by new CNS systems and/or changes in the technical or operational characteristics of existing systems.

[61]http://www1.faa.gov/ats/aaf/asr/library/rfi/6050.19/6050-19.pdf.

[62]*Ibid.*, 57, 58.

[63]*Ibid.*, 38.

[64]http://www.treas.gov/regs/td86-02.htm?IMAGE.X=31\&IMAGE.Y=11.

[65]DO, bureaus, the OIG and TIGTA have adopted the Spectrum XXI software program, or its successor, as the Department's standard automated spectrum management tool, effective May, 2001. The program standardizes frequency and equipment data, frequency nominations, reviewing and processing digital agendas and performing other functions that directly impact the Department, as well as establishes a standard frequency application procedure. Use of Spectrum XXI is in accordance with the Department of the Treasury Spectrum XXI Standard Operating Procedures (SOP). *Ibid.*

[66]The Chief, Wireless Programs Office (WPO):

(1) Manages the radio spectrum interests and functions of the department;
(2) Designates a Treasury representative to the IRAC and its subcommittees as appropriate;
(3) Ensures that the requirements of the NTIA Manual are adhered to within the Department; and
(4) Provides training, policy and procedural guidance for departmental electromagnetic spectrum management.

[67]The Deputy Assistant Secretary (Information Systems) and Chief Information Officer, Bureau Directors, the Inspector General and the Inspector General for Tax Administration, as it relates to their respective bureaus and offices:

(1) Designate a Radio Frequency Coordinator and an alternate to serve as liaison to the Wireless Program Office;
(2) Issue the necessary guidance to meet the provisions of this directive;
(3) Ensure compliance with the requirements of the NTIA Manual and Departmental guidelines and procedures, particularly with respect to Part 8.2.6 and Annex F concerning the Five-Year Review Program;
(4) Submit information pertaining to frequency applications, or modification of existing assignments, as specified in Chapter 9 of the NTIA Manual, to the WPO staff.

[68]Harmful interference to authorized radio operations are reported to the WPO staff via telephone, for appropriate action. Follow-up reports are submitted on TD F70-04.8, "Radio Interference Report" or equivalent form. Interference to U.S. stations by radio stations in Mexico are reported in accordance with procedures outlined in Section 8.2.30 of the NTIA Manual. *Ibid.* 38.

[69]Questions regarding the use of other wireless/radio services, specialized mobile radio service, maritime channels, and airborne radar systems, etc., are coordinated with the Wireless Programs Office staff, for determination of frequency authorization/licensing requirements and to ensure compliance with NTIA and/or FCC rules for such authorization and/or licensing, as may be applicable. *Ibid.*

[70]2001 Telemedicine Report to Congress, U.S. Department of Health and Human Services.

[71]http://www.va.gov/oirm/telecom/freq/.

[72]"In Building Antenna Systems" and "Wireless Communications Engineering Support" provide resources to systems design and engineering support for wireless communications systems.

[73]http://www.va.gov/oirm/telecom/.

[74]Memorandum of Understanding (MoU) between FCC and NTIA on coordination between Federal spectrum management agencies to promote the efficient use of the radio spectrum in the public interest. http://www.fcc.gov/oet/spectrum/.

[75]The International Telecommunication Union (ITU), headquartered in Geneva, Switzerland is the international organization within which governments coordinate global

telecom networks and services. The United States is a member of the ITU. The ITU maintains the [International] Table of Frequency Allocations, which is reproduced in columns 1–3 of the FCC's Table of Frequency Allocations. The FCC and the NTIA assist the U.S. Department of State in developing U.S. proposals to revise the ITU's Table of Frequency Allocations.

[76]http://www.fema.gov/library/civilpg.shtm.

[77]While performing duties as a RACES operator, members may not communicate with amateurs who are not RACES members. Only emergency communications may be transmitted as defined in FCC Rules and Regulations. No amateur radio station shall be operated in the RACES unless it is certified as registered in a disaster service organization.

[78]http://nasa-spectrum.grc.nasa.gov/about/default.asp.

[79]NASA Policy Directive (NPD) 2570.5B, Radio Frequency Spectrum Management, is agency-wide and reflects the legal obligations mandated by the NTIA. This process must start at the conceptual stage of the system design and be strictly followed during each phase. The NASA Radio Frequency (RF) Spectrum Management Manual NHB 2570.6A, provides instruction in obtaining a frequency authorization in compliance with NPD 2570.5B.

[80]http://www.nsf.gov/od/lpa/news/publicat/nsf03009/mps/ast.htm.

[81]http://www.aas.org/~light/pollution_nsf.html.

[82]*Ibid.*, 38.

[83]*Ibid.*

[84]*Ibid.*, 82.

[85]http://www.ibb.gov/ibb_about.html.

[86]*Ibid.*

[87]http://monitor.ibb.gov/.

Chapter

7

Observations on the Present National and Worldwide Spectrum Management and Its Remedies[1]

7.1 What Should We Do Today?

This chapter gives an overview of strategies that scientific users of the spectrum should be aware of today. This chapter is particularly aimed at scientific policy makers and scientific users of radio who have responsibilities in spectrum management or system design when, for example, frequency selection is an issue. Most of the explanations given relate to radio astronomy, but similarities are easily found in other sciences.

7.1.1 The observation

The observation is done at a radio observatory, a "radio astronomy station" in ITU terminology. A radio telescope tracks a celestial radio source for some time, and its receiver collects information on various aspects of the radio waves emitted by this source. These aspects can be the frequency dependence of the intensity (for spectral analysis), broadband/continuum characteristics, narrow band/spectral line characteristics, polarization characteristics, and variability of these aspects as a function of time, frequency, or celestial coordinates. The radio telescope can be a single dish instrument, an antenna array, or an interferometer. The mutual distances between interferometer elements range from a few tens of meters to thousands of kilometers (in the case of Very Long Baseline Interferometry). Since February 12, 1997, the VLBI networks have been extended into space with the Japanese HALCA satellite, which carries a radio telescope for VLBI operations.

The scientific question determines at which frequency the observations are made, with which instrument these observations are made and also the integration time and duration of these observations. The integration and observation duration depend on the sensitivity required to address the scientific issue. Usually, astronomers working at a university pose the scientific question. They are the "customers" of the radio observatory.

It is common practice that initially it is the *radio observatory staff* that is concerned about the quality of the observation in the technical sense to serve the scientific community as a whole in the best way considering state-of-the-art technology. Usually the radio astronomer for whom the observation is done is interested only in high quality data to serve his own individual scientific goal. This is referred to as "scientific quality", which will also depend on the technical and data manipulation skills of the scientist. Radio astronomers are often satisfied with the removal of "bad" data (e.g., owing to harmful interference) to improve this scientific quality. If after this operation the data quality is too low, the observation has to be repeated for reparation. This also holds for interference effects in observations.

At present, hardly any hardware or software tools exist to filter harmful interference adequately from the scientific data without data loss. Furthermore, the feasibility of such a technique depends totally on the characteristics of the instrument, the data processing methodology which is intimately connected with the scientific case, and also on the characteristics of the scientific question (for one question, the harm has less impact on the result than for another question). An example related to the technical characteristics of the instrument is: interference removal techniques applicable to radio interferometers which make use of the fact that the interferometer fringe pattern for a celestial source is different from that for a terrestrial source, cannot be applied to single-dish telescopes.

From a management and technical point of view, one may consider that interference excision techniques based on data editing is an inadequate procedure. Which funding agency will support an instrument, knowing that inevitably a certain amount of data will be lost? Good arguments must be found to defend an investment with intrinsic loss of value because of effects beyond the operator's control and without the option of a replacement under different conditions.

The nature of the scientific activity implies that the question of frequency protection is not only relevant to present day problems but in particular to guarantee the high quality of this activity for future generations of scientists. In order to achieve this, clarity has to exist about a number of questions rising from the simple and "naive" statement:

> *Radio astronomy is only possible when it can use frequency bands which are sufficiently free of interference considered harmful to radio astronomical observations.*

However, before formulating the question of what we should do to make radio astronomy possible to the extent necessary for each scientific case, clarity has to exist about the following:

- The radio astronomy station suffering the harmful interference should have made it clear that the interfering source is external to that station. If not or if the interference is because of insufficient protection built into the telescope system, the radio astronomy station must improve its own instrumentation before starting any further action. This is in fact a key issue, since scientists must make and maintain their instruments in a state-of-the-art condition as much as possible.
- When radio astronomers explain to nonradio astronomers (either in a technical or an astronomical sense) that they are working hard on interference "rejection"/"removal" techniques, they should bear in mind that these techniques should have been incorporated in their systems in any case: such an explanation could invoke the impression in nonradio astronomers that these radio astronomers have not done their work properly.

In all discussions on the protection of scientific research and investigations to reduce the vulnerability of scientific experiments to interference, one should consider that state-of- the-art scientific research is often only possible with equipment which is not commercially available but has been built for a special scientific requirement. Furthermore, quite often such equipment has been built with technology not yet commercially available, since the scientist uses radio frequencies that are as yet commercially inaccessible to the nonscientific user. This holds especially true for radio astronomy. Therefore, each radio astronomy station can be considered to be its own prototype.

Measurements with a scientific instrument must always be taken with the highest possible sensitivity, since it is vital for scientific research to obtain data of the best possible quality. Defective data results in faulty scientific results.

Because radio astronomy and various remote sensing investigations require measurements at such high frequencies for which no equipment is commercially available but has to be individually designed to satisfy the specific scientific requirements, it is in most cases not possible to design and build the instrument according to existing standards and regulations since these are usually not suited to the technical specifications needed for the scientific instrument. If measurements need to be done in a radio frequency range for which the technology is still very much under development, regulations or guidelines will not exist for a long time, and nonscientific users are not yet deploying activities, scientists cannot take these apparently unknown criteria into account in the design and construction of their instruments.

Another issue is that in many radio frequency bands, radio astronomy instruments and other scientific equipment have already existed for a long time—usually for several decades—and came into operation often before the aforementioned technical standards were developed. This implies that many radio astronomy and other scientific instruments do not have adequate capabilities to suppress or remove harmful interference. Studies related to interference-robust equipment are therefore in first instance applicable only to newly designed equipment and can currently be undertaken only for those radio frequency domains for which supporting knowledge and technology already exists, for example, below 20 GHz. It is a waste of effort to develop interference-robust equipment for operations in a frequency range where nobody has any idea how this could be achieved and for which currently no test equipment exists, for example, at frequencies above 100 GHz.

Nevertheless, scientists must develop their instrumentation according to available regulations and technical standards. Administrations and active users must understand that scientists develop their instrumentation to a state-of-the-art condition and understand that everyone is served by obtaining the highest possible quality and the most feasible interference-robust equipment. By its nature, scientific research is continuously developing, and therefore the instrumentation is regularly adjusted to the changing scientific questions and upgraded accordingly.

Given the available instrumentation, scientists in general and radio astronomers in particular must be able to address the following questions raised by the statement given above:

1. What determines whether scientific research is *possible* in this context?
2. What is the criterion for *sufficiently free of interference*?
3. What is considered to be *harmful to a radio astronomical observation*?
4. What is the meaning of *guarantee of data quality* in the context of action to be performed?
5. What is the criterion for *scientific priority* of a specific frequency band?

These questions can be answered in many different ways. Their answers depend on:

1. Knowledge of radiation mechanisms in relation to the physics in and around the celestial radio source. Astrophysics and fundamental astronomy should provide the motivation to claim protection in a specific frequency band.

2. A knowledge of the state-of-the-art and limitations of hardware and software technology to serve matters of instrumentation and data processing for obtaining relevant and adequate information about the object of interest.
3. A knowledge of development trends in both active and passive frequency use and in receiver development (what is harmful today may be harmful tomorrow or not harmful at all).
4. A knowledge of the state-of-the-art and limitations of technology to extract this information from data distorted by, for example, interference by unwanted natural or man-made emissions.
5. The availability and implementation of this knowledge.
6. The view of the priorities in the development of astronomy as a science at both a national and an international level.
7. The research priorities in the different astronomical institutes; these are many different (often incoherent) opinions and interests.
8. The local/national science policy.

Table 7-1 shows the relationship between these five questions and the eight reply parameters.

Given these considerations, we will undertake an attempt to answer the questions just raised:

The *possibility* of radio astronomy. Assuming that radio astronomy at a certain frequency is theoretically possible, the practical possibility of radio astronomy depends entirely on the achievable quality of the

TABLE 7-1 Relations between Aspects and Considerations Relevant to Answer the Five Questions

	1. Possibility of research	2. Interference free	3. Harmful	4. Guaranteed quality	5. Priority
a. Physics	+				+
b. Limitations	+	+	+	+	
c. Trends	+			+	
d. Technology	+	+	+	+	
e. Implementation	+	+	+	+	
f. View on development					+
g. Priorities					+
h. Policy					+

measurements. Radio astronomy is possible when the quality of the observation data is adequate to understand the physical conditions of, inside or surrounding, the celestial radio source. This data quality determines whether the scientist is able to improve his understanding of the physical processes maintaining the structure of this celestial object.

What does *sufficiently free of interference* mean? One of the most difficult questions is how one can identify a signal as interference, since scientific instruments are not usually built to do so. The scientist notices that "something is wrong with his data but he cannot proceed to understand the real cause further." Interference can be manifest as, for example, impulsive interference or an increase of the root mean square (rms) noise in the data (possibly as a function of frequency and polarization). An increase of the rms noise in the data must be distinguished from the system noise: but how? This becomes more complicated when the interference is manifest in the same way as the characteristics of the signal of interest, for example, in spectral line observations.

Radio astronomical measurements are always done at the limit of what is technically achievable. Sensitivity levels can be calculated theoretically, and radio astronomers accept that an observation is sufficiently free of interference when the sensitivity achieved in the final radio astronomical result is at the level of what is theoretically achievable with the instrument used. This criterion may be understood as an ideal that one wants to reach. It should be recognized, however, that in real life, external situations such as anomalous propagation conditions play a role. But by all means, radio astronomy receivers are made state-of-the-art in the sense that the ultimate physical limitations are approached as close as possible. External effects, which have a non-man-made origin, have to be accepted as being unavoidable. For some radio telescopes using receivers cooled at cryogenic temperatures, the sensitivity is determined for a large part by the thermal noise in the system.

These notes imply that it is generally not possible to explain whether an observation is sufficiently free from interference. Therefore, for practical reasons, radio astronomers consider that an observation in which the interference is below a specified threshold, that is, as specified Recommendation ITU-R RA769, it is sufficiently free from interference. However, radio astronomers should have already prepared their case carefully by choosing sites as free as possible from interference. It should furthermore be recommended that when proposing frequency allocations, administrations take into account that it is very difficult for the radio astronomy service to share frequencies with any other service in which direct line-of-sight paths from the transmitters to the observatories are involved. Above 40 MHz sharing may be practicable with services in which the transmitters are not in direct line-of-sight of the observatories, but coordination may be necessary, particularly if the transmitters are of high power: for this reason Recommendation ITU-R RA769 was developed.

What is considered to be *harmful to radio astronomical observations*? As a criterion for the intensity at which an interfering signal is considered harmful, the level which causes an increase of 10 percent in the measurement errors, relative to the errors owing to the system noise alone, is used.[2] This definition implies that in interference calculations the usual practice is to assume that this interference level is the same as that which causes an increase in the receiver output by no more than 10 percent of the rms output fluctuations owing to the system noise. It is assumed that with this criterion the minimal data quality necessary to make the science of radio astronomy possible is guaranteed. This criterion is also used in the determination of the levels of detrimental interference as published in Recommendation ITU-R RA769.

How can freedom from harmful interference to radio astronomical observations be *guaranteed*? The guarantee of freedom from harmful interference can be achieved by regulatory measures by a national administration and by technical measures at the interfering transmitter and at the radio astronomical receiver.

The regulatory measures can be various, such as setting up an exclusion or coordination area around the radio astronomy station, limiting power levels for the transmitter, defining antenna pattern constraints to the transmitter, or even a kind of time-sharing scenario (when it is feasible and acceptable).

The transmitting side of the signal path could be engineered in such a way that unwanted emissions are adequately filtered and the transmitting spectrum is properly shaped to avoid interference. Filtering must be implemented in all cases when the passive service does not operate in the same frequency band as the active service, since when the passive service suffers interference, it is because of spectrum pollution and energy waste, which should be avoided.

The receiving side of the signal path could be subject to adequate filtering if that does not degrade the performance of the system. It should be noted that only when the active and passive applications enjoy an allocation of the same frequency band can arguments be given to look for such filtering (see Sec. 7.5.7).

What is the criterion for *scientific priority* of a specific frequency band? The physical conditions and the characteristics of the object of interest determine the priority of a specific radio frequency band for scientific observation purposes. However, various other arguments and interests also play a role. It is impossible for each institute to work in all fields of scientific research; therefore, choices have to be made. Such choices are based on various arguments, such as the qualities of some particular scientist without which modern radio astronomy is unthinkable (i.e., owing to the work of J. H. Oort and H. C. van de Hulst in the Netherlands in the mid-twentieth century, which led to strong stimulation of hydrogen line research and studies of galaxies), financial reasons (e.g., solar research is in many respects less expensive than millimeter-wave radio astronomy with its study of interstellar molecules), and personal preferences. In addition, the views developed within the scientific community enhance or reduce the noted priorities.

Priorities developed within the scientific community are translated by the responsible national ministers into a national science policy with its own priorities.

In conclusion, the priority of a specific frequency band is primarily scientific, but it must not be forgotten that other elements also play an important role.

7.1.2 Keys to a strategy

We now know that in principle, radio astronomy is possible if it can use frequency bands that are sufficiently free of interference considered harmful to radio astronomical observations. Such a happy statement can also be developed for other sciences using the radio frequency spectrum.

However, the pressure on the radio spectrum from various spectrum users and radiocommunication services shows that much work must be done to translate this possibility into reality. In the past, such a translation appeared to be unnecessary, since scientists could use their radio frequencies without any conflict with other users of the radio spectrum. Of course, the scientists knew that frequency bands used for broadcasting and military applications had to be avoided, but until about 1980 the radio spectrum was relatively free (also for frequencies below 1 GHz).

7.1.3 Education

The new situation of rapidly increasing pressure on radio frequencies indicates the need for adequate and appropriate answers and information in reply to the questions these nonscientific frequency users raise, since these users often do not understand why scientists need to use a particular radio frequency band. But in practice, one cannot simply take all the necessary or wanted time to "educate" or study the problem in a "scientific way" when urgent, adequate, and appropriate action is required: The time scales set by the new situation and the developments of telecommunication technology do not allow for this.

Besides, the nonscientific users of radio frequencies request support by the administrations for their cases and ask for access to radio frequencies to which their radiocommunication service has an allocation (although possibly not used for a long time). These administrations also ask scientists for information and cooperation in coordination issues.

It becomes, therefore, urgent to understand what strategy scientists should follow regarding such requests for information and education. To answer these questions, they face three major situations:

1. Increasingly, scientists are facing degradation of their observations owing to man-made interference and also re-allocations of frequencies, which may be a threat for existing and future frequency bands needed for passive frequency use. Therefore, the scientist must provide a clear opinion and evaluation of the current situation. From that, he must develop unambiguous answers to the questions raised by nonscientific users of the radio spectrum. However, besides the need for answers and education, usually each individual scientific user of the radio spectrum has almost exclusively or primarily knowledge of his own research interests, and he considers therefore the frequency bands that he uses as the most important ones. An example is in the remote sensing field; meteorologists have their own specific interests and priorities, which may be very different from those of aeronomy or atmospheric research. A consequence is that scientists have difficulty in finding a single coherent answer to the questions asked by an administration or the nonscientific user of the radio spectrum. In his response, the scientist must make this clear or refer the questioner to a colleague or organization which is more qualified to reply.
2. At stations of scientific research, not many people are working on frequency management issues. Usually it is not more than one person, and this is generally on part-time or voluntary basis in an environment which hardly understands the issues. (It should be noted that usually none of these people is adequately trained at a university or any other educational institute). The time scale of the learning curve for this job is long: usually on the order of more than 5 years.
3. In administrations and related bodies, the tenure of appointment in a certain position is typically on the order of about 2.5 years. This implies that at this time scale, the "education" and "provision of information" by the scientist has to be repeated regularly as an ongoing process. This tenure is usually short in relation to that of satellite programs (when we consider that the typical schedule of a satellite program is 7 years design phase, 5 years construction phase, 15 years operations phase).

The scientist educating the nonscientific user of radio frequencies or the administration must be aware of a number of issues of the whole spectrum management mechanism as well as the nonscientific applications and needs, plus possible ways and means of solving problems in a constructive manner.:

To this end and as far as reasonably possible the scientist should:

- Know and understand the nature and characteristics of the current and expected developments and problems in his own field.
- Know and understand the developments in active services and passive frequency use, such as technical developments, standardization process, (potential) threats or compatibility problems.
- Know the national/international frequency regulatory process. This knowledge helps significantly in responding in a straightforward manner to the questions and uncertainties of the nonscientific user of the radio spectrum or the administration. This knowledge elevates the discussion level between the parties.
- Know and understand the frequency allocation and management process to avoid mistakes.
- Know and understand the unwritten policies in this process to avoid or escape pitfalls. Therefore, it is important that when one enters this education process one should consult colleagues who already have experienced it.
- Have adequate communication with the national authorities. These could be the ministry responsible for science policy and support, the national administration responsible for frequency management and spectrum engineering, or military authorities, since several frequency bands allocated to radio astronomy are also used by military applications.
- Be aware of the relevant internal communication and decision-making paths within these authorities.
- Participate in decision-making processes (local and regional). This is particularly important in cases of sharing and protection studies or in work in preparation of, for example, ITU-R world radiocommunication conferences.
- On all occasions, present one single opinion/one single voice (local, regional, and global) since fragmentation in opinions, diverging views, and conflicts may have irremediable consequences.
- Indicate priorities for their scientific research (at a local, regional and global scale). This issue may sound logical and evident, but in practice, scientists usually consider their favorite frequency band to be the most important. During frequency management discussions, it is claimed that every frequency band is very important or the most

important for a particular issue; the conclusion can only be that no frequency band is really important.

- Indicate what fraction of science is lost if a frequency band is lost. But note that losing data because of interference other than variable propagation conditions (ITU-R Recommendation RA769) may lead to erroneous conclusions or even cut off a whole new development (see Sec. 7.5). Examples of such degrading effects can be found in meteorology.
- Give the right comment at the right moment to the right body. But is he/she prepared to do so?
- Inform and educate his/her scientific community about the "frequency" world problems. Until today, the scientific community has ignored the developments in the real world and has considered frequency management as a nonscientific activity and therefore a waste of the resources needed to "develop the marvelous world of science." This attitude must change, since scientific frequency management is a fundamental requirement for a scientific institute, just as much as paying the electricity bill.
- Inform and educate directors of institutes about the global frequency problems. Directors must be able to give adequate guidance to their respective institutes, and when political and management issues are discussed, they have to give the right answers. Education of colleagues and directors is also important because a key issue is training of new experts in frequency-management issues.
- Maintain adequate, accurate and appropriate interaction and exchange of information within the scientific community and bodies to actively protect radio frequencies important for research, such as the *Scientific Committee on the Allocation of Frequencies for Radio Astronomy and Space Science* of UNESCO, IUCAF, and *Committee on Radio Astronomy Frequencies* (CRAF).

These are merely keys for a "strategy under construction." However, in order to act adequately, much information, communication, listening and reading is mandatory to decide on the correct criteria. The key problem is to be active and seek solutions to the challenges in cooperation with relevant partners and services. Scientists should not wait until space communication people come to them to ask for their scheduling information, etc.: that will be too late, and furthermore, giving such information is wrong (see Sec. 7.2). The scientist should actively approach administrations and nonscientific users of radio frequencies as soon as he knows that they are planning activities that potentially may be harmful to radio astronomy: this could be seen as an assertive anticipatory attitude. Scientists should develop and maintain a review of the global, regional and local interference pressure as a function of frequency band, radiocommunication service and radio stations on the World Wide Web.

In summary, scientists must generate a general awareness in society of the pressures they meet in the deployment of their research, owing to pressure on radio frequencies. The educational process must start with students in schools.

7.1.4 Communication

In order to acknowledge the reality of an interference or electromagnetic compatibility problem the administrations and active spectrum users need to be informed about the nature and (quantitative) characteristics of notified and expected interference problems. This information has to be transmitted to a community, which usually does not have the slightest idea of passive spectrum use, the nature of a passive service and the characteristics of scientific use of radio frequencies. "Tools" have to be developed to answer this need for specific education. This education is a major element in communication with nonscientific users of the radio spectrum and administration.

Such communication can either be directly with the nonscientific spectrum user or the administration, or through conferences or lecture series. In conferences, one could have sessions dedicated to special subjects relevant to scientific interests. One could also consider lectures during meetings—how fruitful these might be. These presentations should not only be addressed to, be relevant for and of interest to the "already converted." All contacts must be aimed at making "converts." In an aggressive but accurate and informative way, the scientific community has to present its case at such conferences, like conferences on EMC problems, studies on frequency allocation techniques, frequency management meetings, and so forth.

The following communication channels seem important:

- With local administrations to keep them up-to-date on the scientific spectrum issues.
- With regional and global public bodies involved in the regulatory and standardization process to keep them aware of the scientific requirements.
- With organizations of active spectrum users to make them aware of the requirements and needs of scientific usage of radio frequencies.
- With scientific organizations, such as the International Union of Radio Science, URSI, to involve them in support of scientific radio frequency matters.

- With directors of observatories, since they have to represent their institutes and set priorities for the different activities in these institutes.
- With space agencies (U.S., European, Russian, Chinese, Japanese) to keep them up-to-date so that they take the requirements of scientific usage of radio frequencies into consideration in their planning and design activities.
- With industry to make it aware of the specific unusual requirements which scientists must explain to them as being vital for scientific progress.

But it must be noted, the administrations cannot provide a miraculous solution for everything. Furthermore, we should recognize the fact that they are under pressure from those who need access to the spectrum, and in some countries administrations are under the pressure to generate income by different spectrum pricing methods. The administrations are facing new and previously unknown issues related to the various telecommunication developments already discussed; space issues especially contribute significantly to this. Although it is apparent that education of administration officials is important, education of those who need access to the spectrum might be even more important. The process of education, communication and promotion should address key professional and social groups, with emphasis on those who prepare the background to decisions and who make decisions.

7.1.5 Methodology

International scientific organizations. One of the weakest points among scientists relating to the handling of interference problems, influencing frequency management and spectrum engineering is their *individualism*. Nevertheless, scientists have organized themselves into a couple of organizations that work on spectrum management issues. At a global level, the efforts in scientific frequency management are coordinated by the *Scientific Committee on the Allocation of Frequencies for Radio Astronomy and Space Science* of UNESCO, IUCAF. The parent organizations of IUCAF are URSI, the *International Astronomical Union*, IAU, and the *Committee on Space Research*, COSPAR. IUCAF is the channel through which the scientific unions, URSI, IAU and COSPAR provide expertise and information in ITU-R Study Group 7. At a regional level, the radio astronomy community in Europe is organized in CRAF, while in the Americas, the interests of scientific frequency users are covered by the *Committee on Radio Frequencies* of the U.S. National Research Council, CORF, currently representing scientists in the U.S., Canada and Mexico. IUCAF and CRAF are ITU-R sector members and in this position, they collaborate at ITU-R level to take care of the interests of radio astronomy and passive scientific use of radio frequencies.

Coordinated by IUCAF, regional and local committees like CRAF and CORF should continuously put pressure on the international scientific community, for example, the URSI, to stimulate discussions and (where possible) to achieve harmonization of opinion among scientists to form a single clear and concerted view on a specific topic. The scientific unions, such as the URSI, should actively initiate these discussions and trigger studies, the results of which could help IUCAF and the related regional and local committees in their efforts.

If it is seen that the international scientific organizations do not actively deploy these activities sufficiently well, a regional organization such as CRAF should inform IUCAF, which should bring this observation to the attention of the scientific organization concerned, for example, during its General Assembly, by communication to its Executive Council or by input to its information bulletins. This action of IUCAF and regional and local committees such as CRAF and CORF should be a well-coordinated joint effort.

During general assemblies of the scientific unions, IUCAF supported by regional, or local organizations could organize sessions during which the following subjects could be discussed:

- The nature and characteristics of the problem field: development of spectrum demand, development of telecommunication applications and their impact on the radio frequency spectrum.
- Short-term, medium-term and long-term developments.
- How to encourage scientists and the unions to make clear statements on research priorities.
- What is the structure, membership and correspondence of IUCAF and how is the work done and by whom?

Local scientific community. The "scientific individualism" mentioned above is often too clearly visible in the attitude of individual radio astronomers, staff of radio astronomy stations and observatory directors. On many occasions their opinion is required on various topics in relation to the protection of frequencies for radio astronomical research. However, nobody can be a specialist in all fields.

Experience shows that the areas of frequency regulation, protection, and management are very complex, and most people need several years to become familiar with all aspects. Therefore, given the existence and activities of IUCAF, CRAF, and CORF, radio astronomers, staff of radio astronomy stations and directors should coordinate all activity in this field with these bodies. Besides their experience, another important reason is that IUCAF, CRAF, and CORF are well-known bodies to administrations and several international private and public organizations. Working outside these three bodies would confuse and spoil the radio astronomy case.

Furthermore, where IUCAF and organizations like CRAF or CORF make recommendations contrary to the hopes of the individual radio astronomer, radio observatory staff and observatory director, these people should follow the recommendation unless they are able to prove that IUCAF, CRAF or CORF has come to an incorrect conclusion. This methodology does not imply that the mandate of the individual radio astronomer, observatory staff member or observatory director is restricted, but that the bodies who have been created for this job and to which this specialized activity has been referred are able to fulfill their job properly.

Similar cases can be extrapolated for other scientific communities.

Collaborative attitude. Scientists are usually funded by public money, and their interests are outlined in a governmental strategy on research and development. This public aspect and the fact that, with some exceptions, scientific research depends on support from the administration mean that the scientific communities must maintain good relations and interact with these administrations at a national level. At an international level, similar good relations and interactions are required. IUCAF and regional organizations like CRAF are ITU-R sector members. This improved the interaction with the ITU-R and the participation in ITU-R (-related) activities. In Europe, CRAF has developed to become the European radio astronomy discussion partner of the CEPT. Organizations such as IUCAF, CRAF, and CORF participate actively in many working groups, study groups and project teams devoted to spectrum engineering and regulatory issues.

Under the auspices of the ITU-R, much work is done (e.g., in WP7C by remote sensing scientists and in WP7D by radio astronomers) on definitions and criteria for sharing conditions. However, it cannot be assumed that the colleagues in this activity do have complete and sufficient knowledge of specific scientific needs and problems. Organizations such as IUCAF and CRAF should cooperate closely with the chairmen of the relevant ITU commissions/working groups. IUCAF and CRAF should initiate studies in support of the ITU-R working party activities. When necessary and possible, these work programs should be used as subjects for discussions under the auspices of the international scientific community, that is, URSI.

The structure of this activity is basically reactive and addressing issues which arose on this side of the horizon; however, a clear perspective for the mid-term future must be available and regularly be updated. An essential element of such a plan is that it is related to science policy in those countries supporting scientific research. Politically speaking, relations with organizations, for example, in Europe the *European Commission,* should be improved where relevant.

This item must be on the agenda of all scientific users of the radio frequency spectrum.

In addition to good relations with regulatory and political organizations, scientific communities should collaborate actively with space agencies, telecommunication operators, broadcasting organizations and other active users. The aeronautical and space projects may especially cause major problems relating to the process of frequency selection and development of protection tools, where the considerations and requirements of the science service have not been taken into account properly. Protection of science services against space-to-Earth or Earth-to-space transmissions need special attention (depending on the specific scientific interest, of course).

Collaboration with industry should be developed as much as possible. A good example of where such a collaboration can lead is the joint project of the European Space Agency (ESA), the Swiss-Italian company Oerlikon-Contraves, and CRAF to develop adequate filter technology to protect radio astronomy receivers' operation in the frequency range around 95 GHz against harmful interference and physical destruction by signals from a space-based cloud radar system planned to operate in the band 94.0 to 94.1 GHz. The result of this collaboration was that such a filter technology was developed in 1997.

Specific events of harmful interference. Events of harmful interference certainly have an origin. This origin can be *local, regional,* or *global.*[3] When scientists addressing the protection of radio frequencies important for scientific research, that specific local, regional, or global organization must be kept informed of the interference events to be able to give adequate guidance and support to the affected scientific community. This is a major role for CORF, CRAF, and IUCAF and other similar organizations when they come into existence.

Local questions and problems have to be dealt with locally, that is, in bilateral communication with the national, civil, or military administration and the radio observatory involved. In such communication, usually the following information is needed:

- Place, date, and time of interference observed and the frequency band within which the interference occurred.
- Frequency at which the interference occurred.
- Allocation status for the specific frequency band.
- Characteristics (when known) of the interfering signal: strength, direction, modulation (if possible).

All questions and problems need to be communicated to the organization working on spectrum management for scientific research. The information on interference events should continuously be registered, collected, and documented in a Web-accessible database. Organizations such as IUCAF should produce a common public domain and Web-accessible software toolkit for that purpose. If necessary and proper, this organization or IUCAF may provide additional support in necessary communications with the relevant administrations.

Regional questions and problems have to be dealt with regionally. In Europe, CRAF, as the European body concerned with the protection of frequency bands allocated to radio astronomy in Europe, has to communicate with national administrations—CEPT, the EC, and ESA. The required information is the same as that provided to local administrations. It is mandatory that in Europe the radio astronomers follow the line taken by CRAF or improve their scientific case via CRAF only. Such a position holds for other scientific communities and regions as well.

It is also important that regional organizations for the protection of radio astronomy frequencies are established in other regions. Some activity is currently being developed in the Asia-Pacific region, but that needs further attention. CRAF and IUCAF can play a supporting role.

Global questions and problems have to be dealt with globally. IUCAF is the body to take the initiative and to give guidance. However, CRAF and its American sister CORF should provide IUCAF with timely, proper and adequate information. ITU-R members such as IUCAF and CRAF could also inform the ITU-R Radiocommunication Bureau directly about any issue.

In cases of specific events of harmful interference, scientists should follow the following action path:

- If possible the scientific station should determine whether the source of interference is local, regional, or global. In any case, it must be sure that the interference is not internal to the station.
- If the scientific station is connected with an organization dealing with spectrum management and frequency protection issues, this station should bring the event to the attention of this organization. This communication helps to build a database of events which after some analysis enables the organization to address the frequency protection issue adequately to a regional administrative organization or, if necessary and the organization is sector member of the ITU, to the ITU-R Radiocommunication Bureau. An organization such as CRAF or IUCAF may be able to advise and may provide support.
- The scientific station suffering the harmful interference checks the allocation status of the frequency band in which the interference is

experienced. The bases for this are the ITU Radio Regulations and the local national frequency distribution plan. In case the scientific service has no or a secondary allocation in the band, one may discuss the issue with the national administration of the country in which the station is located, but then the case is a local problem which has to be solved locally, that is, with the national administration.

- If the interference is experienced in a frequency band allocated to the science service suffering this interference, the interference event should be well documented and discussed with the national administration. The latter may need an independent confirmation of the harm. This confirmation is usually performed by one of the monitoring stations operated by an administration, for example, the satellite monitoring station of the German administration in Leeheim, Germany. If no adequate monitoring facility is available, the local administration should be asked to request assistance from a facility elsewhere.
- After independent confirmation, the radio astronomer who contacted the administration asks this administration for advice and help. In practice, such an administrative action may imply that a formal statement of complaint is sent to the "operator" whose system is causing interference, or any other adequate action may be taken. If the transmitting source is outside the country, and the location of the source is known, the administration can be requested to inform the administration of the country in which the transmitter is operating. It is also strongly recommended to inform the ITU-R Radiocommunication Bureau about specific interference cases, or request its assistance in solving the problem. The ITU Radio Regulations Board may also be requested to solve the interference problems according to ITU Convention No. 140.

Hopefully, these supportive administrative actions may be sufficient to solve or alleviate the problem.

Work. On a global scale, scientists enjoy only two full-time frequency managers: one located in Washington, D.C. and working for the U.S. National Science Foundation; the other located in Dwingeloo, the Netherlands, being the European frequency manager of the ESF, CRAF. Besides these full-time frequency managers, several other members of the scientific community work on a voluntary basis on frequency management issues (with their own experience and know-how of course), but they are not always available. Given the large variety of issues that require attention, the members of organizations such as IUCAF, CRAF, and CORF should be dedicated to specific tasks on a continuous level. In this way they would build group know-how within the community and support the frequency managers in their work. Using modern means of communication, the physical nonavailability of IUCAF, CRAF, and CORF members could often be alleviated.

Meetings. The number of meetings where a scientist competent to discuss spectrum management issues needs to participate is growing rapidly. A single frequency manager is physically not able to attend all such meetings. Furthermore, in various meetings such as those of the CEPT and the ITU, more than one representative of the scientific community is often needed. This is because of the structure of the meeting when various drafting groups work simultaneously. When required, the frequency manager needs to be assisted by other scientists; the first option is assistance by a colleague of the country in which an event takes place, if available. Experts for specific issues should be available to assist as well.

If studies of specific questions are required, the scientific community and especially its organization dealing with spectrum management and protection issues should organize dedicated workshops to address these.

Agreements between scientists and nonscientific radio spectrum users. In some cases regulatory imperfections exist which must be resolved before completion of a coordination process is possible. To satisfy the needs of various radio spectrum users, private agreements have been made and are being made with the aim of filling regulatory gaps or to solve a problem that administrations cannot solve because of these inadequate regulations or because of political reasons. Reference to such private agreements may be incorporated in administrative regulatory procedures, but such agreements may also develop into precedents to undermine current spectrum regulation and respect for these regulations. This process may be exacerbated by economic and political pressures.

A complicating factor in Europe is that the telecommunication legislation in some European countries does not allow for reference to or attachment of a CEPT Recommendation as an authorization. Other countries may have another different legislation. This implies that bilateral agreements of such private entities with other private legal entities are considered very probably the best way out for the private parties involved.

In some cases, scientists are asked to enter an agreement with nonscientific radio spectrum users. The reasons for such a request can be various, but often such an agreement helps an administration to solve its problem when no adequate regulation for a case exists yet. Such agreements are often an option in coordination issues between different users of the radio spectrum.

If it seems that scientific organizations or stations are urged to come to agreements with active users, it is of major importance that sister organizations and IUCAF are consulted properly and adequately, and that their opinion is taken into account. In all cases one should first investigate whether the requested agreement is necessary. The reason for this is that any agreement may create a precedent and may also put constraints on sister organizations/stations.

It must be noted that in cases of radio frequency matters, a national administration has only the authority to request an agreement from the considered scientific community. Where unexpected problems occur and an administration has no immediate opportunity or method of solving the issue, an agreement could be made between the scientific community and the operator under the condition of full compliance with the ITU Radio Regulations. No private legal entity has the right to ask an agreement for any other purpose. This is particularly apparent where the local, regional, or global use of a frequency band is changed (often reduced for scientific applications): the national government of a sovereign state can make decisions, which set only the political priorities, such as "science or telecom". If the national priority to science is not to be degraded, but the scientific frequency use has been restricted, adequate action is required. In this case, expert organizations like IUCAF, CRAF, or CORF may play an active role in guidance and support.

When agreements cannot be avoided, one should carefully evaluate under which conditions the idea of an agreement can be accepted and when it should be rejected. An agreement should not be accepted for the simple reason that the station involved in this agreement is not interested in that particular frequency band or that it is considered to have local relevance only. A case that is considered to have local relevance only may at a later stage develop into one with global importance. Often such an agreement has severe implications for sister organizations that may be extremely difficult to modify or correct.

If agreements are needed, the framework for these can only be international and national legislation. International legislation is formulated in the ITU Radio Regulations (e.g., its table of frequency allocations and related footnotes and Article S29 that describes the characteristics of radio astronomy as a radiocommunication service and its protection). No agreement must be accepted that does not fully comply with these legal frames. Furthermore, one must avoid the situation in which agreements become a substitute for the international and national regulations because this development degrades the status of international legal structures.

7.2 What Should We Not Do Today?

The issue of agreements mentioned above brings us to the question of what we should not do today. We must be aware of actions to avoid because of the increasing pressure of use on the radio frequency spectrum.

Commercial users of radio frequencies, especially those working in the field of telecommunication by satellite, are eager to enter the radio frequency spectrum as soon as the system they want to operate enters its test phase. Of course they then have to go through the whole process of ITU-R notification and coordination with national administrations, but once this process has been completed and frequency band(s) have been assigned to this system, the design phase of the system has to be completed in such a way that guarantees can be given that other users of that part of the radio spectrum will not suffer from unwanted emissions. If everything runs well, and if in the system design and construction all necessary precautions have been taken to eliminate spectrum pollution, and if all regulations on standards and electromagnetic compatibility have been obeyed, no problems should occur.

However, daily practice often does not comply with this ideal situation. For commercial and competition reasons, industry prefers quick and cheap solutions rather than the implementation of results of thorough technological and scientific research. This industrial practice often results in badly-designed or defective systems (we call a *defective* system any system that technologically avoidably pollutes the radio frequency spectrum and causes harmful interference to another radiocommunication service). The result of this is that on a number of occasions the final system may not be as perfect as anticipated, requiring the operator to enter a coordination process with a possible victim service. Such a process is usually performed under administrative auspices. At that point the administrations face the situation that the new system exists in such an already advanced state that refusing its operations may have an unacceptable impact on the operator. Although in the strict regulatory sense the frequency regulatory authority must refuse operation of any such defective system, political arguments can and, in practice, do overrule such a regulative action.

So far, examples of defective systems which have caused significant and irremediable damage to scientific research using radio frequencies are:

- The Russian system for radionavigation by satellite GLONASS at 1.6 GHz;
- The U.S.-based satellite system for mobile communication by satellite Iridium at 1.6 and 23 GHz;
- The U.S. military satellite TEX at 328 MHz;
- The fixed-satellite system ASTRA-1D operated by a Luxembourg-based operator at 10.7 GHz.

Of these systems, the GLONASS system is being improved to protect passive radio usage: this improvement will be completed by the year 2006.

The Iridium satellite system is a clear example of bad design in which no protection of other radiocommunication services was built in.

The TEX satellite is an example of a system which is no longer operational but which, since it has no OFF-switch built in, cannot be switched off, resulting in harmful interference because of malfunctioning of the system. In order to keep it silent, the company responsible for the system must continue to allocate manpower and a tracking station and keep providing the system with the proper commands.

Another example is the ASTRA-1D satellite, which causes out-of-band emissions far above the acceptable levels of harmful interference in an adjacent frequency band because of the wrong design or malfunctioning of system components. Since the satellite is in a geostationary orbit, it will take a long time before a better, nonpolluting, one will replace it. The issue has been brought to the attention of the German administration, which confirmed the observed interference by observations at its Leeheim Satellite Monitoring Station and, supported by this evidence, also to the attention of the operator. The operator, *Société Européenne des Satellites* in Luxembourg, did improve the system to some extent, but a complete cure of this problem has not yet turned out to be possible.

The system is heavily used in Europe for direct-to-home broadcasting and therefore, political arguments oppose the shutting down of the system in the frequency range it is polluting.

These examples help us to understand the limited applicability of agreements.

The solution reached in the case of the GLONASS problem is the result of an agreement between the GLONASS administration and IUCAF concluded in 1993. The GLONASS system had been causing harmful interference to radio astronomy since it started operations in 1987. Radio astronomy stations were the first to notice this initially military Russian system since it operated partly in a band allocated to the radio astronomy service; in fact the GLONASS designers were not aware they were causing harmful interference to radio astronomy. The agreement between GLONASS and IUCAF foresaw the move of the frequency channels of the GLONASS system outside the radio astronomy frequency band. Furthermore, this agreement also included a time path for this modification.

Since it was too late to start a coordination process for the already operational GLONASS system and since it was highly unlikely that the Russian authorities would switch off the system, GLONASS and IUCAF reached the aforementioned agreement in order to solve the problem in the not too distant future while at the same time allowing both radiocommunication services, that is, GLONASS under radionavigation by satellite on the one hand and radio astronomy on the other hand to continue their operations. It was out of the question to demand filtering of the radio astronomy receivers, as radio astronomy enjoyed a primary allocation in the affected frequency band of 1610.6 to 1613.8 MHz.

We note that the GLONASS-IUCAF agreement has its roots in the ITU Radio Regulations: the adopted modifications are in compliance with the obligation of the GLONASS system not to cause harmful interference to the radio astronomy service, which was initially caused by the space-to-Earth transmissions of this system in the band 1610.6 to 1613.8 MHz.

The situation is different for the TEX satellite; because of malfunctioning of the system, the operator is not able to switch it off or to bring it into a different attitude to prevent the reception of its emission on Earth. The case has been well known since 1992, when Dutch and Indian radio astronomers discovered the interfering source. The satellite operator has been known to the radio astronomy service since 1998 and he has acted according to the regulatory criteria. Radio astronomers will need to wait until the satellite finally "dies"; however, with the necessary continuing effort of the operator, its interference has been effectively removed.

In this specific case scientists did not conclude an agreement with the operator. The administration of the Netherlands informed the ITU-R Radiocommunication Bureau of the problem, which then notified the U.S., Russian and Chinese administrations. The Russian and Chinese administrations responded that the satellite was not theirs. After several years of investigation, the U.S. administration admitted in 1998 that the interfering satellite was a U.S. military satellite, which was supposed to have ended operation in 1991. The United States took the responsibility of acting against the operator, who has since then taken measures to keep the satellite quiet.

Also in the ASTRA 1-D case, it is unlikely that an agreement can be reached which will lead to anything acceptable for scientists. At the time of this writing, the frequency band 10.6 to 10.7 GHz therefore simply cannot be used for scientific research, owing to spectrum pollution from the satellite which operates just above 10.7 GHz. In the regulatory sense, this issue is not yet closed, and discussions at European level are continuing.

7.2.1 Mistakes with agreements

The Iridium lesson. The Iridium satellite system caused a lot of trouble for the radio astronomy service since, despite guarantees to protect radio astronomy, no precautions to realize this have been built in into the system. Furthermore, other defects were noticed in the meantime.

An analysis of the history of the Iridium tragedy shows a number of avoidable mistakes made by scientists at the time, which had a large impact on the coordination process elsewhere.

The situation is as follows: The WRC 1992 allocated the band 1610.6 to 1613.8 MHz to the radio astronomy service on a primary basis. It also allocated the band 1610 to 1626.5 MHz to the Mobile-Satellite Service (MSS), MSS (Earth-to-space) on a primary basis, and MSS (space-to-Earth) got a secondary status in the band 1613.8 to 1626.5 MHz. Furthermore, footnote S5.372 was added to the band 1610 to 1626.5 MHz which says that *harmful interference shall not be caused to stations of the radio astronomy service using the band 1610.6 to 1613.8 MHz by stations of the radiodetermination satellite and mobile-satellite services.* These allocations and the footnote hold for all three ITU-R Regions (see Fig. 6-1 in Chap. 6).

The band 1610 to 1626.5 MHz is used by several MSS operators for the Earth-to-space transmission path. The Iridium Satellite System wanted to use the band 1621.35 to 1626.5 MHz for space-to-Earth transmissions (conforming its U.S. license and allocation). The Iridium Satellite System consisting of 66 satellites in Low Earth Orbits flying at an altitude of 780 km was designed and built by Motorola.

Already soon after the first specifications of the Iridium system became known, the radio astronomy community (represented by IUCAF) discussed with Motorola/Iridium the protection of radio astronomy observations in the frequency band 1610.6 to 1613.8 MHz against possible harmful interference from its MSS stations. By letter of October 9, 1991, to the chairman of IUCAF, Motorola stated: "*Motorola's goal is to share frequencies in a manner that will not interfere with radio astronomy or other MSS/RDSS services.*"

In Europe, CRAF tried to have technical discussions with Iridium on the protection of radio astronomy from unwanted emissions from the space-to-Earth transmissions of the Iridium Satellite System into the frequency band 1610.6 to 1613.8 MHz. However, before 1998, Iridium was not inclined to have such discussions.

In due course of time it became clear that Motorola had not implemented any proper measure in the Iridium system to protect radio astronomy or any other radiocommunication service, as the system design showed. Instead, Motorola proposed some possible operational solutions to radio astronomy stations. However, these would affect radio astronomy observations only without having any impact on Iridium system's operations. Furthermore, Motorola showed that their space-to-Earth transmissions in the band 1621.35 to 1626.5 MHz would cause harmful interference of up to 30 dB above the levels permitted as given in ITU-R Recommendation RA.769, as was eventually confirmed by radio astronomical observations.

At no time, was either Motorola or Iridium willing to have *technical discussions* with IUCAF or CRAF in order to seek a solution for this problem. Iridium only wanted to discuss its view on operational solutions with radio astronomy stations on a site-by-site basis, which soon turned out to be a divide-and-rule policy with respect to radio astronomy stations by this operator.

Mistake 1. The GLONASS-IUCAF agreement complies fully with the ITU Radio Regulations and provides the radio astronomy service with the guarantee that, according to a specified time path, the space-to-Earth transmissions of the GLONASS system will be moved outside the band 1610.6 to 1613.8 MHz in which the radio astronomy service has a primary allocation and that through adequate technical means the harmful interference to the radio astronomy service will be suppressed. The agreements concerning the Iridium system are of a quite different nature: they are meant to "regulate" the harmful interference from spurious emissions from the transmissions of the Iridium system (space-to-Earth). While the GLONASS-IUCAF agreement is made to solve an issue of spectrum pollution, the agreements with Iridium regulate the pollution that the victim service must accept. Therefore, these agreements are fundamentally of a different nature, and the existence of the GLONASS-IUCAF agreement must not be accepted as an argument to conclude the kind of agreement as made subsequently with Iridium.

Mistake 2. In 1994, the U.S. National Radio Astronomy Observatory (NRAO) signed a Memorandum of Understanding (MoU) with Motorola in which NRAO accepted that radio astronomy observations at its stations in the band 1610.6 to 1613.8 MHz would be done, to the greatest extent possible, during low traffic hours of the Iridium system, that is, during 4 night hours. In addition some other obligations of a technical nature were accepted by NRAO. Although an MoU is a common enough kind of agreement, as such it is not legally binding. Furthermore, this MoU put burdens only on radio astronomy while neither Motorola nor Iridium had to accept any constraint or obligation. Nor does the agreement contain any time-path, milestones or an arbitration procedure in case of conflict. In addition, NRAO accepted a nondisclosure agreement concerning technicalities, which could arise from the Iridium case.

The tragic aspect of the NRAO-Motorola MoU is that NRAO entered it in good faith, without apparently being aware of the far-reaching consequences of precedents for sister organizations: Motorola used this MoU by stating publicly outside the U.S., that *"the radio astronomers agreed"*, without explaining that it was reached with one institute in the world only. Had NRAO been aware of the potential implications elsewhere, it would certainly have consulted the radio astronomy community outside the U.S.—which was not done, however. Only if on the other hand, the issue could have been considered as a truly local one with no consequences whatsoever for radio astronomy operations outside the U.S., then NRAO would have been correct in treating it as a local issue. One could conclude that NRAO had simply overlooked the consequences, but even a basic analysis of the Iridium system parameters would have revealed that the footprint of the satellites has such dimensions that its transmissions cannot remain confined within the U.S. alone.

On the issue of the nondisclosure agreement signed, some comments are noted in Sec. 7.2.4.

Mistake 3. In 1997, the U.S. National Astronomy and Ionospheric Center, NAIC, operating the Arecibo (Puerto Rico) Radio Astronomy Station accepted a similar agreement with Iridium which states that for about 8 hours per night the Iridium interference will be below the level allowed for detrimental interference for radio astronomy. It also accepted a nondisclosure agreement with Iridium, like NRAO had done. Furthermore, NAIC accepted that they would notify Iridium when exactly it wanted to do observations in the frequency band 1610.6 to 1613.8 MHz.

On the notification or demonstration-of-need issue, some comments are noted in Sec. 7.2.4.

7.2.2 Consequences of mistakes 2 and 3

These mistakes put pressure on radio astronomers outside the U.S. to enter into similar agreements with Iridium. In Europe, the CEPT and the EC put pressure on CRAF to accept an agreement with Iridium because of the conditions included in the licensing of the Iridium system in European countries to guarantee protection of radio astronomy in Europe.

The guidance of the CEPT resulted in a Europe-wide agreement signed in 1999 between the European Science Foundation on behalf of CRAF and Iridium, obliging Iridium to remain for 7 hours per day and a number of full weekend days below a specified power flux level. The agreement states explicitly that it is legally binding on both the ESF/CRAF and Iridium, as well as to the successor of each of the parties, in the event that this might occur. It does also state that by January 1, 2006, the Iridium system must cease to cause interference above a specified power flux density level. The agreement also foresees an arbitration procedure in the case of any legal dispute. Until now, dispute handling has not been covered by agreements outside Europe.

Throughout the negotiation process towards the ESF-Iridium agreement, CRAF cooperated intensively with IUCAF which gave strong supportive guidance. Elements of this agreement have also been taken into account in Canada and India. In Australia, no agreement between radio astronomers and Iridium has yet been reached.

The events in the U.S. show clearly where failing to consider precautions properly may lead. In Europe, the existence of an MoU between NRAO and Motorola was the argument to make European radio astronomers negotiate their own agreement which, after all, is nothing more than an agreement to time share with the radio waste of a defective satellite system, and thus a precedent was established that such an agreement could be reached regardless of the view of the ITU Radio Regulations. For the NRAO case, the ITU Radio Regulations have limited status, since local legislation has priority over local issues. In Europe, however, the situation was different, since a Europe-wide solution was sought. Therefore, one may question whether an agreement as such was the proper solution, given the legal status of the ITU Radio Regulations.

However, European licensing directives and the difference between national legislation in the individual European countries called for a swift and pragmatic solution to break the deadlock.

Lesson 1. The GLONASS-IUCAF agreement proves that in the case of unforeseen difficulties, the public legal context of the ITU Radio Regulations provides an adequate framework for full cooperation and good faith in how to solve the problem. This attitude is essential for mutual coexistence and cooperation in the use of radio frequencies.

Lesson 2. While the GLONASS-IUCAF agreement was made to *solve* an issue of spectrum pollution, the agreements made with Iridium *regulate the pollution that the victim service must accept*. Therefore, these agreements are fundamentally of a different nature and the GLONASS-IUCAF agreement should never have been used as an argument to force the conclusion of an agreement on another kind of issue, as made in the case of Iridium, or any other future issue.

In both cases, the existing regulations and legislation proved inadequate to solve the issue at hand. Nevertheless, the GLONASS-IUCAF agreement set a precedent and has found its way into national rulemaking, and has in this sense integrated into the national regulatory processes.

The same occurred with the various agreements with Iridium. However, it must be noted that as far as it is currently known, private agreements between radio astronomy communities and Iridium were required to fill an evident regulatory inadequacy regarding the protection of a radiocommunication service from plain pollution, despite the various articles and clauses in the ITU Radio Regulations addressing this issue.

Agreements like that signed between GLONASS and IUCAF comply fully with the public legal regulatory framework. However, if it becomes publicly accepted that victims can be forced to sign agreements accepting pollution regulation, as happened in the Iridium case, and if this becomes common practice for any purpose, this will weaken the public regulatory process and the legal status of an international treaty. Administrations should prevent such agreements rather than suggesting, or even asking for them in order to solve an issue.

Also, the pollution victim should not accept an agreement. For, in doing so, he contributes to the creation of precedents showing that private agreements can be made to regulate pollution, quite separate from the already available legislation, simply to make life easier for the polluter alone.

Lesson 3. Another lesson to be learned from the Iridium story is that individual scientific radio stations must adequately coordinate all their frequency management efforts with the expert organizations working on the issue, for example, IUCAF, CORF, and CRAF. When questions are asked on coordination issues, and their impact transcends the borders of the country where the station is located, agreements between only that specific station and the nonscientific user must be avoided by all means. The reason is that separate or site-by-site agreements generate site-specific or scientist-specific characteristics which may not apply elsewhere, but the most advantageous of which will be used by the nonscientific user in the attempt to reach his goal.

Mistake 4. When it is necessary for scientists to enter into an agreement with a nonscientific entity, the scientists must be careful that the correct legal context and concepts are used in the agreement. The frequency regulation is of a *public legal* nature, while an agreement between a scientific organization and a nonscientific entity arises in a *private legal* context that must comply with the public legal framework. The ITU Radio Regulations have this public legal status, as do national frequency regulations and telecommunication legislation; agreements between private legal entities typically cover private legal issues. Therefore, an agreement must comply with the proper public legal context and may neither modify the legal interpretations for the benefit of one private party nor contain language taken from a public legal text that is not adequately defined in a private legal context.

This issue was raised in the discussions between CRAF and Iridium when Iridium tried to introduce public legal language taken from EC Directives into the anticipated agreement, such as the need to comply with the so-called "proportionality principle." The proportionality principle implies that the lightest possible restrictive regulatory measures should be imposed on the parties to the agreement. From this, Iridium inferred that protection needs to be granted only on demonstrated need. This interpretation was disputed by CRAF as improper legal reasoning. When nonprivate language and concepts find their way into clauses in a legally binding agreement between two private legal entities, the execution of the agreement may well have negative consequences for one of them.

Lesson 4. When it comes to discussions on agreements, scientists must seek legal support. In a region like Europe, this is not easy, since legal questions contain elements of international and national law, both public and private. Scientists must be extremely cautious that their good faith is not abused.

Mistake 5. When an administration asks scientists to enter into an agreement with a nonscientific entity, in their good faith, the scientists may consider that once a number of clauses on operational or technical issues have been agreed upon, the agreement is complete. The mistake is that when the agreement does not explicitly state that it is legally binding, each party still has the freedom to act in any way it likes. In itself the issue of making an agreement legally binding is important, but as in daily practice, conflicts and disputes cannot be avoided, the scientists must not be satisfied with an agreement in which dispute arbitration is not included, despite the fact that the agreement is, evidently, made in good faith.

Lesson 5. Any formal agreement between a scientific and a nonscientific private organization must include a clear statement to make the agreement legally binding as well as clauses regulating dispute resolution. Disputes cannot always be resolved via arbitration, and one has to verify if the agreement contains adequate parameters within which arbitration is possible. Furthermore, when dispute resolution becomes necessary, one has to state under which law this has to be performed, and one has to make clear that adequate jurisprudence related to the subject of the dispute actually exists.

Therefore, also in this case, adequate legal support and guidance is essential. To generalize this conclusion, scientists must be aware of the legal impact of their discussions and cooperation with nonscientific parties. Usually this is not a problem, but scientists must avoid any unwanted difficulty.

7.2.3 Nondisclosure agreements

In discussions with industry and operators, the issue of an agreement not to disclose confidential information to third parties not involved in the discussion is frequently raised. Such confidentiality may relate to technological items or to commercial issues, knowledge of which may be of great importance to each of the parties entering such an agreement. Therefore, nondisclosure agreements are a well-known practice in industrial and commercial collaboration and discussions.

However, by its nature, science has no real use for specific commercial information provided by the other party. This implies that any nondisclosure agreement on commercial issues can only be an empty binding agreement. Only where the scientific question is directly related to a specific technological solution of which the other party has the patent or cannot disclose information for understandable and explained reasons, a nondisclosure agreement could be considered. However, in this case, it must be clearly stated to what exactly the agreement applies, and no vague or unspecified terms must be accepted in the agreement. The root of this position is that the nature of science is different from the characteristics of a commercial-industrial enterprise. Also the scientific culture of freedom of information exchange does not comply with a nondisclosure agreement. Therefore, nondisclosure agreements cannot support the progress of science.

If a scientist accepts a nondisclosure agreement without a clear guarantee that it is not a void agreement, he binds himself to the other party in the sense that he accepts not to disclose information received within the scope of the terms of the agreement. This may have the nasty consequence that the scientist's mouth is sealed while the other party is under no obligation to provide him with even the smallest bit of useful information.

For this reason, *nondisclosure agreements driven by commercial motives must be avoided by all means* unless this would hamper the very progress of the particular scientific context relating to the agreement. A specific case in which a nondisclosure agreement is preferred concerns the coordination of the operations of transmitting stations with radio astronomy stations. The agreement between the operator of these stations and the other (scientific) party could be subject to a nondisclosure agreement if there is a real threat that other operators could abuse elements of this coordination agreement. In any case, great care must be exercised, and warnings should not be underestimated.

7.2.4 Providing scheduling/planning information

Some operators support the interpretation that protection of another radiocommunication service is needed only during actual operations of the victim service.

It must be noted that such an opinion is not supported by the ITU Radio Regulations and the ITU Constitution. The ITU Radio Regulations and regional and national frequency plans have a public legal status, that is, as *allocation* plans they are not dealing with the factual operational status of an application in a radiocommunication service after an allocation has been given on publicly demonstrated need. The protection arguments would be different if these regulations concerned *assignment* planning for specific applications to which specific conditions might apply.

Secondly, if it is accepted that one radiocommunication service is protected only on its demonstrated need, then other questions arise, related to the criteria used for this demonstration and to why the service asking for a demonstration of this need has the mandate to demand this at all. Furthermore, it is far from clear what arbitration process is adequate to resolve conflicts.

Thirdly, if accepted, the difference in allocation status between radiocommunication services becomes arbitrary and dependent on the pressure and power of private (usually commercial) entities. If such a practice is accepted, the status of the regulations is significantly weakened.

Therefore the reply to the question *does scheduling/planning information need to be provided* cannot be answered in the positive. Furthermore, if the answer were to be positive, other questions arise relating to the criteria for the justification to schedule/plan a specific observation or an experiment. Users agreeing to respond to the question run the danger of entering into unsolvable disputes, since the other party is usually not competent in the necessary evaluation. Furthermore, precedents for other users in the same radiocommunication service can be created which are bound to be difficult to remedy at a later stage.

In the vein of the faulty interpretation of protection touched upon above, the U.S. agreements between U.S. radio astronomers and Motorola and Iridium contain clauses stating that radio astronomy observations should be scheduled to avoid peak traffic periods of the Iridium system. The NAIC has even accepted an obligation to show Iridium its observing schedules, to demonstrate to Iridium that the requested observing time is indeed necessary. The European agreement between the ESF and Iridium does not contain such a clause, as it would violate the internal sovereignty of each individual radio astronomy station to do the observations at the time it considers best to achieve the scientific goal.

When nonscientific users ask the scientists to specify when a particular frequency band will be used, the *answer must also be a categorical refusal to respond.* The reasons are as follows:

- Such a request is in conflict with the ITU opinion on protection of a radiocommunication service, as indicated above.
- A notification of demonstrated need is not related to the protection of the notifying radiocommunication service, since the protection concerns the *prevention* of interference detrimental to the victim service by technical and operational measures, which must be taken by the interfering service, and not by *regulation* of the interference.

- No private user of the radio spectrum has the authority to ask another user to show when a particular radio frequency band is needed, since in simple terms, it is none of their business. The background to this is the role of the administration in the coordination process. When a radiocommunication service enjoys the allocation of a particular frequency band, the very fact of this allocation shows that the international community accepted that the need to use this band by that service has been justified sufficiently.
- Such a notification procedure places the notifying radiocommunication service in the victim position and in a secondary status with respect to the service to which the notification is made, since it relieves the interfering service from the obligation to take appropriate measures to prevent interference (e.g., ITU Radio Regulations S0.2 and S0.3).

7.2.5 Monitoring

When harmful interference is experienced in scientific projects, it is important to know what exactly the characteristics of the interfering source(s) are. This information is important background information when the interference issue is discussed with a regulatory authority. In many cases it is not possible for the scientist to derive this important information from the interference noted in the data resulting from his experiment/measurement. In such an event it seems attractive to the operator of the scientific radio station to install a monitoring facility on a vehicle and to drive this around in order to obtain dedicated measurements of the interfering signal.

Warning 1. It must be noted that not every country's national administration will be happy with such a monitoring experiment performed on the initiative of the scientific radio station. On the contrary, even the publication of monitoring data from such an experiment may actually be forbidden by law, as is the case in the Netherlands and in France. When there is a need for such a monitoring experiment, adequate consultation with the administration must be undertaken to find ways to obtain the desired information.

Some scientific stations have installed instruments dedicated to monitoring particular radio frequency domains. It is recommended that the National Regulatory Authority be consulted beforehand in such projects. It has been known on some occasions that this authority has had an active interest in the construction of such a facility, as it is important for a regulatory authority to have adequate knowledge of what is going on in the radio frequency spectrum in its territory. The scientist's special interest in monitoring this information concerns, for instance, scheduling reasons, the establishment of a frequency plan for his experiment.

Warning 2. The scientist must note that the data obtained with such a monitoring facility do not provide information on interference but rather on spectrum occupancy. Therefore, these data are not pertinent to provide support for a specific interference complaint to a regulatory authority, since interference corrupts scientific measurement data and no scientific measurement data are degraded in this monitoring facility. Furthermore, the scientist must be aware of the need to keep the monitoring information to himself or interested colleagues in his institute only, as in many countries it is not allowed to show such information to third parties or to make it public. The reason for this is understandable, since information on spectrum occupancy reveals to other users of the radio spectrum where the spectrum is unused and to what degree. Such information could have high commercial value.

The committee on Radio Astronomy Frequencies maintains a database for spectrum occupancy monitoring data and has a facility to query this database and to manipulate its contents accessible on the World Wide Web under password protection. In order to serve the scientific community and other authorized entities, all astronomical monitoring stations are urged to send the data with a defined data format to CRAF.

7.3 How Are We Operating Scientific Radio Projects after 25 Years?

The way sciences are able to use the radio frequency spectrum after 25 years, depends on parameters such as the development in radio spectrum usage, technological developments on the transmitter and on the receiver side and various other developments such as in the social and economic fields. Before outlining the perspective for scientific radio frequency use after 25 years, we need to make comments on these developments.

The development of science proceeds in an unavoidably intensive relationship and cooperation with other technological developments, especially in the field of the so-called "information technology industry." Advances in information systems and communication networks are driven by digital technology. These developments have brought and are bringing new challenges to the regulatory and legislative regimes and have begun to blur traditional regulatory definitions and jurisdictional boundaries.[4] Universal Access and Interconnection are key factors in the development of competition in the telecommunication industry. This will as far as we understand not change in the next decades but be enhanced.

7.3.1 Spectrum development

Telecommunication developments depend on the possible capacity of communication channels per Hertz and the costs to deploy a service per Hertz. It is a well-known fact that this capacity for satellite-based mobile and fixed services is more than an order of magnitude lower than the capacity for terrestrial systems and even declining. It is therefore not certain that communication by satellite is the future development for telecommunication. This development will heavily depend on the telecommunication infrastructure in the regions where it is applied. In Europe, for instance, the terrestrial wired and wireless telecommunication is very well developed. In other areas of the world this can be much less: for example, in the United States or many remote areas, this Earth-based telecommunication system is less advanced and therefore the interest in space-based systems can be greater in these areas than in Europe.

Recently constructed long distance optical fiber cable systems between, for example, Germany and China, and a newly planned transatlantic high capacity cable system show that space systems are not an apparent route for telecommunication development. The fraction of overseas telephone traffic carried on satellites has declined to an amount of about 30 percent, while in the 1960s a single satellite could have a capacity almost equal to that of all of the RF/coaxial cables laid under the Atlantic Ocean up to that time. In the year 2000, satellites are the prime means of distributing television pictures around the globe, either to cable head ends or directly to subscribers. Satellites are also increasingly being used to provide private data networks for companies (such as banks) that have widely-distributed offices. This service usually involves a central large (hub-) station linked to many very small aperture terminals.[5]

The full blending of broadcasting, Internet, telecommunication and computing facilities and the merging of the Broadcasting-Satellite and Fixed-Satellite Services, indicate that satellite applications are likely to remain for broadcasting since by that, the other functions mentioned can be provided. Geostationary satellite systems are preferred for broadcasting direct to the home because of technological simplicity. Satellite radiocommunication is likely to remain important for navigation and surveying purposes and for safety-of-life services.

From this perspective, it is likely that spectrum can be freed by reducing the allocations to space services, that is, the Mobile-Satellite and Fixed-Satellite Services.

The blending of Fixed and Mobile Service applications will also lead to more efficient spectrum allocation, which is enhanced by increasing the communication capacity per hertz owing to the improvement of transmission and reception techniques.

Other developments are the increasing use of very wide band systems, which may have bandwidths of about 50 percent of the frequency of operation. This shows that the bandwidth requirements per application may change. This development may lead to an increase of incompatibilities between different services, since the bandwidth usage characteristics and requirements are different for different radiocommunication services. Efficient frequency allocation management must be applied to avoid sharing between incompatible services in this regard.

The increasing demand for the aforementioned new broadcasting applications requires, on the basis of current technology, an increase of high-density applications. This development adds to the demand for exclusive use of spectrum because of sharing difficulties for these high-density applications. However, we may not dwell too long on this consideration since technological development will certainly find a solution for this.

Given the nature of scientific research and physics, it is not likely that the spectrum demand for scientific applications will reduce significantly. However, it is not impossible that some scientific application may prefer different frequencies than are now used. That this perspective on spectrum development differs from nonscientific applications is owing to the fact that for most of the non-scientific services the frequency choice is much less critical than for scientific research: the latter must select the radio frequencies because of the scientific question and the laws of nature.

These developments in radio frequency requirements must lead to a reduction of the fragmentation of spectrum allocation.

Furthermore, it can easily be noted that the telecommunication development in developing countries is different from that in the developed countries. The reason for this is apparent: the developing countries will certainly lack money for the implementation of expensive telecommunication infrastructure and systems, while on the other hand areas of very low population density make any such implementation commercially not attractive (not even for satellite-based communication systems). In mid-1999, the number of telephone connections per 1,000 inhabitants in developing countries was about 0.5 percent of that in the developed countries.

A similar ratio is observed for television sets or newspapers, as statistics provided by the World Health Organization and the World Bank show. This imbalance between developing and developed countries is recognized by the ITU, but it is unlikely that within 25 years the discrepancy between these two cultural levels will decrease: on the contrary, the noted gap appears to become bigger and may even change geographically.

Such regional differences in telecommunication development imply that the characteristics of scientific interest in radio frequencies will also show regional differences.

In summary, our perspective on the situation after 25 years is the following:

- Significantly less radio spectrum is allocated to the Mobile-Satellite Service and to the Fixed-Satellite Service.
- The spectrum demand of the Fixed-Satellite Service is significantly reduced.
- Terrestrial services remain well developed, but some are blended, such as the Fixed and the Mobile Services, which implies reduction of spectrum allocated to these services individually with respect to the situation in 2000.
- There are more exclusive radio frequency allocations.
- The fragmentation of the frequency allocations is reduced, leading to a more efficient use of the radio spectrum.
- Regional difference between the requirements and use of radio frequencies remain, which is reflected in the ITU Radio Regulations table of allocations. However, the geographical distinction of radiocommunication regions may have changed, since the economic-political characteristics are intrinsically dynamic.

7.3.2 Technological development

Superficially, one may note that nonscientific users of the radio frequency spectrum are apparently more concerned about the quality of their signal channel from transmitter to receiver than scientists, because of the significant commercial implications. This quality includes the spectrum efficiency and purity of the signal. For commercial companies and industrial enterprises, this quality can be a matter of survival of the organization in the battlefield of hard competition. Scientists on the other hand consider themselves much less under such pressure because their motivation and interest is driven by the scientific question they explore. However, real life does also affect their view on system development, design, construction and operations.

7.3.3 Outside science

Current commercial pressure and hard competition lead to a practice of priority for a specific industry regardless of other users of radio frequencies. This is especially noticeable in the current practice that for space systems operating in the fixed-satellite and mobile-satellite services no mitigation factors are considered, and all the burden of compatibility issues is put on the terrestrial services. The next decades will see more regulations to address electromagnetic compatibility issues. Regulations for space systems will especially require more attention because interference experiences by industry implies that the current regulatory gaps providing relative freedom for the mobile-satellite and fixed-satellite industry need to be filled. The telecommunication industry is developing adequate mitigation techniques to provide a reliable service to the end user. At the time of writing, this aspect has hardly taken hold, but industry cannot avoid developing of its products to be less vulnerable to interference. Such a development may well lead to a product quality with the level of safety margins of, for example, airplanes, because spectrum pollution implies direct impact on the quality of life: pollution means that essential communication systems and applications using radio frequencies become unreliable, and related malfunctioning of equipment may well have great impact on the cultural, health and social aspects of society.

7.3.4 Scientific applications

Scientists and radio astronomers, in particular, intensively study new directions in system design to build instruments which are more interference robust. However, the driving force for this work is not the changing radio environment but the scientific need for systems with much better sensitivity and better angular resolution. The physics of the universe implies that besides research within radio frequency bands allocated to the radio astronomy service, much work must be done outside these bands in an environment with much interference. Research in interference free frequency bands remains a matter of the highest priority to achieve maximum sensitivity and to ensure the adequate calibration of the instrument used.

This view on the development of scientific instruments is not new and in fact has been in daily practice for decades. However, the pressure on the radio spectrum urges scientists to consider questions such as how can one build an instrument and at the same time address the question of how one is supposed to discover what has to be rejected experimentally when building the equipment?

Much research is in progress on new antenna systems for scientific instruments and improvement of calibration technology. However, the research on techniques to suppress RFI in frequency-, time-, space-, and multi-interferer domain is still in its infancy. Methods must be found to answer the following questions:

- How can one determine the characteristics of the interference with an instrument that has been built for a different purpose?
- If one knows the modulation scheme of the interfering source, is one able to remove the interfering signal? If that is the case indeed, how can one determine from the observed characteristics the modulation scheme?

At the time of writing this document, the answers to these questions are still far away.

Before any adequate answers are found to the questions just mentioned, any real research on mitigation techniques is premature. Nevertheless, it is urgent that mitigation techniques for experiments outside radio frequency bands allocated to a science service are developed. However, before doing so, one has to consider what is meant by "mitigation technique."

It is apparent that these considerations are fed by the situation at the time of writing this document and that research is inevitably progressing. However, one conclusion should probably be made in any case: Scientific research faces in the year 2025 an environment with more harmful interference, and scientists do not have the tools to avoid or remove this.

The simple reason for this is that technological development outside the realm of science is also advancing but not in coordination with scientific development. It is impossible to design and build instruments that are robust against all harm, that is, unknown or unanticipated at the time of design and construction.

We believe that the comments on radio astronomy apply to other scientific applications using radio frequencies.

7.3.5 Administrative development or the ratio between public and private sectors

A significant development, which began in the 1990s, was the changing telecommunication legislation to stimulate privatization and competition. To a large extent this process goes hand-in-hand with a process of de-regulation and convergence in regulation, such as the grouping of different industries into one industry with one regulator (e.g., the Malaysia Act of 1999). New legislation has given rise to new, separate telecommunication regulatory agencies. The governing structure of the new separate regulators, despite significant national and regional diversity, seems to point to a new model for telecommunication regulatory bodies.

While in the view of the ITU[6] the increase in regulators and legislative reform is certainly encouraging, new technologies and services are moving faster than the bodies that regulate them. Convergence is not a simple issue for telecommunication regulators. The challenge is to determine ways in which to regulate technologies that are continuously evolving and, more importantly, to determine the role of the regulator in a converged sector. The challenge for regulators is to develop consistent and relevant regulation that does not inhibit the growth of the sector, but rather encourages technological innovation.

This ITU perspective complies with the view on radio frequency management restricted to the commercial, monetary and industrial aspects of the radio frequency spectrum. It is noticeable that many administrations adhere to this point of view. However, one may also consider that the ITU perspective reflects the common view of its members and their governments.

7.3.6 Science in 2025

During the next decades, scientists will increasingly face a view on radio frequency management restricted to the commercial, monetary and industrial aspects of the radio frequency spectrum.

They will have to explain that the radio frequency spectrum is *not* limited to these aspects, although they certainly are of major importance. The intrinsic meaning of "management" is *the sparing or frugal mode of administering scarce resources to serve all humankind.* [7] This view holds certainly in any case for radio spectrum management.

In the world of commercial force and pressures, the interest of passive use of radio frequencies by science services is expected to be low to negligible.

Scientists should not expect more radio frequency bands to be allocated to science services. They must accept that, as with other users of radio frequencies, frequency management—the work to prevent and remove harmful interference—is an expensive and immense effort, which will constitute a much larger part of their work than in the mid-twentieth century. The cooperation between scientists and administrations should be intensified and improved since both parties, the user and the administration, need each other for sound frequency management.

7.4 What Management Tools do Scientists Have?

Scientific users of radio frequencies are private legal entities, which have to obey the law regulating frequency allocations, frequency assignments, and frequency allotments. As private legal entities, scientists have no tools to manage the regulatory aspect of frequency management which belongs in fact to the mandate of the regulatory authority of a sovereign country.

However, this is only a part of the story. The other part is that in radio spectrum issues, the administrations having regulatory responsibility also have the facilities to improve the radio frequency situation and environment for the scientist when the circumstances demand this. These circumstances may vary, from the request to remove radio interference to guidance when the project requires the use of a frequency for which no allocation to the science service exists. In countries with a well-profiled scientific community, administrations are usually very receptive to questions, requests, information, and other communications from the scientific community. This is partly related to the fact that scientific research is usually funded only by public money, and also that scientific frequency usage is rather vulnerable to interference and the impact of pressure by nonscientific radio spectrum users.

Furthermore, in collaboration with the national and regional administrative bodies, scientists also have the opportunity to influence the radio frequency environment. In Europe, CRAF is very active in this respect, and a number of positive results can be noted.

Another aspect the scientist must not forget is that in designing, building, maintaining and operating his equipment, he should take the opportunity to reduce its susceptibility to interference and improve the robustness of the system.

It is highly recommended that in using these tools, scientists cooperate intensively with their expert organizations. In the case of radio astronomy and remote sensing, these are for example, IUCAF, CRAF, and CORF.

Besides these administrative and technical-operational tools, no other management tools are immediately at his disposal; however, when the scientist is using these tools in an optimum way, no more may be needed. One may object that this is not sufficient, and that scientists should improve their political profile and influence at national and international level. Although this is very true, such a political action is not spectrum management as such but rather lobbying for a desired objective. Table 7-2 summarizes the conclusions of this chapter.

7.5 With What Harm Can We Live?

Harmful interference is commonly quantified in terms of the power level, which exceeds a specified level received from a transmitter. However, this parameter is not completely practical in compatibility issues in which interferers with variable time and location must be considered. Separation distances may alleviate this problem, but in coordinating radio astronomy stations, statistical methods may be used. Input for such methodologies requires quantification of, for example, the tolerated percentage of data or time loss as a result of interference. In this chapter, some ideas on tolerated losses are given for radio astronomy and other passive services applicable for frequency bands shared with other services.

Some frequency bands are passive exclusive bands, such as the band 1400 to 1427 MHz, which is allocated to the Earth Exploration-Satellite

TABLE 7-2 Spectrum Management Tools for Scientists

Collaboration with national and international administrative bodies
Improvement of hardware and software of scientific instrumentation
Collaboration with expert organizations, e.g., IUCAF, CRAF, and CORF

Service (passive), Radio Astronomy Service and Space Research Service (passive). In the ITU Radio Regulations, footnote S5.340 is added to such bands, which states that for that specific frequency band all emissions are prohibited. This footnote applies to all purely passive bands in the radio frequency spectrum. It does not state that all *transmissions* are prohibited, which would address an assignment/allocation issue, but rather relates to emissions, which is always related to some transmitting station. The ITU Radio Regulations do not provide a definition for transmission nor an explanation of the difference between emission and transmission. Nevertheless, the first conclusion is that it is forbidden to assign a station in a frequency band to which footnote S5.340 applies. In relation to the assignment of frequencies to stations adjacent to the relevant frequency band, the decision below shall apply in conformity with the ITU Rules of Procedure:[8]

> To resolve cases of harmful interference between services in adjacent bands it was decided that, irrespective of the phenomena at the origin of the interference (out-of-band emission, intermodulation products, etc.), the administration responsible for the emission overlapping a nonallocated band shall use appropriate means to eliminate the interference.

7.5.1 Radio astronomy

An important parameter in sharing the radio spectrum between the radio astronomy service and active services is the percentage of observing time lost to interference. In coordinating radio astronomy stations with active service operations, for example, by using statistical software tools like the Monte Carlo methodology, administrations need a quantification of this parameter. Existing limits to time losses tolerated by various other services in ITU-R Study Group 7 are given in Table 7-3.

TABLE 7-3 Examples of Time Losses Tolerated by Radio Services Other Than Radio Astronomy in ITU-R Study Group 7

Radio services	ITU-R Study Group	Time loss tolerated (%)
Earth Exploration-Satellite Service, 3-D sounding for forecasts	(Rec. ITU-R SA.1029)	0.01
Earth exploration, near-Earth spacecraft	(Rec. ITU-R SA.514)	0.1
Meteorological satellite service	(Rec. ITU-R SA.1161)	0.1
Space operations systems S/N > 20 dB for > 99% of time	(Rec. ITU-R SA.367)	1.0
Broadband passive sensors in spacecraft (looking down)	(Rec. ITU-R SA.1029 and Rec. ITU-R SA.1166)	1.0–5.0

Radio telescopes are designed to operate continuously, following a schedule of observing programs requested by astronomers. As a rule, access to radio telescopes is on a competitive basis, with research proposals often exceeding available telescope time by a factor of two or three. Virtually all radio astronomy installations are operated out of public funds and must be used very efficiently. Some loss of observing time resulting from maintenance or upgrading of hardware or software, however, cannot be avoided.

In order to achieve such results, other services will have to design their individual systems and to control their operations to an appropriate fraction of these figures. Prudence dictates that individual systems exhaust only a fraction of the interference budget, depending on factors related to the actual allocation situation, such as band sharing and the interference potential owing to unwanted emissions from other services.

The advent of radio services using space stations and high-altitude platforms requires some reassessment of the measures by which the radio astronomy service is protected from interference. Frequency sharing with such services is normally impossible, but potentially negative effects upon the radio astronomy service by services in other bands arise in two ways:

- Unwanted emissions falling in bands allocated to the radio astronomy service;
- Intermodulation and departures from system linearity in radio telescope systems owing to strong signals in adjacent bands.

It is generally assumed that the satellite operators will use all practical means to minimize unwanted emissions and radio astronomers all practical methods to minimize sensitivity to signals in adjacent bands. Nevertheless, the second problem will be an important consideration when allocating bands adjacent to or close to bands allocated to the radio astronomy service.

The emissions studied in radio astronomy are Gaussian white noise. The measurements are either broad band measurements of integrated flux and polarization over an entire frequency band allocated to the radio astronomy service, or of spectral structure within that band. In many instances, there is little time-variability in the properties of the emission over time-scales shorter than a day. In order to improve the signal-to-noise ratio, observations usually involve integration of the signal over seconds, hours or even days. However, some emissions consist of short pulses or bursts, where simple integration is not appropriate; the radio emissions from pulsars, often of the Sun and of Jupiter are examples. In such cases it can be very difficult to distinguish the desired astronomical signals from impulsive interference. But also in measurements requiring long integration times, it can be very difficult to distinguish the desired astronomical signal from interference, particularly when it is manifest as a low intensity background with little variation in time.

Whenever data loss is mentioned in this document, it refers to data that has to be discarded, because it is contaminated by interference above the levels of Recommendation ITU-R RA.769 from one or several sources. Data loss may result from loss of part of the observing band, or part of the observing time. Both of these can be expressed as loss of effective observing time.

An observing session may consists of a series of shorter observing units of minutes, duration to more than 12 h depending on the instrument used and on the requirements of the scientific project. After the observing session, these units or fractions of them are weighted for data quality, or deleted if unusable. If the measurements are of a single position in the sky, as in measuring the strength of a weak, discrete radio source, the usable blocks will then be incorporated into an overall average. Lost observing time then translates into a poorer measurement that might be unusable. If mapping is being done, lost observing time results in lost detail on the map and a possible need to redo part or all of it.

There are two aspects to the issue of loss of observing time owing to interference to radio astronomical observations that must be considered. In other radiocommunication services, these are the needs of the customers during a budgeting period. The equivalents for a radio astronomy facility are:

- The amount of time during an observing session that is lost as a result of interference. In the simplest case it is that which is blanked by interference. However, saturation of the receivers and other equipment used to process and record the data obtained can lead to greater loss than just the time interval actually occupied by the interference.
- The amount of available observing time at any observing facility that is negated by interference over an operating year. This may not in all cases be the same as the first point because some interference problems might be avoided through appropriate scheduling of observations and maintenance.

There are two further burdens caused by interference to radio astronomical observations:

- The amount of time and effort spent *after* the observing session exorcising the effects of interference from the data. This often far exceeds the loss of observing time that the interference itself causes during the observations.

- As the amount of interference increases, the probability of weak interference not being identified also increases. In addition, the more processing is needed to make data usable, the greater the risk that erroneous conclusions will be drawn from the data.

7.5.2 Interference owing to variable propagation conditions

We consider the following scenarios of interference owing to variable propagation conditions:

Transmitters beyond the horizon. In cases where the strength of an interfering signal varies as a result of time-varying propagation conditions, a percentage of time must be specified for propagation calculations. A figure of 10 percent is given in Recommendation ITU-R RA.1031-1. However, this does not automatically lead to a 10 percent data loss for radio astronomy observations. Propagation conditions vary episodically, typically over a time-scale of days. It should therefore be noted that over periods of weeks at a time, the period for which data are contaminated by interference might be a few days. These effects occur primarily at longer wavelengths. Hence, periods of data loss can be reduced by dynamic rescheduling to about 1 percent.

Transmitters under line-of-sight. For transmitters under line-of-sight, for example, airborne or space-based transmitters, the variability of propagation conditions is negligible compared to the signal strength, and therefore does not need to be considered.

7.5.3 Interference from transmissions variable in time and location

For the present analysis we identify the consequences of interference from transmissions variable in time and location and quantify the extent to which this harm can be tolerated:

7.5.4 Sharing of a frequency band

Terrestrial transmitters. Radio telescopes are operated continuously. Loss of observing time owing to interference is to be avoided. However, some small loss is inevitable. An example is the emissions from mobile (Earth) stations in the mobile (satellite) services, where the location and activity of individual users cannot be fully controlled. An acceptable maximum level of data loss from such a system is 2 percent.

This sharing scenario is dealt with in Recommendation ITU-R M.1316, which also asks for agreed input parameters. This figure of 2 percent may be used.

Space-based transmitters. Sharing with satellite downlinks is not possible in bands where the radio astronomy service has a primary allocation. Hence there should be no data loss from this source.

Unwanted emissions into a radio astronomy frequency band

Terrestrial transmitters. In coordinating radio astronomy stations with the stations of other services operating in frequency bands outside a radio astronomy band, filtering of transmitters and geographical separation should be considered to suppress unwanted emissions into the radio astronomy band to below the Recommendation ITU-R RA.769-1 threshold levels at the location of a radio astronomy station. However, there is a potential for interference when the beam of a radio telescope is pointed closer than 19° to a terrestrial source. The levels in Recommendation ITU-R RA.769-1 are based on the assumption that the interfering source is at the 0 dBi contour. As can easily be understood, a terrestrial source on the horizon (elevation = 0°) can cause detrimental interference in up to 1.8 percent of the visible hemisphere for a telescope that points within 5° of the horizon. Some sources of interference are known and can be avoided. A practical level of 2 percent data loss can be expected from a terrestrial source on the horizon.

Space-based transmitters. Space-based transmitters have the potential to cause extensive harmful interference to the radio astronomy service through unwanted emissions into a radio astronomy band. It is extraordinarily difficult to reduce the data loss suffered by the radio astronomy service from unwanted space-based transmissions even in primary radio astronomy bands. For example, even a satellite transmitting at the harmful interference limits defined in Recommendation ITU-R RA.769-1 blocks 5.5 percent of the visible hemisphere of the sky. If the orbital elements of the satellite are known, the radio telescope may be scheduled to avoid observations within 19° of the satellite position, and data loss can be reduced. This rescheduling, however, results in loss of flexibility of telescope use and loss of observing time owing to the additional steering time required and may not be possible in all cases. Rescheduling of observations may be possible for geostationary orbit and slow moving satellites, but fast-moving satellites pass through a cone of radius of 19° around the observing direction quickly. In practice, a lower probability of data loss will result. It is expected that the overall effect is approximately 2 percent.

The practical assessment reviewed above leads to an acceptable aggregate data loss to the radio astronomy service of 5 percent from all sources. The existence of multiple overlapping sources of interference is a practical aspect that must be accounted for. To comply with this requirement, 2 percent per individual system is a practical limit.

7.5.5 Other science services

Time losses tolerated by radio astronomy have been indicated in detail in Sec. 7.5.1. The reason for this amount of detail is that at the time of writing this document, the radio astronomy criteria were developed only in 2000 and have not yet been integrated in many publications. The criteria for radio services other than radio astronomy in ITU-R Study Group 7 have been explained already in more documents and are summarized in Table 7-3. It should be noted that the structure of the argument is similar for all services in ITU-R Study Group 7 but with their specific articulations.

7.5.6 Regulating tolerated loss

When a fraction of percentage of time loss is tolerated by a radiocommunication service, this should not be interpreted as if this service is happy to accept a regulation in which this time it is unprotected. Just as with road traffic, to prevent the number of accidents exceeding a certain percentage, less stringent precautions need to be taken than when trying to guarantee absolute safety. Therefore, in regulating a tolerated fraction of time lost because of interference, one could approach the protection by considering that the interfering transmissions remain below the interference threshold applicable to this service all the time, but that this threshold does not guarantee 100 percent protection, since daily practice within the accepted thresholds implies an already accepted fraction of lost time.

For further details on this issue, refer to Recommendations ITU-R RA.1513.

7.5.7 Passive versus active: the sharing problem

According to the terminology of the ITU Radio Regulations, *frequency sharing* occurs when a frequency band is allocated to more than one radiocommunication service, both of which have a primary status in that particular frequency band. When a service has a secondary status in that band, it cannot interfere with the primary service nor claim protection and must accept interference from a primary service unless some conditions have been put on that primary service. A situation of a radiocommunication service with a secondary status in a frequency band, in which other services have a primary status, is not called a sharing situation.

It is apparent that a sharing situation implies a potential threat of harmful interference. This is especially true when active and passive radiocommunication services share the same frequency band. The requirements of the passive service and of the active service are usually very different, which adds to the complicated issue of sharing.

TABLE 7-4 Possible Mutual Relations Between Radiocommunication Services

Relation between radiocommunication services	Sharing situation	Potential threat of harmful interference	Coordination required/recommended
1. Frequency band allocated to more than one service, i.e., these services have a primary allocation.	Yes	Yes	Yes—seek advice of administration.
2. Frequency band is allocated to secondary service.	Yes	Yes, but secondary service has no right to complain.	Yes—support by administration depends on good will of these administrations only.
3. Frequency band is not allocated but used by a Radiocommunication service on the basis of an ITU-R footnote.	No	Yes, but secondary service has no right to complain.	No—support by administration depends on good will of these administrations with respect to the victim radiocommunication service.
4. Adjacent band is used in which the service has no regulatory status.	No	No, but out-of-band or spurious emissions may occur when the systems are not designed properly.	No—but in case of harmful interference from out-of-band or spurious emissions, the victim may submit a formal complaint to the national administration.
5. A frequency band is used in which the service has no regulatory status.	No	Yes—the victim service has to accept this situation and to try to find a solution.	No

Table 7-4 summarizes the possible mutual relations between two radiocommunication services and the potential threat of interference. It must be noted that the sharing concept never applies to protection from out-of-band or spurious emissions. In this case, the issue is rather a *compatibility issue*. Such an issue should not be solved as if it were a sharing case. If this happens, this solution is a pseudo-solution that may relieve the interfering service from taking adequate measures to protect the victim service, and it creates easy precedents and dilutes the public legal nature of the regulations that apply.

In a sharing situation, harmful interference of one radiocommunication service with the other with which the frequency band is shared can be avoided when these services are adequately coordinated. Table 7-5 summarizes the possible coordination methods. Such coordination is usually accomplished by, or under, the auspices of a National Regulatory Authority. In regional coordination issues, usually regional administrative organization plays a role, such as the CEPT for Europe.

In the coordination process, operational or regulatory solutions are sought to enable both services to operate in the frequency band, although under certain restrictive conditions. Such conditions may imply the implementation of the coordination area concept. According to the ITU Radio Regulations, a *coordination area* is the area associated with an Earth station outside which a terrestrial station sharing the same frequency band neither causes nor is subject to interfering emissions greater than those at a permissible level. Another option to solve a sharing problem is to set an *upper spectral power flux density limit* (mask) on the transmissions of the interfering source. It must be noted that this limit must always be an aggregate limit. A third option is *time-sharing.* This means that when two services are incompatible with each other,

TABLE 7-5 Possible Coordination Methods

Coordination method	Impact on interfering service	Impact on victim service
1. Coordination area—applicable only to terrestrial interference situations.	Within the area around the associated Earth station the interfering emissions must not exceed a permissible level. The interfering service must implement the technical means to achieve this.	Operations of the victim service unharmed down to the permissible level of emissions from the interfering service.
2. Maximum transmitter spectral power flux density.	The interfering emissions from the interfering station must not exceed a specified spectral power flux density. The interfering service must implement the technical means to achieve this.	Operations of the victim service unharmed down to the permissible level of emissions from the interfering service.
3. Time sharing between interfering Earth station and the station of the victim service—it should be noted that the time-sharing scenario as such is not defined in the ITU Radio Regulations.	Operational constraints for interfering station.	Operational constraints for station of the victim service.

that is, they cannot operate simultaneously, the regulatory authority specifies when each of them can operate without interference, and when one (or more) of them has to cease operation or accept interference.

The following explanatory comments can be made on the sharing scenarios:

Geographical sharing means that a coordination area is defined and installed in such a way that an Earth station outside this area sharing the same frequency band with the station used by the scientific application neither causes nor is subject to interfering emissions greater than those at a permissible level.

Time-sharing means that time constraints are put on the operations of an interfering Earth station and the station used by the scientific application sharing the same frequency band in such a way that each of the stations operating at the same time as the other station, neither causes nor is subject to interfering emissions greater than those at a permissible level. This means in practice that each of the affected scientific stations will blank its reception at the presence of transmissions by the other station, while at other agreed times the transmissions by the active station are prohibited. It is apparent that time-sharing has the potential for many operational difficulties and certainly cannot be applied widely because of the characteristics of the services involved. Time-sharing attempts between a radio astronomy station at Bologna (Italy) in 1992 and nearby broadcasting transmitters show that when the commercial interests are strong, time-sharing is not a practicable option.

The sharing issue is particularly difficult and complicated for space-to-Earth transmissions that interfere with terrestrial stations. In many cases, the satellite operator considers that when the terrestrial station is outside the footprint of the downlink transmissions, the terrestrial station is protected. This may be true for some systems, but certainly not for those using active antennas, because the transmissions through the side lobes of such systems are beyond control. This implies that a terrestrial station is protected when the space station is not visible from the terrestrial station. Similar arguments apply for high altitude platform stations (HAPS) and aeronautical stations.

Scientists often argue that the number of stations they operate in an affected radio frequency band is small. Apart from the fact that for the radio astronomy service, this statement does not need to be made since it is explicitly explained in the ITU Radio Regulations (Article S29.4). It should be noted that it is an irrelevant criterion in the protection of a radiocommunication service to consider whether the number of stations operating in that service in a specific frequency band is large or small.

However, this qualitative (that is, not-quantified) statement may help to add political weight to some arguments. Furthermore, it must be noted that what is true in one part of the world may not be true in another because of evident geographical conditions and the uneven geographical distributions of the stations in that service. The argument that the number of radio stations in some service is small may also be converted into the undesirable argument that less protection requirements are necessary for that affected radiocommunication service.

7.6 Endnotes

[1]Materials for this chapter draw upon the work of the European Science Foundation (ESF) Committee on Radio Astronomy Frequencies (CRAF), "Science and Spectrum Management," 2002. The author wishes to thank the European Science Foundation for granting permission to reproduce in full the various texts, charts, tables, and figures. A special word of thanks is directed to Dr. Wim van Driel, Observatoire de Paris, CRAF Chairman, and Dr. Titus Spoelstra, CRAF Frequency Manager/Secretary.

[2]ITU, "*Handbook on Radio Astronomy*," ITU-Publications, Geneva, p.16, 1995a.

[3]"*CRAF Handbook for Radio Astronomy*," 2nd ed., European Science Foundation, Strasbourg, p. 93, 1997.

[4]ITU, "Trends in Telecommunications Reform—Convergence and Regulation," ITU-Publication, Geneva, 1999.

[5]Evans, J.V., "*The Past, Present and Future of Satellite Communications*," in: M.A. Stuchly (ed.), Modern Radio Science, Oxford University Press, Oxford, pp. 1–23, 1999.

[6]ITU, *Ibid.* 2, 1999.

[7]Dooyeweerd, H., "*A New Critique of Theoretical Thought*," 4 volumes, 2d Impr., Presbyterian and Reformed Publ., Philadelphia, PA, Vol. 2, p. 66, 1969.

[8]ITU 1998b, Rules of Procedure—approved by the Radio Regulations Board for the application by the Radiocommunication Bureau of the provisions of the Radio Regulations, Regional Agreements, Regulations and Recommendations of World and Regional Radiocommunication Conferences, part A1, section ARS4.

Abbreviations

A

AA/OSF	Associate Administrator/Office of Space Flight
ABU	Asia-Pacific Broadcasting Union
AFFMA	Air Force frequency Management Agency
AII	Asia-Pacific Information Infrastructure
AIPS	Astronomical Information Processing System (Image Processing Software)
ALEXIS	Array of Low Energy X-Ray Imaging Sensors
AM	Amplitude Modulation
AMS	American Mobile Satellite
APEC	Asia Pacific Economic Cooperation
APSR-4	Air-Route Surveillance Radar Model 4
APT	Asia Pacific Telecommunity
ARA	Association of American Railroads
ARES	Amateur Radio Emergency Service
ASBU	Arab States Broadcasting Union
ASM	Army Spectrum Manager
ASR	Airport Surveillance Radars
ASTAP	Asia-Pacific Telecommunity Standardization Program
ASTRA	Europe's leading Direct-to-Home satellite system
ATCRB	Air Traffic Control Beacons
ATS	Advanced Technology Satellite
AVM	Automatic Vehicle Monitoring

B

BR	Bureau of Radiocommunications

C

CA	California
CB	Citizen Band
CCIR	International Radio Consultative Committee, a predecessor organization of the ITU-T
C-E	Communications-Electronic systems

CEPT	Conference of European Posts and Telecommunications Administration
CERP	Comité européen des régulateurs postaux
CH	Coronal Hole
CINC	Commander-in-Chief
CIO	Chief Information Officer
CIP	Communication and Information Policy
CITEL	Inter-American Telecommunications Commission
CNO	Chief of Naval Operations
CNNOC	Commander Naval Network Operations Command
CNS	Communications, navigation, and surveillance
COMSEC	Communications Security
CORF	Committee on Radio Frequencies, U.S. National Research Council
COSPAR	Committee on Space Research
COSPAS	Acronym for the Russian words "Cosmicheskaya Systyema Poiska Avariynich Sudov," which means "Space System for the Search of Vessels in Distress"
CPM	Conference Preparatory Meetings
CME	Coronal Mass Ejections
CRAF	Committee on Radio Astronomy Frequencies
CTIA	Cellular Telecommunications & Internet Association

D

DAB	Digital Audio Broadcasting
DoD or DOD	U.S. Department of Defense
DHTV	Digital High Definition Television
DO	Departmental Offices
DOE	U.S. Department of Energy
DOJ	U.S. Department of Justice
DOS	U.S. Department of State
DM	Degraded Minutes
DM	Departmental Manual
DME	Distance Measuring Equipment
DSCS	Defense Satellite Communication Systems
DSF	Disappearing Filaments
DTV	Digital Television

E

EBU	European Broadcasting Union
EC	European Commission
ECM	Electronic Countermeasure
ECTRA	European Committee for Regulatory Telecommunications Affairs

EEA	European Economic Area
ELF	Extremely Low Frequency
ELT	Emergency Locating Transmitters
EMC	Electromagnetic Compatibility
EMI	Electromagnetic Interference
EPIRB	Emergency Position Indicating Radio Beacon
ERC	European Radiocommunications Committee
ERO	European Radiocommunications Office
ESA	European Space Agency
ESCAP	Economic and Social Commission for Asia and the Pacific
ESF	European Science Foundation
ETO	European Telecommunications Office
ETSI	European Telecommunications Standards Institute
EU	European Union
EUTELSAT	European Telecommunications Satellite Organization
EW	Electronic Warfare

F

FAA	U.S. Federal Aviation Administration
FACA	Federal Advisory Committee Act
FAS	Frequency Assignment Subcommittee
FCC	Federal Communications Commission
FEMA	Federal Emergency Management Agency
FLTSATCOM	U.S. Navy's Fleet Satellite Communications System
FLEWUG	Federal Law Enforcement Wireless Users Group
FM	Frequency Modulation
FN	Footnote
FS	Fixed Service

G

3G	Third Generation
GAO	U.S. Government Accounting Office
GCA	Ground Control Approach
GII	Global Information Infrastructure
GLONASS	Global Orbiting Navigation Satellite System
GMDSS	Global Maritime Distress and Safety System
GMF	Government Master File
GMPCS	Global Mobile Personal Communications System (Universal Mobile Telecommunications System)
GOES	Geo-stationary Operational Environmental Satellite
GNP	Gross National Product
GPR	Ground Penetrating Radar

GRC	NASA Glenn Research Center
GPS	Global Positioning System
GSO	Geosynchronous Satellite Orbit
GSM	Global System for Mobile Communications
GWCS	General Wireless Communications Services

H

HALCA	Highly Advanced Laboratory for Communications and Astronomy
HAPS	High Altitude Platform Stations
HETE	High Energy Transient Experiment
HF	High Frequency
H.R	House/Hearing Report

I

IAP	ITU Conferences and Meetings, including common regional proposals
IARU	International Union of Radio Amateurs
IAU	International Astronomical Union
IBB	International Broadcasting Bureau
ICAO	International Civil Aviation Organization
ICW	Interrupted Continuous Wave
IDWM	ITU Digitized World Map
IFIC	International Frequency Information Circular
IFRB	International Frequency Registration Board
IFF/SIF	Identification Friend or Foe/Selective Identification Feature
IFL	International Frequency List
ILS	Instrument Landing System
IMO	International Maritime Organization
IMT	International Mobile Telephony
INMARSAT	International Maritime Satellite Organization
INTELSAT	International Telecommunications Satellite (Organization)
IRAC	Interdepartmental Radio Advisory Committee
ISB	Independent Sideband
ISM	Industrial, Scientific, and Medical
IRAF	Image Reduction and Analysis Facility (Image Processing Software)
IRAS	Infrared Astronomical Satellite
ITFS	Instructional Television Fixed Services
ITU	International Telecommunications Union
ITU-R	International Telecommunications Union Radio Regulations

IUCAF	Scientific Committee on the Allocation of Frequencies for Radio Astronomy and Space Sciences
IVDS	Interactive Video Data Service

J

JFMO	Joint Frequency Management Office
JTIDS	Joint Tactical Information Distribution System

L

LEO	Low Earth Orbit
LF	Low Frequency
LOS	Line-of-Sight
LMCC	Land Mobile Communications Council
LMR	Land Mobile Radio
LSGAC	Local and State Government Advisory Committee
LUF	Lowest Usable Frequency

M

MBPS	Million Bits Per Second
MDS	Multipoint Distribution System
MF	Medium Frequency
MISTAR	Military Strategic Tactical and Relay Satellite
MMDS	Multichannel Multipoint Distribution Service
MMW	Millimeter Wave
MOD	Modification
MoU or MOU	Memorandum of Understanding
MM	Maritime Mobile
MSS	Mobile Satellite Service
MUF	Maximum Usable Frequency

N

NABA	North American Broadcasters Association
NAIC	National Astronomy and Ionosphere Center
NASA	U.S. National Aeronautical & Space Administration
NASDA	Japanese National Space Development Agency
NATO	North Atlantic Treaty Organization
NCO	National Command Authority
NEAR	Near Earth Asteroid Rendezvous
NEXRAD	Next Generation Weather Radar
NII	National Information Infrastructure
NOAA	National Oceanic and Atmospheric Administration, U.S. Department of Commerce

NOR	Norway
NPD	NASA Policy Directive
NPS	National Park Service
NPRG	National Partnership for Reinventing Government
NRA	National Regulatory Authorities
NRAO	National Radio Astronomy Observatory
NTIA	National Telecommunications and Information Administration
NTIS	National Technical Information Service
NVIS	Near-Vertical-Incidence Sky

O

OAU	Organization of African Unity
OET	Office of Engineering and Technology
OH/IR Stars	Stars with strong hydroxyl (OH) masers and strong infra red (IR) emissions from the shell of warm gas
OIG	Office of the Inspector General
OMB	The White House Office of Management and the Budget
OMCM	Orthogonal Multiple Carrier Modulation
OST	Outer Space Treaty
OTA	Office of Technology Assessment
OTH	Over-the-Horizon

P

PATU	Pan African Telecommunications Union
PCA	Polar Cap Absorption
PCC	Permanent Consultative Committees
PCS	Personal Communications Services
PSWN	Public Safety Wireless Network
PT	Pactor System

R

RA	Radiocommunications Assembly
RACON	Radar Transponder Beacons
RACES	Radio Amateur Civil Emergency Service
RADHAZ	Radiation Hazards
RAS	Radio Astronomy Spectrum
RCS	Radiocommunications Sector
RDSS	Radio Determination Satellite Services
RFA	Radio Frequency Authorization
RF	Radio Frequency
RFI	Radio Frequency Interference
RMS	Root Mean Square
RPV	Remote Piloted Vehicle

S

SARSAT	Search and Rescue Satellite-Aided Tracking
SART	Search and rescue Transponders
SCA	Subsidiary Communications Authorization
SES	Sudden Enhancement of Signal
SES	Severely Errored Seconds
SG	Study Groups
SGLS	Space Ground Link Subsystem
SHF	Super High Frequency
SID	Sudden Ionospheric Disturbance
SMR	Specialized Mobile Radio
SNG	Satellite News Gathering
SOLAS	Safety of Life at Sea
SOP	Standard Operating Procedures
SPA	Sudden Phase Anomaly
SPS	Spectrum Planning Subcommittee
SPASUR	Space Surveillance System (USA)
SSB	Single Sideband
STL	Studio-to-Transmitter Links
SUP	Suppress
SWF	Short-wave Fade

T

TACAN	Tactical Air Navigation
T-CAS	Traffic Alert and Collision Avoidance System
TDRSS	Tracking and Data Relay Satellite System
TELWG	Telecommunications and Information Working Group
TENS	Transcutaneous Electrical Neuromuscular Stimulation
TEX	U.S. Air Force Satellite; Mission: Study ionospheric effects on signal propagation
TG	Task Groups
TIGTA	Treasury Inspector General for Tax Administration
TIROS	Television and Infra Red Observation Satellite
TRANSIT-SAT	Polar Orbiting Satellite
TV	Television

U

UHF	Ultra High Frequency
UMTS	Universal Mobile Telephony System
UNCTAD	United Nations Commission on Trade and Development
UNESCO	United Nations Educational, Scientific, and Cultural Organization

URSI	International Union of Radio Science
USD	U.S. Dollar
USCG	U.S. Coast Guard
USAT	Utilities Spectrum Assessment Taskforce
US&P	U.S. and Possessions
USIA	U.S. Information Agency
UWB	Ultra-wide Band

V

VHF	Very High Frequency
VLBA	Very Long Baseline Antenna
VLBI	Very Long Baseline Interferometry
VLF	Very Low Frequency
VRM	Venus Radar Mapper

W

WBU	World Broadcasting Unions
WHO	World Health Organization
WIPO	World Intellectual Property Organization
WMO	World Meteorological Organization
WP	Working Party
WPO	Wireless Programs Office
WRC	World Radiocommunication Conference
WTO	World Trade Organization
WV	West Virginia

Index

T

ABOUT THE AUTHOR

AMIT K. MAITRA, founder and president of SATLINK Communications, provides services and solutions to the space/defense/intelligence communities. He holds two master's degrees (MPA and MA) in policy making, analysis, and administration from the University of Minnesota, Minneapolis, and graduated summa cum laude from the University of Calcutta, India, with a bachelor's degree in systems engineering and architecture. He was a visiting scholar at the Wharton School, University of Pennsylvania, Philadelphia, and is currently completing the Executive Doctor of Management (EDM) program at Case Western Reserve University, Cleveland, Ohio.